Astronomers' Universe

More information about this series at
http://www.springer.com/series/6960

C.V. Vishveshwara

Universe Unveiled

The Cosmos in My Bubble Bath

 Springer

C.V. Vishveshwara
Bangalore Association for Science Education
Jawaharlal Nehru Planetarium
Bangalore
India

ISSN 1614-659X   ISSN 2197-6651 (electronic)
ISBN 978-3-319-08212-7   ISBN 978-3-319-08213-4 (eBook)
DOI 10.1007/978-3-319-08213-4
Springer Cham Heidelberg New York Dordrecht London

Library of Congress Control Number: 2014953633

Printed on acid-free paper

Springer is part of Springer Science+Business Media (www.springer.com)

To the loving memory of my elder siblings
My sister Rathna
and
My brothers Puttanna and Chandru

Praise for
Einstein's Enigma or Black Holes in My Bubble Bath

Beautifully written and thoroughly entertaining, Vishveshwara's "Einstein's Enigma" provides an authoritative but distinctly original approach to an explanation of basics and subtleties of Einstein's general relativity and of the astrophysics of black holes. I warmly recommend it to beginner and expert alike.

Sir Roger Penrose, *Author of* Shadows of the Mind *and* The Road to Reality

The main dish in this feast is a clear and sound presentation of the science underlying black holes from a distinguished scientist who has been contributing to their study since before they were named. Furthermore this science is presented in a sauce of philosophy, history, literature, gastronomy and imagination from an entertaining personality who needs several alter egos to show all the different ways he can think about his subject. Among the cartoons and drawings are the few lines of optional mathematics which are included for those who like that approach.

Prof. Charles W. Misner, *Co-author of* Gravitation *by Misner, Thorne and Wheeler*

The unusual format and whimsical style of "Einstein's Enigma" should not obscure the fact that this is a serious book, which aims to get across the essentials of the theory of general relativity and some related topics to a readership which is not assumed to be fluent in advanced mathematics. I believe the author, who has a long experience in presenting this kind of material to non-specialist audiences, has succeeded in the task he has set himself; the book will amply repay sustained and diligent reading by even a totally unmathematical reader."

Sir Anthony J. Leggett, *Nobel Laureate in Physics, 2003*

The only thing this intoxicating is Salman Rushdie's Haroun and the Sea of Stories.

Dr. Richard Isaacson, *former Program Director for Gravitational Physics, Physics Division, National Science Foundation, USA*

An innovative way of trying to explain some of the most complicated concepts in physics to beginners and it contains one of the best accounts I've ever read of the (almost Damscene) conversion of a newcomer to the pleasure of a night's observing. As Nietzsche said: 'What would be the joy of a star but for those who behold it?'

David Reid, *BBC Sky at Night*

Vishveshwara's is among the successful presentations of a most sophisticated, important, and beautiful theory of twentieth-century physics—Einstein's general relativity and its most fascinating corollary, the existence of black holes. A book to inform, enrich, and entertain every science-educated reader. Summing Up: Highly recommended. Lower-division undergraduates through professionals.

V. V. Raman, CHOICE

The author is a major specialist in the field of relativity, astrophysics and gravitation, and is currently working in Bangalore, India. According to me, this book deserves a very large audience: it is very different from all the popular science books written on the subject and as such will be appreciated even by the non-physicists. The physicists will probably learn new things, while the specialists in astrophysics and gravitation will certainly enjoy this new and fresh insight in their field.

Fabrice Louche, Physicalia Magazine

The author blends fiction, fantasy, physics and philosophy to tell the story of gravitation theory...and the astrophysics of black holes. He not only succeeds at doing this, he entertains readers with delightful digressions and illustrates key concepts with wonderful cartoons, some purportedly scribbled on paper napkins...

Michelle Press, Scientific American

Imagination is more important than knowledge. For knowledge is limited to all we now know and understand, while imagination embraces the entire world, and all there ever will be to know and understand.

Albert Einstein

Imagination was given to man to compensate him for what he is not; a sense of humour to console him for what he is.

Francis Bacon

Contents

Prologue

Ah, time flies, time flees, time flows. *Time is the culprit*, said the Master who taught us that all things are relative including time. How long has it been since I addressed you? Several years indeed! And some of you are new. At that time, long back, I recounted my discussions with my dear friend George, Professor George Gallagher, the astrophysicist at the University, that is, regarding black holes and described the fantasies induced by my magical bath tub. In fact, all that was presented in a book form with the title, *Einstein's Enigma or Black Holes in My Bubble Bath*. Some of you might have even read it.

What have I been up to during these years? Well, I have been travelling quite a bit. I have wandered all over Europe. I even journeyed for a brief period to the fabled East. To be free like a bird on the wing, a steed on the run, a fish in the blue waters of the ocean! It is exhilarating, it is euphoric. I imbibed many cultures in all the places I visited, learnt their ways of life, and tasted a variety of culinary delights. In the meantime, George too has been away as visiting professor at several universities abroad. Knowing him, he too must have lived life to the hilt enjoying it in every possible way while working assiduously on his research and teaching. I have missed him dearly indeed.

All this will change soon. I am back and so is George. As a matter of fact, I have made an appointment to meet him this evening. We shall be getting together at the lovely little Italian restaurant run by our friend Bruno. Earlier, I have described how we used to meet at this charming place, enjoy good food and drink, and most of all discuss all about black holes. I just can't wait, you know. I am sure history will repeat itself. I feel it in my guts.

On my way to Bruno's, I meet my friends Fernando and Maria along with their daughter Felicia. As usual, they are relaxing in their lounge chairs, sipping red wine, on the sidewalk outside their grocery store. During these years of my absence, Felicia has blossomed into a beautiful young lady from a plump little kid that she was.

'Welcome back, *Señor* Alfie', Fernando gets up and shakes my hand warmly for a long time. 'We missed you, you know. Ah, before I forget, here is the spare key you gave us. Maria and I took good care of your apartment, no? Kept it spic and span, tiptop'.

'Oh, thank you so much, *muchas gracias*'. I am genuinely grateful.

'But there is something funny with your bathtub you know'. Fernando says hesitantly. 'When I scrub it, I feel—how do I put it—I give bath to someone living. *Hombre*, it gives me creeps. What goes, *Señor* Alfie?' Fernando looks at me searchingly.

I keep quiet. I can't possibly tell him the most well-kept secret shared between me and my bathtub, can I?

'Well, off to Bruno's for a good meal?' enquires Fernando. 'I hear that Professor is back too. Why don't you and him join us for food sometime? Maria will make some *tapas*, very tasty, *mucho rico*. And I bring excellent wine from my cousin Alejandro who makes it himself. Give my best to Professor'.

I say goodbye to the three of them and move on. As I said, I can't wait. I am much too excited. Yes, there will be great food and drink again at Bruno's. More than that, it will be wonderful to meet Bruno himself. And, needless to add, to reconnect with George and find out all the fantastic things that must be happening with him. And maybe we will discuss something else as we did about black holes before, who knows?

All right, I shall let you know how it goes this evening. Till then, as I wish you always, may peace be with you.

Chapter 1
Reconnecting

At Bruno's

The Italian restaurant *Benvenuto*, or simply Bruno's as we call it, seemed to have had a facelift in my absence. It looked freshly painted with a number of flowerpots decoratively placed to make the façade most inviting. As I entered the place, I could see that the furniture had been rearranged to make room for more people without cramping the space. All said and done, the interior had not lost its charm and elegance.

Of course, the most important event I was looking forward to was meeting George. What a delight it was to see George beaming at me and waving his hand in welcome. As always, he was seated at our favourite table waiting for me. George was unusually relaxed. He is normally in a highly excited state. As I approached our table, he gave me a warm smile, looking at me steadily as though he wanted to inspect my appearance in every detail.

'Alfie, you have not changed a bit,' said George. 'I mean it unlike some people who say that for the sake of formality.'

'I know you mean it George. And you haven't changed either,' I said returning his steadfast gaze. '*Age cannot wither you, nor custom stale your infinite variety,*' I paraphrased Shakespeare's quotation from *Antony and Cleopatra*.

'What are you talking about?' George was mystified.

'Well, that was Shakespeare's remark about Cleopatra,' I answered.

'Oh, Cleopatra, whose face launched hundred ships. You compare me to her!' George was amused.

'George, it was Helen of Troy whose face launched thousand—not hundred— ships. It was my poetic way of saying that you haven't changed in all these years.'

'Cleopatra, Helen of Troy, who cares? Beauty queens, if you have seen one, you have seen them all!'

'But, George, neither of us has seen a single beauty queen in all our lives,' I remonstrated.

© Springer International Publishing Switzerland 2015
C.V. Vishveshwara, *Universe Unveiled*, Astronomers' Universe,
DOI 10.1007/978-3-319-08213-4_1

Fortunately, this inane exchange came to an end as we saw Bruno emerging out of the kitchen and making his way towards our table, a big smile playing upon his face. How wonderful to see our good old friend Bruno Beltrametti—not just the owner of the restaurant, but chef, headwaiter, entertainer, and philosopher all rolled into one.

'Don't tell me that I haven't changed,' Bruno addressed us. 'We all change; we just don't want to admit it to ourselves, that is all.'

'Who said you haven't changed, Bruno? You have grown younger, pal,' laughed George.

'All right, you win. You always have the last word, don't you?' smiled Bruno. 'You know, honestly this place is not the same without you two. I am so glad you are back. Let us celebrate, food and drink on the house. What would you like?'

'You decide, Bruno,' said George taking the lead.

'Vegetarian food for me,' I added. 'I have become a vegetarian.'

'Since when?' George raised his eyebrows.

'Since a couple of years,' I answered. 'It is good for health—and for conscience too.'

'This man is impossible, suddenly changing over to vegetarianism,' groaned George. 'All right, let me be a vegetarian too for today, Bruno.'

'Why not, I will make some delicious ravioli. Normally, we fill it with crab or lobster meat. But for you vegetarians it will be assorted cheese spiced with herbs. They will be simmered in special creamy mushroom sauce, my own creation. And you can have excellent Frascati wine to go with it,' Bruno went back to the kitchen. Soon enough, a waiter brought two plates of toasted cheese for us to munch on and a bottle of chilled white wine without label to go with it.

'You must have had some gorgeous time, Alfie, judging from the post cards you were sending me from all over the world,' remarked George after having nibbled a bit of the toasted cheese and taken a couple of sips of the wine. 'I couldn't reply since you never mentioned your address.'

'I didn't know my own address either, George, since I was wandering without planning my next step, you know. As you say, it was fantastic, especially visiting the haunts and habitats of the greats like Galileo, Newton, and the ancient Greeks too.'

'You must tell me all about it sometime, Alfie. I can't say I didn't have a nice time either visiting a couple of universities, balancing work and pleasure. I was not weighed down by responsibilities since my three students you met have graduated and are well-placed. I am yet to take some hapless kids to do their doctoral research.'

'You must have had plenty of free time, then.'

'Yes and no. I filled my spare time with writing a book on the Universe.'

'The Universe!' I exclaimed. 'That is wonderful, George. How is it coming?'

'I shall tell you in a minute. First, let us have something to eat,' said George spotting the waiter coming out of the kitchen.

The waiter brought our steaming hot ravioli in sauce, accompanied by side dishes of garnished vegetables. We started eating slowly, enjoying every morsel punctuated by sips of the delicate wine. Food is the soul of happiness, isn't it?

'Ah, the book on the Universe,' George speared a piece of ravioli with his fork and waved it in the air excitedly. 'You know, Alfie, the study of the Universe is a continuously evolving process. I don't mean the evolution of the Universe. Of course, that happens too. But, the entire cosmic journey has progressed along a meandering path from its early beginnings to the modern era. It is the blossoming of human knowledge, human beliefs, human mind, and the science created by the human intellect. Even the early myths were an attempt to explain what was observed in the sky. I want to describe this exciting adventure starting with the ancient times and ending with the most recent findings. Do I have to add that the quest is one without end?'

George seemed to be a little bit tired not so much by having spoken about his book at length as by the cosmic journey he was envisaging. He sat back and took a long pull at his wine.

'That sounds fantastic, George.' I too was excited. 'How far have you progressed with the work?'

George silently bent down, picked up his briefcase, and pulled out a bound volume of typed manuscript.

'Here, I am handing over to you the preliminary draft of the book. No, no, don't open it now. Go home and read it at your leisure and let me have your feedback. We can meet regularly and discuss the contents, style, and presentation. It will help me with further drafts and the final version. What do you say?'

'That is great, George, really great,' I said with feeling. 'What is the title of the book?'

'I haven't decided it yet, Alfie,' answered George. 'You could help me with that too.'

'Aha, a book in search of a title,' I remarked. 'Like Luigi Pirandello's play, *Six Characters in Search of an Author*! The way it happens in the play, maybe the book will also evolve as we keep changing the possible titles.'

'You are absolutely crazy, you know that?' George shook his head. 'Incurably mad!'

Bruno came over and placed on our table another bottle along with three small liqueur glasses.

'Here is *Limoncello* from Amalfi coast, one of my favourite *digestivos*,' announced Bruno as he poured the chilled yellow liqueur. 'By the way, I couldn't help overhearing a bit of your conversation as I moved around, you know. Not eavesdropping for sure. What is this Universe book? Can I understand it?'

Before George could respond, I answered Bruno's query. 'You bet, Bruno. You guys must have forgotten what I told you long ago. The great Rutherford once told the equally great Bohr that a good theory should be understandable even by a barmaid. Since unfortunately I don't see any barmaids around, I must say that a good book should be understandable by a restaurant owner, that too an intelligent one like you, Bruno.'

'Barmaids indeed,' Bruno threw back his head and guffawed. 'Cheers, here is to the new book!' He gulped down his drink and filled his glass again.

George was enjoying every one of his little sips, eyes close, head tilted back a little, smiling in contentment. Bruno and I joined him in silence.

'By the way, you noticed changes we have made around here' Bruno swept his hand to indicate the interior of the restaurant. 'There is going to be a major one, you know. I am hiring a young chef to help me out. He is very good at all sorts of cooking. So, there will be different types of food from different countries once a week. Do come and taste and tell me your opinion. I must now get back to work. Goodbye, my friends, *addio, amici miei*!"

Bruno got up and so did we. We thanked him profusely for the excellent dinner and the exquisite wines. He waved away our thanks as he usually does.

'Don't wait till you have read a lot, Alfie,' said George. 'I would like to discuss each section of the book as you go along. It will be a great help. Do meet me soon. Bye now.'

'It will be an immense pleasure, George,' I told him sincerely. 'I shall call you soon and meet you. Till then, keep well.'

As we parted, I remembered our meeting of long ago when George had promised to tell me all about black holes. Afterwards, regular sessions of discussion had followed that were most rewarding. Now it was going to be the entire Universe no less. I walked home slowly and happily with the stars smiling in the clear sky above.

A Brief Biography of My Bathtub

I stopped short at the blind alley that is situated close to the street on which I live. My memory flooded back to the eventful day when I had inherited my precious bathtub. I told some of you about it in detail, didn't I? Let me briefly recapitulate what had happened then.

In the short, dark stretch of the alley, there were some derelict buildings as well as a couple of warehouses. But that night, I saw to my surprise one of the small buildings sporting the sign, *Al's All-in-One Store* with dim light shining inside. Among several other signs advertising an assortment of items, there was a large one on which was written in bold letters: *Bathtubs for Half Price*! With some trepidation, I had entered the store. I can never forget what had happened next. Allow me to recount my experience in some detail.

As I gazed upon the chaotic variety of merchandise scattered around, I felt a presence at my elbow. The sight that greeted me as I turned around has been indelibly etched on my mind. An elderly gentleman, presumably the storekeeper, had appeared from nowhere. I was captivated by his eyes looking steadily at me, eyes gentle and kind, but not without a mischievous twinkle in them. Time had woven a web of lines on his serene face that was framed by a halo of white, wispy hair.

He spoke softly with a thick accent, which I could not place. It was one of those accents that belonged everywhere and nowhere. 'I am Al at your service,' he said with a warm, friendly smile. 'So, you have come to collect your bathtub, *ja*? *Wunderbar*, this way please, *mein lieber Herr*.' He gestured with his head, his long, unruly hair waving in the air. I had entered the store with no intention of buying anything, let alone a bathtub. But, as though mesmerized, I had followed Al.

It was a large gallery that held a number of bathtubs of different shapes, sizes, and colours. Al pointed to three of them claiming that they were of great historical value. The first one, which was truly magnificent, had belonged to some ancient Egyptian queen according to Al. Perhaps to Nefertiti or maybe Cleopatra, he said. Moreover, its design and decoration reflected the cosmic order believed by the ancient Egyptians. The Egyptian myth envisaged the sky goddess Nut, decked with shining stars, arching over her reclining husband Seb, the Earth god. And their son Shu, who controlled the winds, knelt between them in the sky. During the day, the Sun, the god of gods Amon-Ra, sailed in his divine barge along Nut's body. Each night he died and entered Amenti, the nether world, to be broken up into myriad stars. At dawn, he was reborn to repeat the perpetual cycle of birth and death.

The bathtub, enormous and shaped like the Sun's barge, was made of black marble. Semiprecious stones of different kinds, like turquoise and lapis lazuli, as well as flakes of gold and silver, were inlaid into the dark background creating a continuous panorama of the sky, the Sun, and the sparkling stars. It was indeed breathtaking.

Next, Al showed me his favourite bathtub. Simple and elegant, it had been carved out of pure, white marble and was elliptical in shape. Engraved on one side was a spiral and on the other the picture of a pole balanced on a conical peg.

'The owner of the bathtub discovered the spiral you see, *mein lieber Herr*,' explained Al. He claimed that, if he were to be given a long enough lever and a place to hover around in space, he could lift the entire Earth.' And then he announced jubilantly, 'His name was Archimedes!'

'From this very bathtub here,' Al continued after pausing dramatically, 'Archimedes ran stark naked, shouting *Eureka*. Imagine our scientists following his example in their haste to announce their results! What a sight it would be!' Al roared with laughter tears in his eyes. What a contrast it was between his soft speech and this bellowing laughter that echoed from wall to wall of the store!

The third bathtub Al showed me was essentially a crude rectangular box made of mere sandstone, chipped here and there. Unadorned, it looked more like a coffin than a bathtub. Jean Paul Marat, one of the leaders of the French Revolution who suffered from some terrible skin ailment he had contracted while hiding in the sewers of Paris, apparently soaked himself in lukewarm medicated water that filled this bathtub. A woman named Charlotte Corday had stabbed him to death during one of his soak sessions. The bathtub had proved indeed to be Marat's coffin so to speak.

'I am told that, if you look carefully, you can still find faint bloodstains in the bathtub,' said Al. 'Gives you the creeps,' Al seemed to shiver at the thought.

Al took out his pipe from his pocket, filled it with tobacco from the pouch he had extracted from another pocket, and lighted it. As he puffed in contentment, he spoke about the benefits of taking a bath, but confessed, 'Personally, I hate baths myself. That follows in the tradition of the great Kepler who bathed only once in his lifetime, that too at the nagging of his wife. It nearly killed him. So, I try to avoid taking a bath although I extol its virtues.' He added, his eyes shining, 'As you very well know, sir, there is nothing nobler than to preach what one does not practice.'

Al bent down to tie his shoelace. His shoes were scuffed and he wore no socks beneath his rumpled trousers. Straightening up, he smoothened the sweatshirt he was wearing. As he escorted me along, he said, 'Pardon me, the small black box over there is not a bathtub. It is my violin case.' Al roared with laughter again.

Finally, we reached our destination. Exclaiming, 'Ah, now for your bathtub!' he pulled off with a flourish the tattered cloth that covered the bathtub. I could not believe my eyes. This bathtub, as he called it, was nothing more than a kitchen sink! It was beyond my comprehension how anyone but an infant could get into it.

Al was amused by the incredulous expression on my face. 'Half price, half size,' he laughed and then assured me, 'Take my word for it. You will find no difficulty getting into your bathtub that will adjust itself to your dimensions. Dimensions are relative you know.'

Al fell silent for a moment and spoke slowly with a seriousness he had not displayed so far.

I remember every word of Al that followed as if I heard them a moment ago: *What is more important is the fact that the bathtub is a magical one. It is filled with myth, math, science, philosophy, art, literature, and above all dreams, not to mention your bath water.*

Al told me that the heavy bathtub would be delivered home along with instructions to install it in five easy steps. 'Oh, I almost forgot. Along with the bathtub, you get a free sample of our special bubble-bath additive.' He produced a plastic bag filled with perfectly spherical black beads. They were black in a strange manner, reflecting no light at all as if they totally absorbed light, but quite pretty in a peculiar way.

'Rest assured that your newly acquired possession will give wondrous moments you could never have imagined. Goodbye now,' said Al as he walked me to the door and held it open for me. As I watched him, I wondered whether I had met him before. No, that was impossible. Did he resemble someone whose description I had read somewhere? Or someone in a photograph I had seen? My mind seemed to have become fuzzy and I was confused. I caught myself meandering through the maze of my memories as I realized that Al was regarding me with a mysterious, knowing smile. As I was about to leave, he said gently, 'We shall meet again, my friend.'

As I started walking home, I turned back. The alley was now plunged in absolute darkness. There was no light in the shop and I could see no signs either. Had I been dreaming? The shop, the shopkeeper, and everything that had befallen me, was it all my imagination? The whole episode was a bit scary.

As I trudged up to my apartment, tired and confused, I once again wondered whether I had imagined the whole episode involving Al and his store. But there it was at my doorstep, in all its three-dimensional reality, my bathtub neatly packed. Surprisingly enough, it was quite light as I carried it inside and installed it easily. I could not wait to take my first bath in my bathtub or kitchen sink, or whatever it was. I filled it with hot water, stirred in half a spoonful of the bubble-bath mixture and eased myself in. Surprisingly, I could comfortably fit into the bathtub: either it had expanded or I had shrunk. As Al had said, dimensions were relative. Maybe the mass of the bathtub was also relative since it had proved to be quite light.

Again, as I recall what followed, every moment of my new experience is summoned up from my memory. The bubble bath was incredibly soothing. The vapours rising from the bathwater seemed to steep into my mind blending aware-ness, thoughts, and imagination into a flowing stream. I was surrounded by count-less bubbles, multicoloured spheres that glistened and trembled, each carrying within it a dark speck that appeared to grow a little as it absorbed the vapour in its vicinity and the light that fell on it. Slowly, words I had heard that evening while talking to George came to my mind: black holes swallowing up matter and energy and growing in mass and size. Was this happening in my own bathtub? Were black holes swarming in my bubble bath?

The bubbles were swirling all around me massaging my body, gently tugging at me. As I luxuriated in the fantastic bubble bath, my eyes grew heavy and I drifted into a supremely blissful slumber.

That is ancient history. But, the past leads to the future passing through the present, does it not? Well, once again today, returning from Bruno's I climb up the stairs to my apartment. I cannot wait to start reading the draft of George's book about the Universe. I sit back comfortably in my old but cosy chair, switch on the lamp next to me, open the manuscript, and begin my own cosmic journey.

Chapter 2
Ancient Times

As George had told me, the book began with astronomy of antiquity. The ancient astronomers from all over the world—the Greeks, Indians, Arabs, and Chinese—had been avid stargazers. Night after night, they could see the fixed patterns of the stars in the night-sky: the *Constellations* as we know them. And the astronomers associated the groups of stars with their own myths and beliefs, stories and legends. Beasts, heroes, heroines, gods and demons resided in those star patterns. Whereas the stories helped identify the star-groups, the cultures that gave birth to those stories were perpetuated through those very stars. Take for instance *Orion*, one of the most prominent constellations in the sky, named after the great celestial hunter of Greek mythology. His story involved the battle between him and a giant scorpion that eventually became the constellation *Scorpio*. Indians have a fascinating story for the same group of stars as well. One of their demigods, *Prajapathi*, took the shape of a stag and chased his own daughter *Rohini*, who had assumed the form of a lovely, young doe. The gods, outraged by Prajapathi's incestuous lust, complained to *Lord Shiva* who, assuming the form of a great hunter, shot down Prajapathi with an arrow from his bow and rescued Rohini. A remarkable story, is it not? The constellation, in both Greek and Indian cultures, is associated with a hunter. Is there some deep inter-cultural relationship at work here, I wondered. Incidentally, Lord Shiva, as *Maha Vyadha* or the Great Hunter, is identified with *Sirius*, the brightest star in the sky. In some accounts, the hunter, who kills Prajapathi, is called *Lubdhaka*. And Rohini happens to be the star *Aldebaran*, as it is known in the west. Furthermore, the three stars in Orion's belt are identified with Shiva's arrow.

As I read the fascinating account of the constellations, I remembered my recent wanderings in the lands far away. I could vividly visualize the ancient monuments of Greece that bore glowing testimony to a glorious era. I remembered the magnificent amphitheatres where the Greek plays had been performed a long time ago, plays by the great playwrights like Aeschylus, Sophocles, Euripides, and Aristophanes. In India, among other things, I visited many beautiful temples, some of them decorated with intricate carvings of breathtaking beauty. And there were others that were famous for imposing sculptures of immense proportions. In the

© Springer International Publishing Switzerland 2015
C.V. Vishveshwara, *Universe Unveiled*, Astronomers' Universe,
DOI 10.1007/978-3-319-08213-4_2

middle-east, I had gazed upon the endless stretches of desert sands, at once awesome and frightening.

I felt tired. The evening had been exciting. I closed the book, marking the page I had just finished with the sheet of paper on which I usually make my own notes. I got up and stretched, thinking of going to bed. Then I suddenly remembered my bathtub. Of course, I needed a good, hot bath to melt away my fatigue. I walked over to my bathroom and started filling my bathtub. But what about the fabulous bubble-bath mixture? Had I exhausted my supply? I didn't remember. I looked inside the cupboard where I had stashed away the sachets containing that magical ingredient. To my surprise, there were still quite a few of them left. But the mixture looked quite different from what I had used earlier. It did not consist of black beads as before, but was in the form of luminous little disks. Without thinking about this difference in appearance, I opened one of the sachets and was about to add a spoonful of the content to the bathwater. I stopped short. Had I heard a faint, eerie sound emanating from the bathtub? How could it be? I was startled.

I heard it again, a low gurgle followed by what seemed like the sound of clearing the throat. Good heavens, I had completely forgotten that my bathtub could speak!

'Hmm, we had completely forgotten that both of us can speak, had we?' There was again the low sound of resonant laughter. 'How could you, boss?'

I felt really ashamed of myself.

'Oh, well, let it pass,' resumed my bathtub. 'All these years I have been waiting for you, boss. Why did you have to ask your friend down the street to wash me regularly? He scrubbed me so hard! Fortunately, we bathtubs do not bleed. But then, he seemed to be afraid of me. That is funny.'

I remembered what Fernando had told me. Whenever he touched my bathtub, he used to feel that it seemed to be human in some uncanny manner. Maybe he had been a bit frightened by the experience, who knows?

'That was perceptive of him to think of me as human, I must admit grudgingly. Anyway, I am so glad that you are back. I am delighted that you still remember me. I would be the happiest—well, whatever I happen to be—if you recall what I used to be in my previous life.'

I was not sure about what my bathtub was referring to. What did it mean by the phrases, *whatever I happen to be* and *my previous life*?

'Surely you have not forgotten, boss!' There was a hint of reproach in my bathtub's voice. 'I was a kitchen sink in my Master's house. That house was known by one of the most famous addresses: One Hundred and Twelve, Mercer Street, Princeton, New Jersey, United States of America, no need perhaps to add the World and the Universe too. Don't forget what you were told about me: *What is more important is the fact that the bathtub is a magical one. It is filled with myth, math, science, philosophy, art, literature, and above all dreams.* How did I acquire all those things, boss? Well, I absorbed all that from the discussions and dialogues the Master used to hold with his visitors, even from his musings and his unexpressed thoughts, didn't I?'

Oh, yes, whatever my bathtub had told me long ago came back to me in vivid detail.

'Ah, that is better. By the way, boss, why do you keep referring to me as *my* bathtub? Sounds very possessive, you know. Why don't you call me KSBT— Kitchen Sink Bath Tub? Or how about combining the letters into *KhaSBaTh*? It refers to bath and, moreover, the word means *confidential talk* in Hindi, the Indian language that you have learnt a little bit. Confidential talk, hush-hush, *gupchup*, that is what we do all the time, don't we? How about that?'

Crazy, but wonderful, the name *Khasbath*, I thought.

'Well, I don't want to hold up the great pleasure you are about to have. But, fear not, we shall have a long chat one of these days.'

My bathtub, sorry Khasbath I mean, fell silent. I stirred a spoonful of the bubble bath mixture into the bathwater and gently lowered myself in. I could feel the water lifting me up. Well, I knew that the Archimedes Principle was at work: *The upward buoyant force that is exerted on a body immersed in a fluid, whether fully or partially submerged, is equal to the weight of the fluid that the body displaces.* Needless to add, the body in question belonged to me and the fluid was my bathwater. Among the profusion of the bubbles swarming around me, I could now see not black specks as earlier, but scintillating little sparks tinged with different colours. On close observation, I realized many of them formed definite patterns. I could not concentrate long, as I felt intensely drowsy. My eyes slowly closed and I drifted away into a strange state of mind that seemed to hover on the verge of total nothingness.

The Celestial Theatre

I slowly opened my eyes. I could see countless stars shining away wherever I looked. Those stars differed from one another in their brightness. If one looked carefully, they seemed to be tinged with varying colours as well. As I knew already, many of the stars appeared to form definite patterns. Ah, the constellations that convey amazing stories, I thought. Stories! I loved those stories. But where there are stories, we need a good storyteller, don't we?

Out of pure, empty space, the figure of a man materialized. First just the outline became visible. And then all aspects of his person filled in with complete clarity. I knew from my past experience that this would happen time and again, whenever a new character appeared on the scene.

He wore a snow-white robe held by an ivory clasp at one of his shoulders. From his apparel, I could conclude that he was probably from ancient Greece, perhaps a scholar from that bygone era. His black beard, streaked here and there with a few strands of silvery grey, was well trimmed and his eyes gleamed with good humour.

He stood in a studied pose holding a book in one hand with the other extended as though he was ready to speak ceremoniously.

'My dear ladies and gentlemen, you who have gathered here in such large numbers,' the Greek addressed an imaginary audience in a booming, theatrical voice, surveying the non-existent crowd. Realizing that there was hardly any audience at all, except for a lone listener, he cleared his throat and directed his words towards me. I watched him with fascination. From time to time, he would forget that he had only a single listener and revert to speaking to the large, imaginary gathering.

'Ah, there you are, my learned guest, you who have come from afar to listen to me and watch our little theatrical performance,' he began. 'Please allow me to introduce myself with all modesty in spite of my considerable reputation as a playwright. My name is Aristophanes, a redundant piece of information since that name, I believe, is known far and wide.' Aristophanes paused dramatically for effect and added, 'You may call me *AR* for short. Saves time and energy!'

Aristophanes the great Greek playwright! I knew about him well and had read a couple of his plays. He lived around fifth and fourth century BC and was famous as the Father of Comedy and the Prince of Ancient Comedy.

'As you perhaps already know, I hail from ancient Greece that was the source of all knowledge—history, philosophy, science, art, music—the list is endless.' Holding up his hand next to his mouth as though guarding his words from spreading out, he said quite loudly, 'A hyperbole, an exaggeration, sir, if you ask my opinion. Ah, you are puzzled by my gesture, are you not? Well, it is known as a stage whisper and I employed it for your benefit since I am an adept at stagecraft. In other words, I am a playwright or a writer of plays and I produce my plays on stage for people to watch and enjoy.'

He looked at his captive audience of one steadily.

'Perhaps our honoured guest knows that I wrote a play entitled *The Frogs*. It was about both playwrights and playwrongs, so to speak, quite hilarious if I may say so myself in all modesty.'

I had of course read the play *The Frogs* by Aristophanes, which was indeed hilarious. I also knew who the targeted 'playwrong' was—it was Euripides!

'Enough of ancient literary history,' said Aristophanes after a momentary pause. 'Now up with the curtain and on with the show.' He swept his hand across the entire firmament. As he did so, a multitude of figures gradually appeared against the constellations in the background. Naturally, there were not only men and women, but also beasts, demons, and inanimate objects.

'Well, my honoured guest, many constellations tell heroic tales involving the figures you see outlined against the star patterns,' Aristophanes went on. 'For instance, focus your esteemed attention on the wondrous pattern over there with three brilliant stars in the middle that can never be missed among their myriad companions. That star pattern is called the constellation *Orion*.

'How so the name Orion? Answer me not for I shall enlighten you regarding the origin of the name of the constellation and the tale that goes with it. *Tale* and not a *tail*, mind you, although in fact a tail does figure in the tale, as you will discover soon.' Aristophanes drew himself up to his full height preparing for his narration. After a momentary pause he went on.

'There is but little charm in telling a story just with words. After all, I am a playwright, am I not? Should I not present to you some dramatic display worthy of your erudite appreciation? Yes, we shall proceed to do so without further delay,' commented Aristophanes. 'Tonight, we shall stage one of those dramas enacted night after night on the heavenly vault. In order to supplement my narration that I have created for this special occasion, I have summoned my able minstrels, namely the chorus of my play *The Frogs*, who, with their mellifluous voices, will sing for us from time to time.'

As Aristophanes gestured with his hand, a chorus of frogs of different ages and sizes materialized. They were dressed in togas somewhat similar to the robe worn by Aristophanes. Together they bowed to me formally, but then grinned, giggled, and winked at me most unceremoniously, while some made croaking noises that were far from mellifluous as had been promised by Aristophanes.

Woes of Orion

'Now for the dramatic presentation of Orion's tale!' announced Aristophanes with a flourish. And the chorus of frogs made sounds, or noises, that were supposed to simulate the blare of trumpets and the beat of drums. As Aristophanes recounted the story, the characters involved, who had positioned themselves against the constellation, enacted their roles accordingly, punctuated by the choral rendering of the frogs. I began to enjoy this novel show immensely.

'Orion was the great celestial hunter with enormous strength,' began Aristophanes as concentrated starlight focused on the figure of Orion who struck a heroic pose. 'Regard him, if you please, holding aloft his enormous club in one hand and a lion's skin in the other as his shield.' Aristophanes added in a stage whisper, 'A lion's skin for a shield? What could it protect him from except perhaps pebbles pelted by a child?'

Then the frogs sang in unison.

Chorus:
Brekekekex ko-ax ko-ax
Brekekekex ko-ax ko-ax
Oh, oh, Orion, hunter in the sky
You squashed a mighty dragon? Could you squat a fly?
You own a club, eh? A night club you run?

Are you a bouncer? That must be fun!
Holding up a lion's skin, a lion that hardly bites
Cut it up into pieces, man, and wear them as tights
Oh, oh, Orion, hunter in the sky.
Brekekekex ko-ax ko-ax
Brekekekex ko-ax ko-ax

'Irreverent, nasty little creatures, these frogs,' Aristophanes shook his head suppressing his smile. 'The refrain—*Brekekekex ko-ax ko-ax*—is supposed to mimic the croaking of the frogs. I included it in my play *The Frogs.*'

The frogs giggled, winked, hopped around and finally settled down again.

'You must have heard the saying that brawn and brains do not go together,' Aristophanes continued his story. 'Well, that is precisely what Orion managed to prove. He boasted that he was so strong that he could kill—destroy, exterminate, annihilate—all the animals on the Earth. To whom did he boast, you may venture to ask and I shall oblige you with the answer to your unexpressed query. To Artemis of all people, I mean of all gods and goddesses. A goddess Artemis certainly was, is, and will be, and that too a goddess protecting animals! She was alarmed, if you permit an understatement, and conveyed her anxiety in proper alarmed language to her colleague in the Greek pantheon, Gaea, the Earth goddess, who too was alarmed beyond measure. What would happen if Orion in his eternal folly—for follies tend to be eternal—carried out his threat in a moment of madness? There would be no animals, insects and such left on the Earth anymore.'

Chorus
Brekekekex ko-ax ko-ax
Brekekekex ko-ax ko-ax
Oh, oh, Orion, aren't you full of brawn?
Where is your brain, man, gone with the dawn?
Why boast your head off to Artemis?
The goddess of beasts, that arty-miss
You can kill off all animals, did you say?
Wiping them off the Earth a child's play?
Do you want to live or do you want to die?
Oh, oh, Orion, hunter in the sky.
Brekekekex ko-ax ko-ax
Brekekekex ko-ax ko-ax

'To continue. In order to avoid such a calamity, Gaea sent an enormous scorpion with the specific intention of having Orion eliminated once and for all. Once and for all? How many times can one eliminate another, I ask you. Do not trouble yourself to answer, for it was but a rhetorical question. Be that as it may, the celestial hunter and his equally celestial antagonist were engaged in mortal combat. Although Orion was able to wound the scorpion with a terrible blow from his mighty club, the struggle was short lived. Despite all the strenuous efforts of Orion, the scorpion got better of him, whipped up its tail, and killed him with its deathly sting.'

After a momentary pause to let the story sink in, Aristophanes added parenthetically, 'Ah, I told you early enough that a *tail* figures in the *tale*, did I not?'

Chorus
Brekekekex ko-ax ko-ax
Brekekekex ko-ax ko-ax
Brethren of the Chorus, you see what I see?
A steel-armoured scorpion larger than a tree!
Six sturdy legs, all multiple-jointed
Enormous pincers knife-sharp pointed
The hunter and the arachnid, watch their mortal fight
There is no room here for any cowardly flight
Club comes down, pincers snap
One false move—a dire mishap
Even the great gods shudder with fright
Although safe in their Empyrean height
Oh, dear, Orion trips and slips
His body is now in the pincers' grips
The tail lashes out and stings his hips
Dead as a Dodo, splattered like a fly,
Oh, oh, Orion, hunter in the sky.
Brekekekex ko-ax ko-ax
Brekekekex ko-ax ko-ax

'Ah, so we come to the end of our story,' announced Aristophanes. 'But how can we conclude our play on a sad note? Would not the audience clamour for their money back? No, sir, it would not do. Here then comes *Ophiuchus*, the heavenly healer, stepping forth holding a serpent in his arms. Ah, that erudite serpent happened to be the mentor of Ophiuchus. The slithery scholar had taught Ophiuchus the ancient secrets of healing. From the satchel that hung by his side, Ophiuchus now took out a vial containing the extract of some magical herbs and applied it to the two antagonists—one dead and the other wounded. The two sprang up full of life at once, bowed all around, and withdrew.

'All this time, the immortal gods had been watching the mortal combat with utmost amusement, for the misery of others is their inexhaustible source of entertainment. In their infinite wisdom which we mortals cannot fathom, those gods placed the two celestial combatants, Orion and Scorpio, on the opposite ends of the sky dome as constellations, so that those two would never be seen together. That is the end of our story. My chorus and I are indeed greatly honoured by your kind attention and are indebted to you till eternity yawns and turns over in her sleep.'

I applauded vigorously as Aristophanes and the frogs bowed low repeatedly.

Aristophanes swept his hand across the celestial characters who had taken part in the drama. They had gone back to their respective constellations and were about to fade away. They too bowed in unison. The constellations had slightly moved from their original locations. But the stars making them up had maintained their relative positions as before.

'Aha, my distinguished guest has noticed that the star patterns have moved while we were staging our play,' commented Aristophanes. 'Oh, yes, stars rise and set just like the Sun and the Moon. So they have moved, but their relative positions are unaltered. In other words, the constellations have remained the same. I shall not attempt to describe these celestial phenomena any further, for I am but a mere story-teller. But, explanations shall be given, rest assured, my honoured visitor, not by me but by another eminent Greek.'

Wisdom of the West

As Aristophanes spoke, another figure had slowly materialized next to him. He too was dressed very much like Aristophanes. All ancient Greeks seemed to be similarly clothed for that matter. Tall and well built, he had dark curly hair. His keen, observant eyes focussed sharply on me.

'Be pleased to meet Aristarchus of Samos, one of the greatest astronomers, or the studier of stars, our country has produced,' proclaimed Aristophanes. 'A man younger than me by some hundred and odd years, he will relate to you the mysteries of the stars. I take leave of you, but fear not, we shall meet again. Farewell for now!'

With those words, Aristophanes slowly faded away in the same manner as he had appeared.

Now it was the turn of Aristarchus to hold forth on the starry stage. He smiled and held out his hands in a gesture of welcome.

'I greet you in the name of all the glitter, glory, and grandeur of the celestial sphere and its denizens, my friend from far off land and far off future,' said Aristarchus warmly and bowed.

I bowed in return instinctively.

'As you have been informed by my playwright predecessor, my name is Aristarchus, an avid stargazer of ancient times,' Aristarchus began. 'You may call me *AR* for short,' he added with a smile.

I was amused. Here was another *AR* of antiquity.

'Did I say the *celestial sphere*? I must explain what is meant by that term, must I not?' said Aristarchus. 'When you look up at the night sky, you feel that a dark inverted bowl, carved out of some black crystalline material, is fixed above you. Permanently stuck inside the bowl in definite patterns, or constellations, are the glittering stars. As the bowl turns about the pole, with the polestar or *Polaris* close by, the stars go round in their courses. Obviously, constellations too rise and set, since that is what the stars do. But, the relative positions of the stars making up the constellation are fixed. This is what you noticed at the end of the theatrical production of Aristophanes. We, who dwell in the northern part of the Earth, see only the hemisphere above us. And those who inhabit the southern regions of the globe see the other half of the celestial sphere with its own constellations. Perhaps, my friend, someday you will make your way to the southern hemisphere of our Earth and enjoy those unseen star-patterns.'

Yes, that would be great, I felt.

'Many Greeks before me had studied the night sky very carefully and thought about things happening up there,' said Aristarchus looking heavenwards. 'For instance, more than three hundred years before I was born, there was this great man by the name of Thales of Miletus. He watched the sky all the time. Even when walking on the streets. While doing so, it seems he fell into an open well.'

'He did not want to be rescued even then,' Aristarchus went on. 'That would have disturbed his watching the sky. He knew a lot including how the Sun seemed to move in a circle around the Earth, how long it took for the Sun to go round this way. Thales thought that all things around us were basically made up of water.'

'Aha, that is because he fell into a well with water in it,' I thought irreverently.

'Could be, could be,' Aristarchus laughed. 'But, seriously, that was an important step Thales had taken as you can very well imagine. In the centuries that followed, people have wondered whether all things are made up basically of some common ingredients. Again, my friend, you will learn all about it in good time. As for me, I shall tell you, as we go on, about some of the other great Greeks who studied the sky. But now the time has come to talk about the Sun and the Moon, the dearest companions of the Earth.'

Aristarchus paused before continuing.

'I have no doubt that you know whatever I am going to tell you, since astronomy has advanced enormously since our times,' reflected Aristarchus. 'Yet, I am confident that you will agree with me that it is important to understand the evolution of any subject over the past centuries. By simple observations, one can find out that the Moon circles the Earth in twenty-nine days, always presenting the same face to us on the Earth. Why? Is the Moon ashamed of its other side? No, because Moon's period of rotation happens to be exactly equal to its orbital period. Round it off to thirty days, and you have the month as a measure of time. Then again, the Sun takes a year comprising three hundred and thirty-five days to go around the Earth. As the Moon shines by reflecting Sun's light, we see different phases of the Moon, do we not?'

Aristarchus went on to describe the different phases of the Moon in detail. I knew it all, of course. But, it was so nice hearing it from this venerable astronomer of centuries ago.

Aristarchus fell silent. He must have been thinking of the past.

'The philosopher Protagoras wrote, *Man is the measure of all things*. What did he mean by that? He meant that there is no truth but that which individuals deem to be the truth. He got into trouble because of such a radical idea, what a pity! Time and again, people have to overcome the prejudices of those who fail to accept anything new. Yes, man is the measure of all things. But man is also the *measurer* of all things, is he not? So, long ago, I wondered how far the Moon and the Sun were from the Earth. Wondering is not enough I said to myself, we must find this out by actual measurement. And I did manage to devise a simple method to measure the distances of our two neighbours relative to the size of our own Earth.'

I knew that the so-called *simple methods* are invariably brilliant ones too. I had no doubt that this was the case with the measurements of Aristarchus.

'Well, what was my result? My standard of measurement was the size of our own Earth. And, I found out that the Moon's distance is ten times the size of the Earth and the Sun's distance about some two hundred times.'

Aristarchus stroked his beard for a while and went on.

'I am told that other stargazers, in ages after mine, have made similar measurements by different and more sophisticated methods. And they have found the distances of the Moon and the Sun to be respectively about thirty and twelve thousand times the size of the Earth. Ah, that is progress.'

'But the one who plants the seed must be given due credit for the fruits that has resulted,' I felt.

Aristarchus seemed quite pleased with my unspoken words.

'That is not all, my companion in curiosity,' continued Aristarchus. 'How big are the Sun and the Moon compared to the Earth? I wondered again. To repeat what I told you earlier, wondering is not enough, I said to myself, we must find out. And I did manage to devise a method, again simple enough, to measure the relative sizes of our two neighbours. What did I find? I discovered that our dear Earth is about three times larger than its follower, the Moon. And the Sun is seven times larger than the Earth.'

This must have been the first time such a measurement had been attempted and, I was sure, that the method must have been ingenious as well.

'As in the previous case, others who came after me have discovered that my estimate of the Moon's size is not bad at all. Our Earth is about four times larger than the Moon. But the Sun, on the other hand, is actually a little more than a hundred times the size of the Earth. The Sun is quite large compared to our Earth. I am confident that you agree with me.'

Oh yes, I agreed readily.

Aristarchus fell silent again for a while, musing about his findings no doubt. Then he spoke up, his eyes shining. 'Yes indeed, even seven times the size of the

Earth, as I believed, was quite large for the Sun, if you ask my opinion. That set me thinking. Of any two bodies, the heavier one always tends to be sluggish and stays at the same place, while the lighter one could run around more easily. Then the Sun must be at the centre with the Earth going around it. And not the other way round as everyone thought. Does it not make sense to you?'

'Oh, yes, it does, it does,' I wanted to say.

'I wish my friends were like you. But, nobody took me seriously,' Aristarchus seemed to be a little sad. 'Everyone continued to believe that the Earth was at the centre. Only after two thousand years my idea was proved to be correct.'

'By the way, we know that the Sun is far enough from us. But, I knew that the stars were much, much farther. Our story does not end there. In fact, it is just the beginning,' Aristarchus pointed his finger at the sky. 'Look at the night sky. Apart from the Sun and the Moon, there are five other heavenly bodies that are very special.'

Five objects brightened up to show their locations. Unlike the stars, they did not twinkle.

'They move against the backdrop of the fixed stars,' explained Aristarchus. 'They are the *wanderers* or the *planets* as they are called. They rise and set like all other objects in the sky. Furthermore, they too appear to go around the Earth at different speeds completing their full circles in different periods of time.'

The planets started moving majestically against the backdrop of the stars in the night sky.

'As you know, those planets bear the names Mercury, Venus, Mars, Jupiter, and Saturn. Apart from these planets, strange heavenly bodies with tails visit us from time to time, the *comets* as they are called.'

Aristarchus thought for a moment and then went on. 'In addition to rising and setting, the heavenly bodies—the planets, the Sun and the Moon—go round the Earth, do they not? They follow their own *orbits*. The great thinker Plato said that the path of each of these bodies, or its orbit, must be perfect. Why? That is because they are heavenly entities and do not belong to the imperfect Earth. A circle is the perfect curve, each point on it being at the same distance from the centre. Therefore, the orbits of these celestial objects had to be circles. Then again, being perfect, each of the heavenly bodies was supposed to move with constant speed around the Earth, never faster never slower. That is what Plato said and, of course, everyone believed it without doubt.'

Aristarchus nodded and stroked his beard a few times.

'But, you may ask how these vagabonds of the skies—for that matter the Sun and the Moon as well—are supported in their eternal journey,' continued Aristarchus. 'Ah, that is a good question, my learned friend, even though you did not ask me. Yes, those celestial travellers cannot just hang in empty space, can they? So, what is the answer to your unspoken query? The answer came from the great Master, the Master of Masters, Pythagoras, who lived about two hundred years before my time and was, like me, from Samos, my beloved hometown. I am sure you have come across the work of Pythagoras some time or the other. He studied so much, asked so many questions, found out so many answers! Pythagoras was fond of numbers and asserted that everything we know is based on numbers. He was deeply interested in music and related it to numbers.'

Aristarchus pondered for a moment or two before proceeding.

'Coming back to the planets, according to Pythagoras the answer to the question regarding their support was that each one of them was fixed in a crystal sphere.'

Aristarchus moved his hand in a circle and opened his fist. As if by magic, a beautiful piece of material in a perfect geometric form lay in his palm, scintillating with the reflections of the surrounding stars.

'Here is an example of such a crystal,' demonstrated Aristarchus. 'Once again, just as the circle is the perfect curve, the sphere has the perfect surface with all points on it being at the same distance from the centre. But the crystal sphere in which a planet is fixed is much more beautiful than the crystal I have here, so exquisite that it is transparent and you cannot see it at all! And the spheres turn carrying the Sun, the Moon, and the planets along. That is not all. I told you that Pythagoras, the Master, knew all about music too, did I not? Yes, he said that as the crystal spheres turned, they made beautiful music. Each sphere produced a different pleasing sound. These different sounds did not clash with one another. They did not fight but were at peace. This was *celestial harmony*, the *music of the spheres*.'

'Yes, the music of the spheres imbued with celestial harmony was indeed divine. Just as you could not see the transparent crystal spheres, you could not hear the music made by them either. You had to be very, very special for that. Only the Master was that special. Only the Master could hear the music of the spheres, that is what people said,' explained Aristarchus. And added in a whisper, 'The Master never denied what people believed!'

That must have been quite disappointing to the ordinary inhabitants of ancient Greece I surmised.

'How do we know that Pythagoras was right, although no one doubted his word?' asked Aristarchus. 'Let me tell you how. There was this fantastic scholar who was born some ninety years before me. He thought of practically everything— how objects fall, how fire rises, how wind blows, how plants grow, and so on. He thought deeply and extensively and that is how he discovered so many things. His name was Aristotle. Well, we could say he was the greatest *AR* according to people for a long, long time.'

Oh, here comes yet another *AR* I thought. How is it that so many names began with those letters!

'This wise man too agreed with Pythagoras that the Universe is made up of these crystal spheres carrying the planets,' said Aristarchus. 'What about the stars then, you may ask. Yes, it was agreed that they too were stuck in an immense crystal sphere that was beyond all the others and they too turned.'

Aristarchus was silent for a while before continuing.

'Yes, they were both great men no doubt, Pythagoras and Aristotle,' reflected Aristarchus meditatively. 'But should we take their word for everything without examining their ideas carefully? As I told you earlier, I wondered about the possibility of the bulky Sun relaxing at the centre. Perhaps, the Earth itself turned about an axis so that all other bodies we see in the sky seemed to travel as they do, rising and setting. Again, as I said before, it was a long, long wait, my friend, before all this became clear.'

All of a sudden, an ancient Greek similar in looks to Aristarchus streaked totally naked shouting some word in Greek repeatedly.

'There goes the craziest *AR* in Greece!' exclaimed Aristarchus hitting his forehead with his open palm. 'That Archimedes is running stark naked straight from his bathtub shouting *Eureka* again and again. The word means *I have found it*. What has he found out? Maybe that the bathwater is too hot!' He laughed heartily before adding, 'Actually, Archimedes is a brilliant scientist. Still, I hope he covers himself up before he gets into trouble with the Greek police.'

After this momentary interlude of mirth, Aristarchus fell silent as he seemed to be absorbed in his own thoughts. I too waited quietly turning around in my mind all that I had heard from him.

Wisdom of the East

Indian Continuation

'*Bho, maha yavana khagola vigyani, vandanam!*'

I was startled by these incomprehensible words of an unknown tongue intoned by a soft voice in an almost musical cadence. Aristarchus too looked up sharply in surprise. Both of us realized that a strange figure had appeared while we were absorbed in our own thoughts. He was wearing a pure white length of cloth, fastened around his waist and folded in pleats reaching to his knees covering each leg separately. Another length of equally white cloth covered his torso loosely. The colour of his body—a lustrous shade of dark brown—stood out in contrast to his spotless white clothes. His chest and shoulders were powerfully built. Long wavy hair, dark as night without stars, cascaded down to his shoulders. He smiled at us rather with his eyes than with his lips.

I knew that there was an unseen automatic translation mechanism that converted all languages into my own. Probably it was having a problem converting an exotic language simultaneously into two languages, namely my own and ancient Greek. Ah, I intuitively realized that the problem had been fixed. Instantly, the words we had missed were repeated again.

'O, great Greek astronomer, my salutations to you,' the stranger said politely. He spoke like a poet in a flowery style. 'Yes, I speak the most ancient tongue, namely *Sanskrit*, the unparalleled divine language of my glorious country *Bharathavarsh* or India as it has come to be called in modern times.'

As the stranger spoke, an ethereal backdrop of a magnificent temple with beautiful carvings on its façade appeared symbolising his culture. I fondly remembered the fantastic temples I had seen during my visit to that fabulous country.

'With your permission, O my esteemed friends, kindly allow me to inform you of my name and calling,' said the stranger with folded hands, his torso slightly bent forward in an attitude of respect. 'I too am an astronomer, a devout stargazer like

you, my great Greek companion. I answer to the humble name of Aryabhata.' Then
he added with a smile, '*AR* for short!'

Oh, no, no, not another *AR*, I thought. At this rate, Astronomy will have to be
called *AR*stronomy! Well, I was eager to hear what this new *AR* had to say.

'O my Greek brother of the celestial lore, during my lifetime on this Earth, seven
to eight hundred years after your departure from your earthly abode, I too pondered
the mysteries inherent to the motion of the stars and the planets,' said Aryabhata.
'And, just as you speculated that the Sun, being ponderous, rested at the centre,
while our Earth, light and lively as it is, circumambulated the former, I too came to
the conclusion that the phenomenon of the rising and the setting of the celestial
objects was only an appearance created by the rotation of the Earth itself, which we
inhabit and hence have made our seat of observation of the heavenly activities.'

It took some initial effort on my part to follow the long, winding structure of
Aryabhata's sentences. But soon, I was enjoying the style of his speech.

'I recorded all my findings in verse, in the form of poetry,' Aryabhata went
on. 'After all, the firmament is filled with poetry, is it not? Stars and planets are the
letters, their patterns are the words, and together they sing the eternal song that
follows the rhythm of their movement. I wrote in Sanskrit, but the verse in
translation reads thus:

As from a sailing boat one sees
The backward glide of the line of trees
So do the heavens appear to be churning,
Whereas, in truth, it is the Earth that is turning.'

'Oh, that is beautiful, what a lovely analogy!' exclaimed Aristarchus as he
clapped his hands in delight. 'And above all, it is a brilliant idea that explains
how we see the motion of the celestial bodies because of our own rotation.'

Aryabhata bowed low with folded hands in great humility acknowledging the
admiration expressed by Aristarchus.

'Just as in your case, illustrious seeker of Truth from the West, no one cared for
my novel idea,' sighed Aryabahata. 'Even some of the great scholars who came
after me criticized me bitterly.'

'Oh yes, my friend and brother, your people have given this world so much
knowledge,' continued Aryabhata. 'So have we, we from the land of the sacred
river *Ganga*, from times immemorial. Your Master, Pythagoras, offered so much
learning. He unravelled the mysteries of numbers. It is well known through the ages
that he discovered a most profound truth about certain type of *tricona*, the triangle.
It does not diminish his greatness, if I were to state the fact that our people had
known this truth termed as the *Pythagorean Theorem* earlier. Even before your
Master graced our Earth, two of our own scholars, *Apasthambha* and *Bhaudayana*,
made use of the theorem repeatedly in their tract entitled *Sulvasutra*, or the *Rules of
the String*, which was used in the construction of altars for our sacred rituals. Ah,
then again, the numbers, those glorious entities that gods themselves love! Forget
not, O scholar from the West, *we* were the ones who presented to the world of

learning the supreme concept of *shunya* or *zero* that wholly transformed the mode of counting.'

Aristarchus nodded, acknowledging this fact.

'Zero used to be represented by a *bindu*, a mere point. It used to get lost among other numbers. It so happens that I, the humblest among the humble, introduced the small circle as the symbol for zero,' added Aryabhata in a low voice with shy hesitancy.

'Your people I am told, O seeker of Truth, knew only the numbers up to ten thousand, one followed by four zeroes,' continued Aryabhata. 'On the other hand, our ancestors had envisaged numbers up to one followed by fifty-three zeroes, each of the important numbers among them bearing a particular name, which even the great Buddha, who had enlightened this world before your Master appeared on it, knew well. Only the invention of zero could have brought to fore such large numbers, all of which have yet to be used, but will be used one day, in the realm of the stars.'

Aryabhata paused before adding, 'And we did so much in developing other branches of mathematics, *rekhaganitha* or geometry and *beejaganitha* or algebra.

Aristarchus seemed to be quite impressed with Aryabhata's narration of the ancient Indian contributions and bowed his head honouring the greatness of all this achievement.

'Just as those who preceded me had discovered marvellous truths, so did many who came after me,' said Aryabhata. 'And those wonderful findings flowed west-wards just as the sunlight, emanating first in the east, travels towards the west. But, you will agree with me, my friend, if I proclaim that light is light and it dispels darkness wherever it might have originated from, east or west.'

'I agree with you absolutely, my illustrious neighbour,' said a deep, resonant voice, again surprising me as well as the two astronomers engaged in their erudite dialogue.

The newcomer who had appeared on the scene was powerfully built like Aryabhata. But unlike the latter, he wore a loose white robe, a turban covering his head. His tanned skin was leathery having apparently withstood rough weather. He had a dark beard and his perfect white teeth flashed like lightning within a dark cloud whenever he spoke. The backdrop for him was a single tent in the midst of a desert, its golden sands stretching far and wide reaching the horizon in all directions.

'*Aghayoone Mohtaram*, O Revered Sirs, I too am a keen observer of the stars,' said the stranger bowing deeply. 'I am from Persia and my name is Omar Khayyam. Kindly call me Omar, which is quite endearing.'

No *AR* this time, I sighed. What a relief!

'Allow me to point out something you have inadvertently missed, my friend,' said Omar smiling at me with amusement. 'My name Omar ends with *AR*, does it not? In our country we read and write from right to left. So whereas the *AR* appears at the end of my name according to you, it is at the beginning by our convention. So, I too am an *AR* as every stargazer ought to be.'

All three of us, including the distinguished astronomers from ancient Greece and India, were delighted by this strange revelation.

'As I was saying, my esteemed friends, knowledge flowed from Hindustan, or India as it is called nowadays, to the west by way of our country, or *Arabia* as our region is popularly known. Yes, it was just as in the case of silk and spices, but immeasurably more precious than all such commodities,' said Omar. Looking straight at Aristarchus he added, 'So much knowledge you and your brethren had gathered over the ages but, alas, to be ignored by many of your own countrymen who had no appreciation for it. On the other hand, we kept it safely with us, did we not? We translated it into our own language, enriched it, and then transmitted it back to the west when the time came.'

Aristarchus bowed, as before, in acknowledgement of this fact.

'Yes, we preserved knowledge gathered from both east and west, like wine in their casks improving with age,' remarked Omar. 'But we added our own ingredients to improve it. Our people held the profession of stargazing in highest esteem. The greatest Arabian astronomer, Mohammad ibn Jâbir ibn Sinân Abu-'Abdallâh al-Battâni, or al-Battâni for short, declared that the science of the stars comes immediately after religion, as it is the noblest and the most perfect of all the sciences, adorning the mind and sharpening the intellect, and that it tends to recognize God's oneness and the highest divine wisdom and power. We, the Arabs, charted the stars and followed the movements of the planets. We gave names to many of the stars in the sky, names that have remained in vogue even up to the modern times long after our own days.'

Omar Khayyam regarded meditatively the lone tent in the background planted in the endless desert sand, swaying in the wind. Once again, I was reminded of the breathtaking spectacle of the fiery deserts I had seen in my recent wanderings.

'I was but a humble tent maker once. But then, I wanted to stitch the tent of science and turned my eyes towards the stars. As many of my brethren did, I too was deeply interested in the reckoning of time and worked on improving the methods of keeping count of the days and hours in a year. I drank deep from the wells of nature and was richly rewarded for it.'

'Omar Khayyam is a modest man. In his time, he was hailed as the King of the Wise,' proclaimed Aryabhata.

Omar Khayyam spoke again. 'Amongst the many phenomena we studied, we the Arabs and our neighbours from Hindustan, eclipses were of paramount interest.'

'Yes, we assiduously studied eclipses—those most fascinating celestial occurrences, which turn out to be the most frightening experience for some uninitiated people,' concurred Aryabhata. 'Every culture has some story, some myth that tries to explain the event. But we know the true cause of the eclipse, do we not? When the Earth's shadow falls on the Moon, that celestial body is eclipsed, while the eclipse of the Sun occurs as a result of the Moon blocking its light from reaching the Earth.'

'Myths, stories, they are interesting no doubt, perhaps even more than reality. But let us not forget poetry,' mused Omar Khayyam as his eyes became distant and dreamy. 'Ah, I came to be known to the world as a poet and not as one who pursued the noble science of the stars. In my *Rubaiyat*, I sang of man and God, of fate and destiny, of learning and wisdom, and, above all, of wine, the symbol of love for the Divine. I did not write my science in poetry unlike my dear friend and esteemed neighbour here. But the poetry of the stars always filled my mind. After all, just as our friend from Hindustan described, there is sublime poetry written on the firmament by the stars and the planets.'

Oblivious of the presence of his listeners, Omar Khayyam recited to himself some lines he had improvised spontaneously.

Ah, my love, raise your glass of wine
To capture the ruby-red shine
 Of the Sun, turning into a dying ember
 As darkness descends like peaceful slumber
Bonding forever your heart with mine.

Then again, Omar Khayyam sang another verse he had composed perhaps inspired by this rare confluence of three different streams of ancient cultures. We listened in rapturous fascination.

Here on the endless sands of time I lie
Watching the black velvet of the sky
 Would I not give my entire soul
 To gaze upon night's inverted bowl
As the celestial fireflies traverse by.

dervishe

A hush had fallen on us listeners as the soft sounds of Khayyam's recitation filled the space around us. Even the stars seemed to be listening in silence.

All of a sudden, seven men, who resembled Omar Khayyam in their features, appeared on the scene. Like Khayyam, they wore long robes and turbans, but of different colours and textures. While whirling vigorously, they started circling around Khayyam at different speeds. The one closest to Khayyam, the man who was smallest in size among them, wore a robe that glowed softly. He whirled and circled in such a way as to face Khayyam all the time. The second one's clothes were dazzlingly bright. Other robes were tinged with different colours. For instance, the dress of the last but one among the men, who was also the largest among them, was painted with bands of different colours. And the last one had a circular piece of clothe stitched to his waist that rose up and formed a ring when he spun around.

'They are holy men, *dervishes* as they are called,' explained Omar Khayyam. 'They represent the Moon, the Sun, and the other planets. They are from the future and, consequently, their robes indicate the aspects of the celestial bodies that were destined to be discovered much later than my own time.'

As I watched this fantastic spectacle, all the characters in front of me slowly receded, presumably towards their own countries, growing smaller and smaller in appearance and finally disappeared altogether.

I gazed at the stars and the star patterns. The planets that were travelling across the night sky were not aimless wanderers after all. Were they not moving in circles, the most perfect paths they could follow as Aristarchus had described?

I gazed at the stars with concentration and listened intently. Did I see the faint glimmer of those lovely crystal spheres? Did I hear the faint harmony wafting from them? I wondered.

'It is all simply beautiful,' I whispered to myself. 'And beautifully simple too!'

The Luminous Loops

I heard the sound of someone clearing his throat and saying, 'Pardon me. To tell you the truth, it is not really all that simple, I mean the motion of those crafty planets.'

The sound had preceded its source, which slowly materialized, namely another person with white beard who wore an orange striped robe and a red skull cap.

'Ah, I happen to be in the same profession as Aristarchus, studying the stars I mean,' announced the newcomer. 'He lived some four hundred and odd years before me. I must tell you my name, must I not, before proceeding any further? It is Claudius Ptolemaeus, or Ptolemy of Alexandria as I am commonly known. As I was saying, the motion of the planets is not all that simple, as you have been led to believe. Watch!'

One of the planets that was moving smoothly seemed to hesitate, slow down, come to a momentary halt, proceed backwards, and then turn around, and finally move forward as before. Yes, it did move in a complicated manner making loops along its orbit. This was true of the other planets as well. I was really bewildered and shook my head in disbelief.

'Oh, yes, it *is* baffling, I must say,' admitted Ptolemy. 'How does one explain this most peculiar behaviour of the planets, their *retrograde motion* as it is called? How does one reproduce their motion using purely circles? Why only circles? As Aristarchus told you, the venerable philosopher Plato had declared that all heavenly motions must be circular with constant speed. Why? Because, the circle is the perfect figure in geometry and therefore in nature. So then, the peculiar paths of the heavenly planets have to be mimicked by means of circles and nothing but circles. We have a down-to-earth problem on hand, do we not? That means we are obliged to go down to the Earth.'

Amusement Park

Slowly, gently, Ptolemy and I descended to the Earth. I could see beneath me a scene of confused activity, which became progressively clear as we approached the ground. There was a big crowd made up of many families. Parents were carrying little children, while older ones ran around excitedly with their senior siblings laughing and chasing them.

'Well, what do you know, we have landed in an amusement park,' exclaimed Ptolemy. 'It is wonderful to see so many happy people at one place, is it not?'

It was indeed a happy place with many events taking place all around us.

There was a little crowd gathered around a man in pink-striped clothes and a long cap. He was throwing three different coloured balls into the air continuously and catching them in turn.

Ah, that was the juggler, wonderfully skilful at his act.

'Yes, indeed,' agreed Ptolemy and added with a sly smile, 'In our profession, some of us juggle facts to suit the hypothesis. Ah, are you confused? Forget what I said, it is of no importance.'

A young girl was selling colourful trinkets made of glass. As a child, I used to love such trinkets, especially the marbles like the ones she had. Although the marbles were perfectly spherical and smooth, one could see varying textures within them.

We heard squeals of laughter coming from an adjoining enclosure. There was a tightly stretched trampoline made of some highly elastic material on which some children were jumping up and down. Whenever they went up high enough they screamed and yelled with pleasure. In a corner sat a merry toddler in the middle of the dimple his weight had created in the trampoline. He would throw a ball making it roll down the dimple straight towards the centre. Or if the child flicked the ball side-wards, it would circle round first and then spiral down to the middle.

'Look, look, there is the balloon man,' Ptolemy pointed excitedly. The ancient Greek seemed to be enjoying himself amidst the modern amusements.

A few yards from the trampoline stood a man with a bag full of balloons with dots of many different colours printed on them. As he inflated the balloons, the coloured dots moved away from one another. It was a pretty sight.

There were many other interesting attractions. But, Ptolemy and I moved on. Looming before us was a huge, upright, slowly revolving wheel—the Ferris wheel. Seats were dangling all along its rim. Passengers in each seat, mostly a couple or two children, were safely strapped in. As the big wheel went up higher and higher, the children squealed with excitement.

'Wonderful, we can now illustrate how the planets move along their paths in the sky,' said Ptolemy rubbing his hands with glee. 'Take for instance one of the points on the rim of the big wheel going round in a perfect circle. It is like one of the planets going round the Earth, Earth being thought of as located near the centre of the wheel. At least that is what one believed to begin with. Let us get an idea of how the planet might have felt moving in such a path. Let us get on to the wheel then.'

We jumped on to one of the seats at the ground level while the wheel turned. I had not taken a ride in an amusement park in ages. This was so exciting! Slowly, we rose. Up, up, and up we went towards the sky. 'Wheeeee!' shouted the children around us in exhilaration.

'Smooth sailing, is it not?' asked Ptolemy. 'This is how a planet was supposed to be moving. But wait, let me show you how it actually travels in its loopy path.'

Till now, our seat had been just hanging down from the rim of the big wheel. Now all of a sudden, it started rotating fast around the pivot from which it was suspended. However, the pivot itself continued to revolve as before slowly with the big wheel.

'Ooh, this is wild,' I stifled a scream reliving my childhood experience—excited, frightened, and rapturous—all at the same time. 'Everything has turned topsy, turvy now; ah, it is all right again; oh, no, upside-down again...' I could not even look at Ptolemy who was also probably in the same state as I was for all I knew.

After a while, the seat stopped rotating and I heaved a sigh of relief. Is this how the planets felt while travelling around the Earth, I wondered.

'Thanks to the heavenly decree, planets, being inanimate, do not feel anything,' remarked Ptolemy. And then he added with a shrug, 'Although some thinkers believed that they are in fact endowed with life.'

We got off our seats when we reached the ground, but the wheel moved on.

'Now, let us see how the planet's motion looks like to an observer on the Earth,' said Ptolemy. Pointing at the top of the wheel, he added, 'Concentrate on the blue light placed on the empty seat over there.'

As if by magic, a bright blue lamp had appeared on one of the seats that happened to be empty. At once, the arm by which the seat was hanging began rotating. To my amazement, the light seemed to be moving not in a perfect circle but in a path made up of loops just like the planet I had observed in the sky. In fact, I could see this clearly since a blue streak was being temporarily left behind in the lamp's wake.

Fantastic, I thought.

'Yes indeed,' affirmed Ptolemy. 'By adjusting the size and the speed of the big wheel, or the *deferent* as it is called, and that of the smaller circle, namely the *epicycle*, the motion of any of the planets can be reproduced.'

This was indeed an extraordinarily ingenious mechanism that had been invented to mimic planetary motion!

'Now you know how the loopy paths of the planets can be recreated using epicycles,' said Ptolemy. A smile of satisfaction spread over his face as he regarded the giant wheel turning in the background.

'Well, my time has come,' announced Ptolemy after a while. 'I must bid you farewell, for you are destined to learn more about the planets and the stars. It will be a long journey, my friend, but believe me, an extraordinarily exciting one.'

Gradually, light faded and so did Ptolemy, smiling and waving goodbye. I was once again surrounded by countless stars glittering in the night sky.

I opened my eyes that were still heavy with the dreamy visions that I had seen. The bath water had become lukewarm. Most of the bubbles had evaporated too. The remaining ones disappeared as did the scintillating specks they harboured, just as stars melt away with the first light of dawn.

The whole experience, which I had not had in a long time, was inducing in me a feeling of intense exhilaration. But all good things have to come to an end, as the saying goes. Well, I slowly got up, dried myself, went to bed and soon fell into a dreamless sleep.

Chapter 3
The Founders and the Foundations

I was to meet George in his office. As always, it was a pleasure to walk across the university campus. The Sun was playing gently upon the trees and shrubs of the lovely little park where George and I had some wonderful moments discussing the works of great scientists like Isaac Newton a few years back. The central part of the campus, *The Quad*, was quiet with very few students hanging around, since most of them were attending their classes. In the evenings, this place would be transformed, buzzing with activities that ranged from throwing Frisbees to playing impromptu music and staging skits. I made my way to the old physics building where George has his office. I knocked on the heavy wooden door, turned the polished brass knob, and entered.

George's office had hardly changed since my last visit despite the passage of a few years in time. His huge ancient wooden desk where he worked, two chairs in front meant for visitors but piled up with journals, the adjoining table overflowing with typed and photocopied articles, notes and what not, mostly 'what not'— *organized disorder* as George called it—it was all there. As usual, some equations had been scribbled on the blackboard, but George was not staring at them. I remembered clearly, as if it were just yesterday, what George had told me about staring at equations: *You have to give the third degree to mathematical equations, you know. Stare long and hard at them. Sooner or later, they will break down and confess their secrets to you.* But now, George was seated at one of the comfortable chairs arranged around a small coffee table, which was always kept comparatively tidy. He seemed quite relaxed, thumbing through one of the massive volumes lying on the table. Next to those volumes was also the manuscript of George's book on the Universe. George motioned me to the chair next to his.

'How come you are not giving the third degree to the equations today?' I smiled shaking hands with George.

'Of course, I did, Alfie,' chuckled George. 'They have already confessed. That's why they look so subdued and sad.'

'Let me tell you something, George. Maybe you know it already. Writers have attributed to the English philosopher Francis Bacon of the sixteenth and seventeenth century the idea of having Nature on the rack and torturing her to reveal her secrets.'

'Aha, what do you know? Great minds run in the same channel!'

'On the other hand, I heard an Indian scholar quoting from an ancient text of his country that Nature is like a shy maiden. If you try to look at her closely, she may hide and disappear altogether.'

'That is lovely, Alfie,' nodded George. 'Profound and true in many ways too.'

'All right, so much for philosophy,' I said. 'I wanted to say something about the manuscript of your book on the Universe. No attempt at flattery, George, but it is enormously interesting. I just finished reading about the constellations, and the ancient astronomers both from the east and the west. Engrossing! I wish you had written more about those times, it is all so interesting.'

'Well, when you have to compress the history of the Universe spanning some thirteen billion years into a mere two hundred and odd pages, you have to economize on the contents, Alfie. So, we have to move on to the next era.'

George indicated the volumes lying on the table. He held out the one he had been leafing through with obvious reverence. It bore the title *De revolutionibus celestium orbium*.

'Here is the book that changed man's worldview completely,' announced George. 'If I remember right, we had discussed the Copernican revolution in some detail long ago when we talked about black holes, hadn't we Alfie?'

'Right you are,' I acknowledged.

'Before Copernicus appeared on the scene, perhaps the last word regarding the geocentric model of the solar system had been written by Ptolemy. He described his epicycles in his book *Almagest*. Ingenious explanation for retrograde motion of the planets, but much too complicated. You must have read it all in the manuscript I gave you. Here is the relevant drawing anyway.'

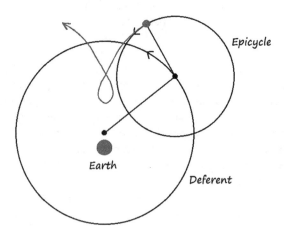

'Yes, indeed,' I said. I almost told him how exciting it was actually to visualize the epicycles through the Ferris wheel in the amusement park. That too guided by Ptolemy no less, who had invented the epicycles. Luckily I checked myself in time. Nevertheless, I explained to George how the Ferris wheel could reproduce the retrograde motion of planets.

'That is absolutely fantastic, Alfie,' George was quite impressed. 'I must include this analogy in the manuscript. Did you make it up?'

'Well, I wish I had thought of it, George,' I admitted. 'You can find it in Arthur Koestler's book, *Sleepwalkers*.'

'Long ago, I told you about the remark that had been made by my illustrious namesake, who was probably my ancestor too,' I reminded George.

'Your illustrious namesake and ancestor? Who was that?' George was puzzled.

'You know that my name Alfie stands for Alfonso L. Sabio. My illustrious namesake was Alfonso el Sabio, Alfonso the Wise, the thirteenth-century King of Castile and Leon in Spain. When he was shown Ptolemy's complex model of the planetary system replete with epicycles and all, he had commented: *Had I been present at the Creation, I would have recommended something simpler to the Lord Almighty!*'

George laughed long and loud. 'King Alfonso was immensely wise and perfectly right, although I very much doubt that he was your ancestor. Probably, Lord Almighty read his mind and must have said: *Let Copernicus be, and all was light*! The Sun at the centre illuminated all the planets.'

'At the same time, Copernicus had achieved great simplicity, hadn't he?' I added. 'I always remember the beautiful saying in Latin. *Simplex sigillum veri*: The simple is the seal of the true; *Pulchritudo splendor veritatis*: Beauty is the splendour of truth.'

'Oh, yes, that is what Copernicus had discovered. Simplicity, beauty, and truth. Let me say a few words about Copernicus and his *De Revolutionibus*, which is considered to be one of the most unreadable books of all time. You can read the details in the manuscript I have given you.'

I noticed something very characteristic of George. He would never say, *my book on the Universe*, but only *the manuscript I have given you*. It was not false modesty, but genuine lack of pride and possessiveness.

'As you know, Copernicus belonged to the Church, he held the position of a Canon with quite a bit of administrative duties,' George went on. 'During his education at Padua, Bologna, and Ferrara, he had studied a number of subjects including astronomy, medicine, and law.'

'Oh, Padua,' I recalled. 'You know, George, I visited that lovely city during my wanderings. The Senate Hall of the University is a veritable place of intellectual pilgrimage. They have covered the walls with plaques bearing the names of the distinguished scholars who had studied at the University: Copernicus, William Harvey, who discovered blood-circulation, so on and so forth. Also, they have preserved the podium from which Galileo gave his lectures.'

'Please Alfie, stop making me jealous of you, telling me all these fantastic adventures of yours,' remonstrated George. 'Well, coming back to Copernicus, he had to do quite a bit of administration as the Canon at the Cathedral in the city of Frauenburg or Frombork, its original Polish name. He had to manage finances, which he wrote about. In addition, he had to take care of the protection of the city from the marauding bands of Teutonic Knights, a rapacious and unruly bunch. Isolating himself in the tower adjoining the Cathedral, he worked away on his astronomical observations using the instruments of his own make. And, of course, he formulated his heliocentric model.'

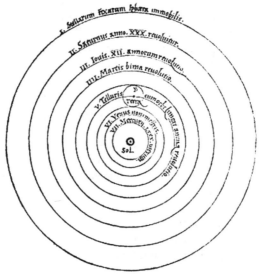

NICOLAI   COPERNICI

net,in quo terram cum orbe lunari tanquam epicyclo contineri diximus. Quinto loco Venus nono menſe reducitur.; Sextum deniq̃ locum Mercurius tenet,octuaginta dierum ſpacio circū currens,In medio uero omnium reſidet Sol, Quis enim in hoc

pulcherrimo templolampadem hanc in alio uel meliori loco po neret,quàm unde totum ſimul poſsit illuminare? Siquidem non inepte quidam lucernam mundi,alij mentem, alij rectorem uo‑ cant. Trimegiſtus uiſibilem Deum,Sophoclis Electra intuentē omnia. Ita profecto tanquam in ſolio re gali Sol reſidens circum agentem gubernat Aſtrorum familiam. Tellus quoq̃ minime fraudatur lunari miniſterio , ſed ut Ariſtoteles de animalibus ait,maximā Luna cū terra cognationē habet.Concipit interea à Soleterra , & impregnatur annuo partu. Inuenimus igitur ſub
                                                                            hac

'Well, we know that Ptolemy invented his ingenious epicycles to account for the apparent retrograde motions of the planets. How did the heliocentric model solve that problem?'

'Ah, that is simplicity itself. Let me explain. Consider the Earth and the planet Mars, for instance. Being farther away from the Sun, Mars revolves at a slower speed than the Earth. As I move along sitting on the Earth, I do not feel my own motion, do I? Let us say that at some point of time I see Mars ahead of me. As I overtake that planet, I find that it is now following me. This is possible only if that planet turned round from being in front of me and went behind me, reversing its motion. A leader has now become a follower! Obviously, the planet must have followed a loop in its course. Of course, this is all apparent motion. The number of loops contained in the orbit of a planet depends on the speed of the planet relative to the Earth. Absolutely simple, isn't it?'

I fully agreed. 'Even in politics, one has to make a u-turn to change from being a leader to becoming a follower and *vice versa*,' I added.

'Indeed, but fortunately planets don't have any political ambitions,' laughed George. 'Well, you can convince yourself that what I told you is true in the case of the inner planets too, Mercury and Venus, which revolve faster than the Earth. Let me show you the diagram that illustrates retrograde motion of a planet as seen from the Earth. It is within the frame work of the heliocentric model.'

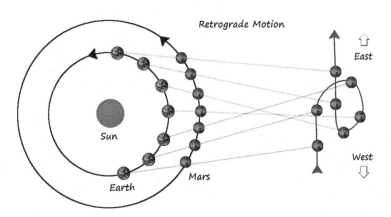

George contemplated the tome in front of him, *De Revolutionibus.*

'After all these centuries since Copernicus, we take his heliocentric model for granted. We are so used to it,' reflected George. 'But tell me, do we really see or feel the Earth and the planets going around the Sun? To give up the age old, universally accepted geocentric model takes courage. To think of the ponderous Earth spinning and moving around the Sun takes imagination. It was a phenomenal leap from appearance to reality!'

'Yes, that reminds me of how Galileo described it in his book, *Dialogues on the Two Chief World Systems, Ptolemaic and Copernican.* Salviati, one of the three characters engaged in the dialogues, who represents Galileo, says,

There is no limit to my astonishment when I reflect that Aristarchus and Copernicus were able to make reason conquer sense, that in defiance to the latter, the former became the mistress of their belief.'

'Beautiful, Alfie, simply beautiful,' said George with feeling. 'Let me quote for you another wonderful passage from one of the most distinguished astronomers of our own times, Sir Fred Hoyle, from his essay on the life and work of Copernicus.'

George read out from his manuscript.

'...The rest of the story is quickly told. The book appeared in March 1543, but a copy did not reach Copernicus until May. He was then on his deathbed, and his actual death occurred within hours of the book's arrival.

To the citizens of Frauenberg it must have seemed as if an infirm old man of some past distinction had passed from their midst. They were not to know that drums were already rolling with the noise of distant thunder. His fellow-canons, as they laid him in his grave, were not to know that Copernicus had detonated an overwhelming explosion of human knowledge, an explosion still with us in our own day, and whose eventual outcome we cannot yet foresee.'

It was indeed a profound and lyrical tribute to Copernicus, quite moving at the same time. 'George, what does Hoyle mean by his statement, *an explosion still with us in our own day, and whose eventual outcome we cannot yet foresee?*' I enquired.

'Alfie, Copernicus had shown that the Earth is not special within the solar system, hadn't he? As we shall see, our solar system is nothing exceptional within the Milky Way. Then again, our Milky Way is just an ordinary member belonging to the realm of the galaxies. And all galaxies are equal. This is known as the *Copernican Principle* based on which cosmological models are built. On the other hand, we have seen only a part of the Universe as it has been in the past. Is this Principle then really valid all over space and at all times? We do not know. And that is the *eventual outcome we cannot yet foresee.*'

As it happens often enough, George sat back and closed his eyes. Perhaps he was far away in the past or far away in the future. I waited patiently, as before, for him to return to the present.

'We meet Kepler then,' said George opening his eyes slowly. 'You know, while Copernicus actually became a Canon in the Church, Kepler too wanted to study theology and become a Lutheran minister. But because of his limited financial means, he was forced to learn and teach astronomy for a living. Then he switched to astronomy altogether. Religion's loss was astronomy's gain. And what a gain!'

'Well, I have read Kepler's life in some detail,' I said. 'Such a dismal beginning to such a magnificent career! Abject poverty, a big family run by an uncaring mother who was often absent from home chasing a drunken mercenary of a husband who beat her on the rare occasions he visited the family. Then again, Kepler was an unimaginable hypochondriac who suffered from all sorts of maladies—both real and imagined. Poor man, it seems he tried everything he could think of to cure himself. That is not all. He was a perpetual exile, moving from place to place, holding petty jobs that paid pittance.'

'Amidst all this, the man goes ahead and discovers his celebrated three laws of planetary motion,' remarked George. 'It was a tremendous job, beginning with the accurate observations of Tycho Brahe, his master. Well, Brahe is considered to be one of the greatest observational astronomers of all time. He made extremely accurate and meticulous recordings of the sky. Also, Tycho recorded the occurrence of a *nova*, appearance of a new star—actually the flaring up of a star—in the constellation Cassiopeia. So, the celestial realm was not immutable after all as Aristotle had proclaimed. By the way, the *nova* was actually a *supernova*, the stupendous explosion of a star, the remnant of which has been named *Cas A*. We shall talk about supernovae when we come to the stars and their evolution.'

'Oh yes, you had told me about Tycho's *nova* and the remnant *Cas A*.' I couldn't help smiling to myself as I remembered my bathtub experience after discussing these matters once upon a time.

'Why are you smiling so mysteriously like that, Alfie?' George was puzzled and maybe a bit annoyed. 'I often suspect you are leading a secret life on the side and hiding something from me, you know. All right, let us pursue Kepler further. Kepler had to sift through Tycho's detailed and complex data, fit them into a theoretical model, trace the orbit of the planet Mars, and finally generalize his findings into his three laws. An account of this arduous task has been given with great appreciation by—guess who—none other than Albert Einstein himself in his collection of essays, *Ideas and Opinions*. We shall come back to it later.'

'I never realized that Kepler's laws involved so much hard work, George,' I admitted.

'Well, it did. First of all, Kepler found out that the planetary orbits were ellipses, not circles. I guess you know your ellipse well enough.'

'Let us see. It is like an elongated circle with two foci instead of a single centre. Draw a straight line from one of the foci to any point on the periphery of the ellipse. Then connect that point to the other focus by another straight line. The total length of these straight line segments remains a constant. In fact, that is how one can draw the ellipse in the first place. On a sheet of paper, stick in two pins to mark the foci. Place a loop of string around them. If you keep the string taut with a pencil and draw, what you get is an ellipse. Now...'

'Enough,' George held up his hand. 'I am convinced that you know enough of an ellipse. With his first law that made the planetary orbit an ellipse with the Sun placed at one of the foci, Kepler had removed the epicycles even Copernicus had to use in his model.'

'I thought those little wheels on wheels of Ptolemy had been banished by Copernicus.'

'No, Alfie, they had remained as vestiges of the Platonic ideal of having only circles as the perfect paths of the planets going round at constant speeds. Of course, the orbit can be a circle in some special case when the two foci coincide. Now, the second law tells you that the area swept in equal time intervals by a straight line joining the Sun at the focus to the planet moving along the ellipse is the same. As you can easily see that this dictates the varying speed of the planet as it goes around the Sun: fastest near the *perihelion* or the point closest to the Sun and slowest at the farthest point or the *aphelion*. Got it?'

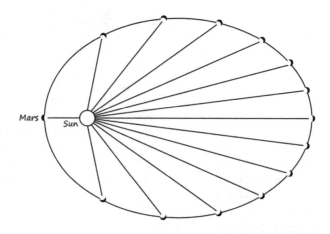

'Got it,' I affirmed obediently.

'Kepler set down all his findings in his book, *Astronomia Nova*. Now for his third law, the law that relates the period of the planet to its distance from the Sun,' announced George. 'Let me make it precise. You know what the *period* is: it is the time taken by the planet to make one complete revolution around the Sun. Now, the largest diameter of the elliptical orbit is called the *major axis* and half of that the *semimajor axis*. Actually, this is the average distance of the planet from the Sun. It is also half the length of the string you used in drawing the ellipse in the first place. Now, Kepler's third law states that the square of the period is proportional to the cube of the semimajor axis. Simple, isn't it? The larger the orbit, the longer the period or slower the planetary motion. Do you know how Kepler arrived at his third law? Kepler firmly believed in the celestial harmony of planetary motion. He tried to assign a different musical note to each of the planetary orbits. In this process he arrived at his third law. Kepler wrote another book *Harmonices Mundi* or *The Harmony of the Worlds* in which he recorded his new discovery along with the musical notes.'

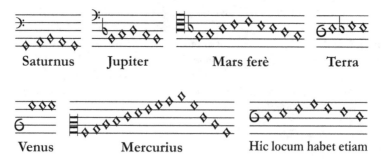

'Well, well, well, what do you know,' I exclaimed. 'Pythagoras and Aristotle were right after all. The music of the spheres does exist! Anyway, Kepler had indirectly proved the heliocentric model with all this, hadn't he?'

'True, but there is more to it than that,' said George. 'Think, three laws, just three of them, govern the motion of the planets, comets, Moons—our own as well as those of other planets—and man-made satellites. Until Kepler came upon the scene, only measurements had been made and models had been built to account for the planetary motions. But now, mathematical laws had been introduced for the first time. The age of general laws describing nature had dawned. Isn't that great? And it was Kepler who had taken the first step in that direction. As a matter of fact, Immanuel Kant, the philosopher, hailed Kepler as *the most acute thinker ever born!* Well, we shall return to Kant later on when we talk about the galaxies. Finally, let me quote the last statement of Einstein from his essay on Kepler that I mentioned earlier: *Kepler's marvellous achievement is a particularly fine example of the truth that knowledge cannot spring from experience alone but only from the comparison of the inventions of the intellect with observed facts.* Ah, that is the ultimate harmony: blending together of experience, observation, knowledge, and intellect—the very foundation of science.'

After a long pause, George took a deep breath. 'Are you ready? We are now going to meet that fantastic man, the great scientist, Galileo Galilei!'

'Not just a great scientist, George, but an accomplished musician, a fine writer, and a gifted artist,' I added.

'He was all that?' George seemed to be surprised. 'I only know his magnificent scientific accomplishments.'

'Well, his father Vincenzo was a musician. Not only did Galileo learn music from his father, but he is supposed to have excelled him, especially in playing the harpsichord,' I elaborated. 'Then again, his writings are testimony to the elegance of his style, composition, and eloquence. I have read that his work, in terms of richness of human content and power of style, set an example showing the path for future Italian prose. And only an artist of high calibre could have made the sketches of astronomical objects as he did. Don't you agree that his drawing of the Moon is so realistic that it rivals the modern photographs?'

'Maybe I should read more about Galileo,' said George. 'But for the moment, let me just enumerate his profound discoveries. Incidentally, we discussed these things when we were talking about black holes, didn't we? But, it doesn't hurt to recapitulate. First of all, he was a staunch adherent of the ideas put forward by Copernicus. Galileo's findings went a long way in supporting the heliocentric model, you know. His little telescope did wonders. With it he could open up whole new vistas into the celestial realm. He showed that there were mountains on the Moon. Amazingly enough, he estimated their heights by his ingenious method. He observed the spots on the Sun and showed that their motion revealed Sun's rotation. Simply incredible!'

'So, there was nothing heavenly about the Sun and the Moon, they were just ordinary denizens of the sky like our own Earth,' I remarked.

'You are right,' George gave a slight nod. 'And Galileo's discovery of the four Moons of Jupiter! That was an extremely important step to show that the Earth was not the only centre of revolution as those who held on to the geocentric model believed, Alfie. All celestial bodies were similar in nature, there was nothing special about the Earth. Then, why shouldn't the Earth itself be going around the Sun? In short, all this work of Galileo did so much to firm up the foundations of the Copernican model. Not only that! He pioneered the method of combining experiment and observation with theoretical explanation. No wonder then that Einstein hailed him as the father of modern science.'

'So, with the heliocentric model firmly established, man lost his central position in the cosmos,' I added.

'Oh, yes, as the Earth moved, the crystal spheres holding the planets were shattered as well as the *primum mobile* that was host to the stars. The *Empyrean*, the abode of gods or God beyond the *primum mobile*, too vanished.'

'Is it surprising then that the Church frowned upon Copernicus and felt that Galileo's *Diologues* was sacrilegious—worse still, dangerous! Perhaps, they found it absolutely necessary to bring him to trial. And make him recant under the threat of torture for their own safety,' I observed. 'Galileo spent his last days practically under house arrest. It was so tragic!'

'Yet, despite his growing blindness, he managed to write his pioneering work *Discourses Concerning Two New Sciences* that embodied his experiments performed over more than three decades,' George pointed out.

'Ah, many of those experiments were carried out using his inclined planes, weren't they, George?'

'Oh yes, as I tell my students, those inclined planes diluted gravity allowing Galileo to carry out his experiments with ease.'

George regarded me with a smile.

'I clearly remembered what you had told me during our discussions of years ago like the one we are having now, Alfie. I have included it in the manuscript and your name will appear in the acknowledgement for that as well as for so many other things you have told me.'

'What are you referring to, George?' I was curious to know what he had in mind. At the same time, I was touched by his comment about my conveying some facts to him.

'The lovely and profound quotation from the French philosopher, Henri Bergson: *Newtonian physics descended from heaven to Earth along the inclined plane of Galileo!*'

After a momentary pause, George announced as he always does when opening a new chapter, 'That means we are ready for Sir Isaac Newton!'

George rose from his chair. 'We need some additional energy to tackle that colossus, don't we? Well, down the corridor they have installed a new coffee machine. Not bad at all, you can even get a nice cup of cappuccino. And you can pick up some good cookies from another wending machine too.'

After we returned from our energising expedition, we sat back to talk about Newton's monumental work.

'Alfie, I am sure you remember the famous saying of Isaac Newton's about the seashore,' began George.

'Of course, George, I know the exact words of that quotation,' I replied and repeated those words. '*I do not know what I may appear to the world, but to myself I seem to have been only like a boy playing on the seashore, and diverting myself now and then finding a smoother pebble or a prettier shell than ordinary, whilst the great ocean of truth lay all undiscovered before me.*'

'Exactly! I have always envied your photographic memory, Alfie,' acknowledged George. 'But to us, ordinary mortals, Newton's work itself seems like an ocean. How can we measure its extent with our little teacups or fathom its depth with our puny rulers? Well, for the moment, let me just summarize Newton's unparalleled achievements focusing especially on gravitation.'

I nodded and waited in anticipation.

'Let me just go over some of the pioneering discoveries Newton made,' continued George. 'You know that the University of Cambridge was closed because of plague for two years.'

'Of course, I do,' I replied and repeated the well-known fact. 'It was during the years 1665 and 1666, *anni mirabilis*, miracle years of Isaac Newton.'

'You are right,' acknowledged George. 'In those two years, Newton established firmly the foundations of mechanics, optics, calculus—or fluxions in his terminology—and the crowning glory—gravitation. He compared the fall of the apple, the acceleration due to gravity on the Earth that is, with the orbital acceleration of the Moon going round our planet Earth. Both of them were produced by the gravitational force exerted by the Earth, weren't they? Newton realized that if the gravitational pull was assumed to decrease inversely as the square of the distance from the centre of the Earth, these two accelerations could be explained simultaneously.'

'Ah, the apple and the Moon together gave birth to gravitation!' I remarked.

'Now, he generalized it: The gravitational pull between any two objects was proportional to the product of the masses of those two bodies and decreased inversely proportional to the square of the distance between them, the *inverse square law of gravitation*. That was a stroke of genius. And the age of Universal Laws that were valid everywhere and at all times had dawned. A simple formula describing the gravitational force, Alfie, but it contained within it the entire solar system. Newton could now explain not only the fabulous observations of Galileo, but also account for Kepler's glorious laws of planetary motion. It was now clear why the Moon, the planets, and the comets moved the way they did.'

George took a deep breath.

'That was not all. Our Earth spins like a top, doesn't it? But a top does not always spin with its axis fixed in one direction, does it?'

'No, sir, the top not only spins about an axis, but the axis itself goes round in a circle.'

'In other words, the top undergoes *precession*, and so does the Earth,' George said. 'So, the Earth's axis goes round a circle.'

I thought about George's statement for a moment.

'That means *Polaris*, the Pole Star, won't remain fixed at one place on the celestial sphere then, since it does not exactly coincide with the pole.' This was news to me.

'Right you are. The period of precession of Earth's axis is some twenty-six thousand years. Even the ancient Greeks knew this. Sorry, Alfie, you are far behind them. In any case, no one knew why this happened. But now, Newton could explain this precession on the basis of the gravitational attraction of the Sun and the Moon on our Earth.'

'Ah, the Greeks, I don't mind following them, but I am not sure I can follow Newton's mathematical explanations,' I responded.

'Well, I don't think there are many who are able to do that, experts included,' assured George. 'Well, let us move on. Newton turned his attention to the shape of the Earth. Gravity pulls the material contents of the Earth inwards. On the other hand, the effect of its rotation is to throw the material outwards. This effect is zero at the poles and is maximum at the equator. So, what is the net result?'

'The Earth, like some people, develops an equatorial bulge,' I grinned, glancing at George's incipient paunch.

'Alfie, you don't have to rub it in,' George wagged an admonishing finger with mock seriousness. 'Like gravity, I do try to pull my tummy in. But the bulge is not due to rotation, it is because of good food, as you know very well. Oh well, coming down to Earth, it has the shape of an oblate spheroid. Newton could derive the ratio of the distance from the centre to the pole and the equatorial radius, which agreed with observation.'

'What next?' My question was not an expression of impatience, rather it stemmed out of curiosity. And George knew it.

'And, to top it all, Newton could clear the mystery that surrounded the occurrence of tides. The true cause underlying it had eluded even Galileo. He had offered a clever but erroneous explanation of the phenomenon, you know. According to him, high tide occurred when the direction of Earth's rotation was along its orbital motion and, conversely, low tide happened when the two directions were opposite to each other. That meant that each of the two tides appeared only once in twenty-four hours. What an irony, Alfie! Galileo must have known this was not true at all. After all, he must have witnessed tides all the time in Venice. Maybe, sometimes cleverness triumphs over correctness! Newton showed that the tides were caused by the differential attraction experienced by the ocean as it is attracted by the Moon. Let me explain this in a little more detail. On the side facing the Moon, the water bulges out since the gravitational pull is higher nearer to the Moon than at the surface of the Earth. On the opposite side, there is again a bulge. The water tries to move away as the force of gravity is less on the outer regions than near the Earth. So, Alfie, we have now two bulges in the ocean, one facing the Moon and the other on the opposite side of the Earth. At these two regions where there is a bulge, there is high tide. In between, low tide occurs. As the Earth turns, those areas on the ocean facing the Moon or situated on the opposite side experience a high tide. As those areas move away, they will have a low tide. The end result is that each of the two tides—high and low—occurs twice in twenty-four hours. As before, Newton was the first one to explain all this on the basis of his law of gravitation.'

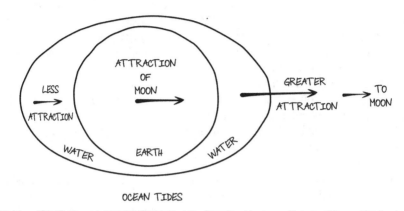

OCEAN TIDES

George paused dramatically before announcing, 'All this and much more were immortalised in the form of the greatest book of science ever written.'

George held up for a moment another large book lying on the table as though it were sacred scripture before handing it over to me. I too opened it with reverence and looked at the title: *Philosophiae Naturalis Principia Mathematica*, Mathematical Principles of Natural Philosophy or simply the *Principia*. As I turned the pages quickly, I could see the formidable contents, both the written material and the accompanying complicated diagrams.

'The three books of the *Principia* are the treasure trove of Nature!' proclaimed George. 'The first book deals with motion and gravitation—all occurring in empty space. The second book considers these motions in a resisting medium such as air or water. And now, look at the title of the third book.' *3 vols.*

I could see the title of the third book printed in large letters, *System of the World*!

'It is not an empty boast, Alfie,' remarked George. 'The contents of the third book more than justify the title. Newton incorporates into the body of his *System of the World* practically everything he had discovered in the realm of mechanics and gravitation. That amounts to everything that was known in his day about heaven and Earth. What more is there to say?'

We both sat in comfortable silence for a while relishing our discussion of Newton's unprecedented accomplishments revealing the workings of Nature. If I thought that it was the end of the matter, I was mistaken.

'You know, Alfie, one of the fantastic applications of Newtonian physics in the planetary realm happened to take place long after his death,' George picked up the thread of his thoughts. 'This was the discovery of the planet Neptune. Earlier, the German-English astronomer, William Herschel, had discovered the planet Uranus in the late eighteenth century. We shall be talking about Herschel, his discovery of

Uranus, and his two amazing relatives—his sister Caroline and son John—when we reach the Milky Way in our discussions. Eventually, the orbit of Uranus was calculated on the basis of detailed observations, taking into account the gravitational pull exerted on it by Jupiter and Saturn. Still, the observed orbit differed from the predicted one slightly. What transpired thereafter is quite tortuous, but interesting. You could read it all in the manuscript, Alfie. Two mathematicians, John Couch Adams in England and Urbain Jean Joseph Le Verrier in France, predicted independently the existence of a planet hitherto unobserved, responsible for the orbital vagaries of Uranus. Furthermore, they pin-pointed its exact position, enabling the astronomers to observe it. And, lo and behold, there it was, the blue planet Neptune, the god of the seas. In all this, let us not forget that it was a triumph for Newton's laws of motion and gravitation. Those four laws had formed the foundation for the detailed computations.'

George completed the list of the planets. 'After Neptune, the last of the planets, Pluto, was discovered. Now, they have relegated its position to that of a *Dwarf Planet*.'

'Why is that, George?' I knew about Pluto becoming a Dwarf Planet, but not the exact reason for this actions of the astronomers.

'Well, there are a couple of logical reasons for that,' explained George. 'All the other planets from Mercury to Neptune have their orbits more or less in a plane around the Sun. The theory is that they all evolved from a flat rotating gas cloud. The central region condensed to become the Sun. The outer part first broke into rings, which in turn shrank, and the rings were transformed into the planets. On the contrary, Pluto's orbit around the Sun is not at all confined to the plane of the other orbits. That celestial object seems to have been captured into the solar system from outside. Recently, more such small celestial objects beyond Pluto have been discovered once again with orbits like Pluto's. And for all you know, more and more of those will be detected. So, for these reasons Pluto and the other similar members of the solar system have been christened the Dwarf Planets.'

'In a way, all this belongs to the continuation of the past history of the planets,' observed George before posing the question, 'Coming back to Newton, where does he stand in our modern times?'

'I know, I know, Newton is very much with us in our classrooms,' I replied.

'Confined to our classrooms? You must be joking, Alfie,' protested George. 'What about our present-day space age? Every other day, satellites are sent up to orbit the Earth, aren't they? How is this done? Newton tells you how. Take a look at this diagram from *System of the World* of the *Principia*.'

George opened to a particular page he had marked with a flag and showed me a picture.

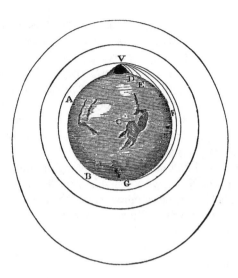

'Let me read the relevant parts of the text for you,' said George. 'Here we go: *...VD, VE, VF the curved lines which a body would describe, if projected in an horizontal direction from the top of an high mountain successively with more and more velocity; the body projected with a less velocity describes the lesser arc, and with greater velocity the greater arc, and augmenting the velocity, it goes farther and farther; if the velocity was still more and more augmented, it would reach at last quite beyond the circumference of the Earth, and return to the mountain from which it was projected... its velocity, when it returns to the mountain, will be no less than when it was at first; and, retaining the same velocity, it will describe the same curve over and over.*'

George looked up to make sure I was paying attention to his reading. And I was hanging on every word of his.

'Here is more,' continued George. '*But if we imagine bodies to be projected in the directions of lines parallel to the horizon from greater heights, those bodies, according to their different velocity, and the different force of gravity in different heights, will describe arcs either concentric with the Earth, or variously eccentric, and go on revolving through the heavens in those orbits just as the planets do in their orbits.* There, Newton has launched an artificial satellite around the Earth or, for that matter, around any other planet!'

I sat silent, simply amazed at what I had just heard.

'Does Newtonian physics end there, Alfie? No way!' George was quite animated. 'What about the paths of the rockets that are launched? What about the orbits of the spacecraft that are being sent routinely to the Moon and even to the other planets? All these require extraordinarily complicated calculations done on sophisticated computers. But what underlies all this? Three laws of motion and one law of gravitation! Can you beat that? Yes, yes, yes, Newton lives on!'

With that, George closed the *Principia* ever so gently as though it was a sensitive, living organism.

I realized that George was quite exhausted with all the explanations he had been giving and with his own emotional involvement with the subject. That was obvious. So, I took over to give him some respite.

'Newton was a strange and complex personality, wasn't he, George?' I commented. 'He was a highly religious person and left behind a large body of religious writings. Then again, after his career as a scientist, he became the Master of the Mint and brought about some important reforms in coinage. Furthermore, he dabbled in alchemy quite a bit trying to find the Philosophers Stone that could transmute base metals into gold, didn't he?'

'Yes indeed, but he didn't have to. He had dug up immeasurable quantities of gold from the depths of Nature as no one had done before,' reflected George.

'One last thing before we close our session,' continued George. 'Let me read a quotation from the manuscript telling us what Einstein, the only scientist who stands alongside Newton, had to say about the latter: *Nature was to him an open book, whose letters he could read without effort. The conceptions which he used to reduce the material of experience to order seemed to flow spontaneously from experience itself, from the beautiful experiments which he ranged in order like playthings and describes with an affectionate wealth of detail. In one person, he combined the experimenter, the theorist, the mechanic and, not least, the artist in exposition. He stands before us strong, certain, and alone; his joy in creation and his minute precision are evident in every word and every figure.*'

That was indeed an unparalleled compliment paid by one great scientist to another one of the past. Both George and I sat in silence. I was sure that he was thinking about those two remarkable men, as I did. After a moment or two, George spoke up.

'You know, Alfie, we spent a lot of time today talking about what seems to be ancient history. But, believe me, those were the days when the very foundations of modern science were being laid. Without those firm foundations, how could we have ever built the superstructures of astronomy and cosmology? Take Tycho Brahe for instance. He made his extraordinarily accurate observations without any optical aids. That accuracy is what we aim for even today—the limits of exact data for building up correct models. Only because of Tycho's exceptionally accurate observations, Kepler was able to hit upon his elliptical orbit for Mars, a radical departure from the circular ideal. What about Galileo? He not only opened up uncharted vistas in astronomy with his little telescope, but combined observation with theory, a blend that is the very basis of modern science. Of course, beneath it all was the revolutionary leap Copernicus had taken from appearance to reality, the unprecedented event in human history. And, finally, Newton! The age of Universal Laws had dawned with him. Yes, Alfie, whatever we are going to say about the Universe, we have to first understand and remember whatever those fabulous founders had achieved.'

There was no need for me to say anything. George knew that I could grasp the importance of the work of the pioneers in the exploration of the cosmos.

George rose and stretched. I too got up.

'It has been a long day, Alfie, but wonderful as always. We shall meet soon. Until then, take care, my friend.'

I strolled slowly across the university campus enjoying the serenity of that place of learning. As I reached home, I could feel the after-effect of my inspiring discussions with George that lingered on.

Chapter 4
Casanova Calls

When I returned home after meeting George, I opened his book on the Universe. This appears to have become a compulsive but most enjoyable habit of mine. I wanted to read up whatever we had discussed: essentially about the foundations that had been laid by those great men of science, namely Copernicus, the duo Tycho Brahe and Kepler, Galileo, and finally Isaac Newton. I vividly recalled the words George had spoken about them in his animated way towards the end of our discussion. I vaguely remembered an old Chinese saying that went something like this: *You hear and learn; you see and understand; you read and remember.* I had learnt by listening to George; I hope I had understood by looking at the material George had shown me; and now, I wanted to remember it all by going over George's book. Needless to add, it is so exciting to learn from George's excellent book that not only explains in a clear and simple manner the ideas, concepts, and the detailed mechanisms of science, but also sketches the lives and backgrounds of the pioneers. I must tell you the surprising fact that I had discovered in the writing of the celebrated thirteenth-century Italian poet Dante Alighieri. Years ago, I had unearthed some verses from his *L'Inferno* or *Hell* that closely resembled ideas inherent to black-hole physics. Now in George's book I found reference to Dante's concept of geocentric Universe along with a nice illustration that showed hell at the centre of the Earth surrounded by the planetary spheres and *primum mobile*.

© Springer International Publishing Switzerland 2015
C.V. Vishveshwara, *Universe Unveiled*, Astronomers' Universe,
DOI 10.1007/978-3-319-08213-4_4

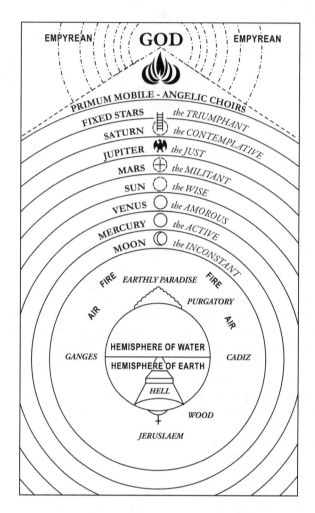

That is not all. Dante must have been quite progressive in advocating the exploration of new intellectual pathways, which was achieved centuries later by Copernicus and others. Let me quote Dante: *For what fruit would he bear who should demonstrate once again some theorem of Euclid; who should strive to expound a new felicity, which Aristotle has already expounded; who should undertake again the apology of old age, which Cicero has pleaded? Naught at all, but rather would such wearisome superfluity provoke disgust.* Isn't this true of all discoveries in science?

After I finish reading George's book and making suitable notes, I always have this urge—well, you know what that is—to have a soothing bath before going to bed. I filled my bathtub with comfortably hot water and took out from the cupboard one of the sachets containing the bubble-bath mixture. When I was about to add a spoonful of the mix, my bathtub started to vibrate. I knew at once what was coming and was quite happy about it.

'This is Khasbath speaking,' announced the bathtub in its peculiar gurgling, throaty voice. Or should I say sound rather than voice? It doesn't matter, does it?

'Khasbath! I like the name you gave me, Boss,' said the bathtub.

What, I gave the name? He—sorry, it—had christened itself like Napoleon crowning himself the Emperor.

'Thanks for comparing me to Napoleon, boss. But, I am not sure I like the controversy regarding who named whom. Let that rest in peace. So, you had your dose of astronomical adventure guided by the Professor. How do I know? Well, remember I can read your mind like, like. . . no, I can't say like my palm, can I?' My bathtub—sorry again, Khasbath I mean—gave its characteristic vibrant laughter. 'I believe you remember very well that is how, by reading minds, I learnt so much when I was a kitchen sink in my Master's home. All, right, go ahead and add the bubble bath mixture. Relax and reflect; ruminate and ratiocinate: aha, you like the sound of those words, don't you? I do. And I shall learn more reading your mind.'

Khasbath fell silent. I added the bubble bath mixture to the bath water. Immediately, the bubbles, accompanied by the sparkling specks, appeared all over, massaging me, warming me. It was a joy to behold them. On closer scrutiny, I could detect here and there extremely minute spheres circling around some of the shining specks. I couldn't help imagining that I was indeed observing, right within my bathtub, miniature planetary systems held in thrall by their presiding stars. As I tried to focus on those tiny orbs in revolution, my eyes became heavy and I drifted into the usual indescribable state that transcended all earthly experience.

I felt myself floating totally weightless. And I was able to navigate freely like a swimmer in water that offered no resistance. But I was now suspended amidst countless stars that greeted me wherever my eyes happened to turn. Then it happened—the familiar and unearthly occurrence of an ethereal figure rising out of pure space, perhaps made up of starlight and nothing else, yet assuming almost material existence. At the same time, I could hear the melodious sound of haunting music wafting from afar.

'I ardently hope that you remember me, *signore*.' Oh Yes, I had heard that deep voice long ago. I had seen this distinguished personage attired fashionably in harmoniously coordinated shades of grey and black. I had seen his countenance— his hollow cheeks, lips pressed together, ends slightly curled up in a sardonic smile, and his translucent eyes that gazed at you steadily as if vivisecting every detail of your innermost thoughts. How could I ever forget this fantastic personality from the past?

'Ah, what an honour to be remembered by *signore!* Yes, it is none other than Giovanni Jacopo Casanova, sometimes referred to as also Giacomo Girolamo Casanova, forever your servant.' He made a deep formal bow. Casanova closed his eyes listening to the beautiful music in the background. 'You hear that sublime music that always accompanies me? As you perhaps know, it was composed by my dear friend, Wolfgang Amadeus Mozart, who had been baptised as Johannes Chrysostomus Wolfgangus Theophilus Mozart, a mouthful of a name in my humble opinion. Poor boy, he tried to take lessons from me in the fine art of winning over delectable damsels, which ignorant individuals condemn as seduction. This, he claimed, was in preparation for composing his opera *Don Giovanni*. Don Giovanni, he and I share not only a common name, but also our common interest.'

Of course, I was aware of the strange interaction that had taken place between Casanova and Mozart.

After a moment or two of quiet thought, Casanova continued, 'And I am sure that you remember the unusual reason for my appearing in the celestial realm: my illustrious name that happens to be the combination of *Cas A*, the remnant of an exploded star and its original appearance that was recorded as a *nova* by that King of Astronomers, Tycho Brahe. Those two terms add up to my name *CasAnova*, do they not? As I remarked at the time of our first meeting, fate has a perverse sense of humour to think of me as the embodiment of an exploded star. Although in my own time I did shine like a star, I never exploded in any manner whatsoever.'

I could vividly recall how Casanova had escorted me to show and describe some of the constellations as well as to visit some famous astronomers of the past.

'That was a glorious journey, was it not, *signore*?' I was not surprised that Casanova had read my thought. It did not feel that it was an invasion of privacy at all. 'We made bridges out of pure space and flew across time. We wandered all over the Universe and visited some great scientists of yesteryears. In addition, although I did not guide you at that juncture, you had the rare good fortune of being able to listen to the greatest poet our country has produced, namely Dante Alighieri, the creator of the immortal poem *Divine Comedy*, who recited to you from *Inferno* those verses that resembled the ideas embodied in the subject you were pursuing at that time. In your recent studies, you have come across Dante's notion of the structure of the Universe, have you not? Where does Dante place the hell in the scheme of things, *signore*? At the centre of the Earth, does he not? Well, perhaps that was true in Dante's days. But, in my considered opinion, in your own time hell has moved to the surface of the Earth!'

Casanova covered his mouth with his lace handkerchief and permitted himself a long, controlled laugh before announcing, 'Now, the time has come again for new adventures, new explorations, and new flights of fancy. Prepare yourself, *signore*, for such an experience as no one in this world has ever had.'

Chapter 5
The Canon and His Cosmos

Casanova and I drifted along and came to an old city with a massive wall going around it.

'We have reached the town of Frombork in Poland or Frauenberg as it is also called,' announced Casanova. 'That wall, the fortification, was built to protect the people of the town, as also the Cathedral standing on the hill, from the marauding Teutonic Knights. This is like a pious place of pilgrimage for those who possess keen interest in astronomy, nay in human thought itself, *signore*. For this is where the great Copernicus worked as a Canon of the Church.'

I could easily discern Casanova's intense reverence for Copernicus, which I too felt.

'As you very well know, Copernicus was well trained in astronomy, medicine, law, and financial matters. As the Canon of Frombork, he took care of the administration of the Church as well, which included the fortification we see.'

We had reached the base of a tall tower that was part of the fortification. We lifted off and rose gently to an open window at the top of the tower and sailed through it. We were now in a large room with high roof, rather dark and minimally furnished. Next to the window through which we had entered, stood a heavy wooden desk piled up with sheets of paper, diagrams and numbers written all over them. In the background stood strange instruments of different shapes and sizes. I wondered what they were.

On the other side of the room was a high chair occupied by an elderly man of serious countenance, dressed in a black robe. He was listening to one of the three persons sitting opposite to him. The speaker seemed to be quite a rich man, judging from his clothes, but looked quite ill.

'O, Nicolaus Copernicus, I have come from afar seeking your help. You, only you, can save me from the dread disease I am suffering from,' wailed the man. 'Your fame as a Doctor has spread all over Europe. You have healed the rich and the poor alike, saved many lives from the epidemic that swept across our province. O, Nicolaus Copernicus, be generous and . . .'

© Springer International Publishing Switzerland 2015
C.V. Vishveshwara, *Universe Unveiled*, Astronomers' Universe,
DOI 10.1007/978-3-319-08213-4_5

'Enough,' Copernicus held up his hand stopping the speaker from going on any further. 'I must inform you that I feel a strong revulsion towards any familiarity and towards frivolous, pointless conversation. Please be kind enough to state the symptoms of your malady and I shall try my best to help you.'

The poor man went about dolefully relating all the problems he had with his health, while Copernicus listened with concentration that appeared to be his inherent trait. Only when his visitor had stopped telling him the list of his woes did he speak up.

'I shall give you a prescription for an Imperial Pill, a universal remedy, which may be taken any time and has curative effect on every disease.' said Copernicus and repeated aloud the prescription he jotted down on a sheet of paper. 'Take two ounces of Armenian clay, a half ounce of cinnamon, two drachmas of tormentil root, dittany, red sandalwood, a drachma of ivory and iron shavings, two scruples of ash and rust, one drachma each of lemon peel and pearls; add one scruple each of pulverized emerald, red hyacinth and sapphire; one drachma of bone from a deer's heart; sea locusts, powdered horn of a unicorn, red coral, gold and silver foil—all one scruple each; then add half a pound of sugar, or the quantity which one usually buys for one Hungarian ducat's worth. And prepare the pills out of those ingredients.'

As the visitor rose to leave expressing his gratitude profusely, Copernicus said softly, 'God willing, it will help you!' and added under his breath as if he was talking to himself, 'Those who inherit diseases are rarely cured of them, and will be wise to endure their suffering in patience.' He then jotted down his prescription on the margin of the open book in front of him.

'Did you notice that book, *signore*?' whispered Casanova with amusement. 'It is a copy of the venerable Euclid's Geometry!'

I looked and Casanova was right. I was amazed.

Having attended to his patient seeking medical help who left the room walking backwards with great reverence, Copernicus turned his attention to the other two visitors. One of them was an elderly person who looked distinguished in his elegant flowing robe made of expensive material. Presumably, he was a high official of the state. The other visitor was young and energetic; he wore crisp, well-fitting clothes that set off his hard muscles. The former spoke with utmost respect.

'Canon of Frombork, my humble greetings to you,' he began and went on to describe how the affairs of the Church were being conducted as Copernicus nodded his approval from time to time.

'How about our financial position? I had outlined some necessary measures to reform the coinage of the country. Have they been implemented?' asked Copernicus.

'Our finances are good and the coinage has been improved along the lines you had suggested, Canon,' replied the official.

'Very well then, we must take utmost care when dealing with the finances of the state since it is my considered opinion, nay, a principle in fact, that bad or debased money will drive out the good,' commented Copernicus. After a moment of thought, he took out a thick volume and handed it over to the official. 'This work of mine dealing with certain aspects of economy and finance should interest you and, hopefully, its reading would prove beneficial in your work.'

As the official respectfully accepted the book, I was able to read the title, *Monete cudende ratio*.

'That means *On the Minting of Coin*,' Casanova informed me and added with a smile, 'I realize that you did not know that the great stargazer, the author of the monumental *De Revolutionibus*, also wrote learned treatises on economics. Very few do.'

Again, my astonishment knew no bounds.

Copernicus now turned to the younger of the two visitors and addressed him, 'Yes, Commander, what have you to report?'

'Canon of Frombork, my humble salutations,' the Commander of the army bowed low. 'I am happy to report that all is well with the defence of Frombork. The fortification is in good repair and our soldiers are ever vigilant.' He proceeded to give the details regarding the army and its various activities aimed at the defence of the country in general and Frombork in particular. 'Most importantly, we do not have to fear an invasion from the Teutonic Knights any longer ever since their shameful defeat culminating in their Grand Master rendering homage to our King, all thanks to your advice, Canon.'

Copernicus waved away the compliment paid by the Commander. He nodded with dignity signifying that the audience was over. The two visitors left after bowing repeatedly to Copernicus, which he acknowledged by raising his hand. Copernicus remained motionless in deep thought for a moment or two and then walked over to the desk by the window taking unhurried steps. He sighed deeply as he sat down and regarded the pile of papers in front of him. He pulled out from beneath the pile a large book and opened it. At the top of the front page was written the title of the yet to be finished tome—*De Revolutionibus Orbium Coelestium*.

'Ah, the book of books, *On the Revolutions of the Celestial Orbs*,' exclaimed Casanova in a low voice.

'At last back to the study of stars, the only and the greatest pleasure I have *in remotissimo angulo terrae*, in this remotest corner of the Earth.'

Copernicus slowly raised his eyes and looked at us, his unexpected visitors, and smiled—a rare event indeed, which lighted up the gloomy room. He spoke in a serious but sincere fashion. I could feel that here was a modest man who was oblivious of his own importance.

'It will be a great pleasure to express my thoughts and ideas in regard to the stars, observing which and learning from them is the noblest human endeavour,' said Copernicus. He regarded us with interest. 'I do not mind imparting precious knowledge to the two of you truth seekers, which I would not pour it into the muddy waters of the human mind in general. Also, I believe that mathematics is for mathematicians.'

'Yes, I made many careful observations of the stars and planets,' continued Copernicus. 'I believe you have already noticed the observational tools here which I fabricated myself.' Copernicus pointed at the strange instruments I had seen and wondered about. 'For instance, I made a *triquetrum* in order to check the positions of the stars.'

The triquetrum or the 'cross-bow' was twelve feet high consisting of three wooden bars and dominated the other instruments.

'Let us not get into the details of my observations. I must admit that the Greeks, my predecessors, possessed more accurate instruments, especially Ptolemy who made sufficiently precise measurements on which my calculations have been predominantly based. Most important to me are the ideas that dictate the structure of our world, our Universe. I shall explain them to you as best as I can.'

Copernicus paused to collect his thoughts and then proceeded to describe his cosmic structure comprising the Earth and the other celestial bodies.

Casanova whispered as before, 'Most of the time, Copernicus will be repeating in his own style what he has written in that big, scholarly book of his *Revolutionibus*. Just as well, since the book is considered to be one of the most unreadable tomes ever written, as you already know. Sometimes, he may repeat long passages. It is always good to know how such great men not only thought, but also wrote. I am sure *signore* agrees with me.' He added, 'I have a feeling that he will be relating only the descriptive parts that are comprehensible to ordinary mortals like us.'

'First of all, we must think of what place the Earth occupies in the Universe,' began Copernicus. 'Without this knowledge no certain computations can be made for the phenomena occurring in the heavens. To be sure, the great majority of writers agree that the Earth is at rest in the centre of the Universe, so that they consider unbelievable and even ridiculous to suppose that this is not so. Still, when one weighs the matter carefully, it will be seen that this question is not settled yet, and for that reason is by no means to be considered unimportant. Now the Earth is the place from which we observe the revolution of the heavens and where it is displayed to our eyes. Therefore, if the Earth should possess any motion, this would be noticeable in everything that is situated outside of it, but in the opposite direction, just as if everything were travelling past the Earth. And of this nature is, above all, the daily revolution. For this motion seems to embrace the whole world, in fact everything that is outside of the Earth, with the single exception of the Earth itself. But if one should admit that the heavens possess none of this motion, but that the Earth rotates from west to east and if one should consider this seriously with respect to the seeming rising and setting of the Sun, of the Moon and the stars, then one would find that this is actually true.'

That is what the remarkable Indian astronomer, Aryabhata, had conjectured centuries before Copernicus. The motion of everything in the sky that happens daily is due to the turning of the Earth. And he had described it through his lovely verse offering the analogy of how the trees seem to move backwards while we are going forward in a boat.

'Now, is the Universe finite or infinite?' Copernicus asked rhetorically. 'We shall leave that question to the quarrels of wise men. For us remains the certainty that the Earth is finite and is bounded by a spherical surface. Why should we hesitate to grant it a motion, natural and corresponding to its form, rather than assume that the whole Universe, whose boundary is not known and cannot be known, moves?'

Copernicus paused several moments and asked emphatically, 'Why are we not willing to acknowledge that the *appearance* of a daily revolution belongs to the heavens, its *actuality* to the Earth?'

'*Appearance* and *actuality*!' Casanova nodded his head a couple of times. 'Think, *signore*, the great man himself asserts the difference, the difference that was to be emphasised again by another great astronomer, my revered compatriot, Galileo Galilei.'

'Since nothing stands in the way of the mobility of the Earth, I believe we must now investigate whether it also has several motions, so that it can be considered one of the planets,' said Copernicus seriously. 'Let us think then of the annual motion of the planets. That the Earth is not the centre of all the revolutions is proved by the irregular motions of the planets, and their varying distances from the Earth, which cannot be explained as concentric circles with the Earth at the centre.'

'Now, on the other hand, let us admit the motionlessness of the Sun and transfer the annual revolution from the Sun to the Earth,' proposed Copernicus. 'Then one can see that the distances to the planets from the Earth do vary.'

That was a simple, but most important, argument. If the Earth stood still and the planets were going in circles around it, obviously the distance from the Earth to any of the planets had to remain constant. On the other hand, if the Earth too were moving in a circle around the Sun, as well as the planet at a different speed compared to the Earth, then the planet's separation from the Earth had to be varying. Simple ideas are often profound!

'Having the Sun immobile at the centre with the Earth revolving around it will also account for the halting and backward motion of the planets,' announced Copernicus. 'Those peculiar motions are not those of the planets at all but they appear to do so because of the motion of the Earth.'

I clearly remembered how George had explained the retrograde motion of the planets on the basis of the Copernican heliocentric model of the solar system.

Copernicus now spoke up. From the tone of his voice it was apparent that he was making a most significant statement that must have appeared in his monumental book. It was the summing up of his heliocentric Universe. 'I say that the Sun remains forever immobile. Whatever apparent movement belongs to it can be verified as due to the mobility of the Earth. Planets move in circular paths of sufficiently manifest dimensions compared to the distance between the Sun and the Earth. But the Universe is such that these distances, as compared to the distance to the fixed stars, are imperceptible.'

Copernicus rose from his high chair with great dignity. I felt that he was about to quote something of utmost importance from his own magnificent work. In a commanding voice that echoed from every wall of the august chamber in which he had discovered the true nature of the cosmos, he proclaimed:

In the middle of all sits the Sun enthroned. In this beautiful temple could we place this luminary in any better position from which he can illuminate the whole at once? The Sun sits as upon a royal throne ruling his children, the planets, which circle round him.

Majestically Copernicus closed his book, *De Revolutionibus*.

'This work has lain concealed in my abode for thirty-six long years,' said Copernicus solemnly. 'The moment has finally come for it to see the light of day.'

Copernicus seemed to have forgotten our presence altogether in his deep contemplation. He walked to the open window and gazed out at the night sky filled with glittering stars. Ever so slowly, the night sky seemed to seep into the room and engulf Copernicus, who gradually merged with the starry background.

'We have come a long way, have we not *signore*?' reflected Casanova. 'Once upon a time, demons and dragons, heroes and hydras inhabited the constellations. Then came the crystal spheres of Aristotle only to be replaced by Ptolemy's complex epicycles. And now, at long last, Copernicus has established his own cosmos with the Sun enthroned at the centre and his children, the planets, circling him in loving obedience. What a glorious revelation, truth shining forth like the central fire!'

Casanova bowed his head as though he was paying his homage to Copernicus. Then, raising his head, he said, 'Let us move on, *signore*, to visit two celebrated astronomers who, at the moment, happen to be engaged in deep discussion of high matters.'

Chapter 6
Hven and the Heavens

The Dwarf and the Giant

We were now in a spacious room, the walls decorated with several paintings and heavy curtains hanging by the windows and doors. At a large wooden table, sat the two astronomers, Casanova had just mentioned. They were absorbed in animated discussion. The older one on the right was dressed in the finest clothes fashioned out of silk and velvet. An intricately designed medallion hung from a golden chain that he wore around his neck. Curled up at his feet lay a dog yawning from time to time, bored as it was with the conversation going on over its head. In striking contrast to the older astronomer, the younger one on the left wore modest clothes crumpled with constant use. He had a pale look as though suffering from some illness or the other constantly. Nevertheless, his eyes were bright and penetrating, reflecting an exceptionally keen intellect.

All of a sudden, I was gripped with a strong feeling of *déjà vu*. Had I not visited this place earlier and listened to the two men in conversation before me? I felt that even the description of the scene that arose in my mind was the same as what I had written earlier.

I realized that Casanova was regarding me with an amused smile. 'Yes, *signore*, mind is a strange entity that often does not distinguish between the past, present, and future. The greatest scientist to grace this world after Isaac Newton, who appeared in your own time, was right when he said that time is a persisting illusion. In the present context, let me acknowledge your innermost feeling that we have been here in the past. And, most likely, you will hear and witness exactly the same things as before. But such summoning up of memories of the past should prove most instructive, not to mention its high entertainment value.' I was reassured by his words and concentrated on what was happening in front of me.

In the background one could see a variety of astronomical instruments somewhat similar to the ones that Copernicus had possessed. Casanova, who seemed to be familiar with astronomical instruments, pointed them out to me: *quadrants* ranging

© Springer International Publishing Switzerland 2015
C.V. Vishveshwara, *Universe Unveiled*, Astronomers' Universe,
DOI 10.1007/978-3-319-08213-4_6

from a radius of 16 inches to seven feet, *sextants*, complicated spherical devices like *astrolabes*, and several other aids used for astronomical measurements.

While I was gazing upon all this with great fascination, I was suddenly drawn to an extraordinary sight. Seated beneath the table was a dwarf chattering away to himself. At that moment he too noticed us and came over.

'Oh, ho, ho, what on Earth do we have here?' exclaimed the dwarf making strange faces. 'Visitors from the future! Hey doggie boy, look, we are honoured by the visit of two distinguished guests.'

The dog yawned, opening its mouth wide, and most offensively thrust out its long red tongue. The dwarf kicked at it muttering, 'Stupid mutt,' and the dog yelped in protest.

'Oh, yes, your honours, welcome to Uraniborg, the Palace of Astronomy, situated on the island of Hven outside Copenhagen, dedicated to Urania, the Muse of Astronomy,' the dwarf bowed so low that his cap fell off. 'Ah, forgive me, I forgot to introduce myself, your honours. I am Jeppe, the pet dwarf of the nobleman you see before you on the right, the King of Observers, Tycho Brahe.' Jeppe pretended to hold up an imaginary trumpet and made a strange noise apparently to mimic its blaring heraldic sound. I was reminded of similar sounds made by the chorus of frogs that had accompanied Aristophanes long ago.

'Tell me, how many astronomers keep pet dwarfs, eh, eh?,' questioned Jeppe. 'None, your honours, none! But, my Lordship does, you know why? Because he takes my advice on the affairs of Hven, which he happens to administer. You know why, let me repeat the question. Because I am blessed with prescience, knowledge of things before they happen, and I can foretell the future.' Jeppe held his open palm against his mouth and said in a conspiratorial, but loud, whisper, 'At least that is what the man believes!' Jeppe rolled on the floor laughing uncontrollably.

Getting up quickly, Jeppe continued. 'Please allow me to relate to you of an occasion on which my master sought my highly valued advice. And to give him due credit, he followed my advice too. On the said occasion, when the peasants of Hven had been insubordinate, I counselled that they be assembled and offered all the beer they could consume instead of punishing them. And it worked, it worked! Alcoholic amelioration of administrative ailments I call it.' He winked at me and added insolently, 'I ardently hope that your vocabulary is good enough to understand what I just said.'

Jeppe drew himself up with pride as tall as he could, after having established his importance to his employer. I watched him with fascination not unmixed with a bit of uncomfortable feeling, while Casanova looked on, his customary amused smile curling up the corners of his lips. I then looked away from Jeppe and focussed my attention on Tycho Brahe and his magnificent attire. My attention was diverted by something strange about his nose, which had a dull sheen to it.

'I perceive that your honour has noticed my Lordship's nose,' observed Jeppe with ill-concealed glee. 'It glows strangely, for it has been constructed out of gold, silver, and bee-wax—an admixture of two noble metals with the ignoble produce of a lowly insect, I must say. My employer designed it himself with great ingenuity. Whatever happened to the original nose made of flesh and blood, you may ask. Ask away, your honour, and I shall oblige you with the correct answer. The God-given olfactory organ was severely damaged in a duel my master fought with his third cousin. A duel fought over the amorous attentions of a dainty damsel perhaps or to put it rather crudely a fight over a woman?'

'Alas, no such romantic happening,' Jeppe sighed. 'Swords were crossed on account of an argument that broke out at the dinner table as to who knew more mathematics, my master or his third cousin. Can there be anything more absurd, I ask you. Allow me to hasten to add that your honour need not strain his brains for an answer, because the question was purely rhetorical, which means, I am told by linguistic experts, a question to which the answer is obvious and therefore need not be given. In short, a query tantamount to waste of time!'

Jeppe was out of breath now, having chattered away with hardly a pause. He held up his forefinger to his lips and hushed me saying, 'Why is your honour talking so much? What did you say? It is I who has been chattering away? No, your honour, I only echo the words and thoughts of others. Be that as it may, let us turn our attention to my Lordship, a great man, though between you and me, no man is great to his own dwarf.'

After catching his breath and collecting his thoughts, Jeppe spoke on again. 'Ah, I have told you so much about my master who is a nobleman. What do we mean by a nobleman, your honour? A man whose parents, parents' parents, parents' parents' parents and so on were all high and mighty, rich and famous, and, mark my words, often lazy and useless. Of course, not so with my Lordship, although he came from a noble family. But forget him for a moment and take a good look at the younger man sitting next to him.'

Jeppe waited for a while apparently to allow me time enough to study the man on the left. 'He is another accomplished astronomer as well, who answers to the name of Johannes Kepler. He came originally from Germany and became master Tycho Brahe's assistant, but never worked here at Uranisborg, but only later on in Prague. But as you know, space and time can get smeared out in this world of imagination, of fantasy, of make-believe. So you see them together here at Hven. Once, our master, the great Dane—no, no, I don't mean the Great Dane, a type of dog, heaven forbid or should I say Hven forbid—used to call Johannes a mad dog. An irony involving dogs, eh, your honour? Anyway, now the master realizes his assistant's worth.'

Kepler

'Please allow me to tell you briefly about Kepler's background,' Jeppe continued. 'He too must have come from a noble family, did you say? My, my, your honour has a warped sense of humour I must admit. What a contrast between the families of the two men! Please do not take my word for whatever I say, no scandalmonger I am, no, sir, never. Kepler himself has bared it all in his chronicles. His father was, according to his beloved son, vicious, inflexible, quarrelsome, a wanderer, and beat his wife to boot—perhaps with his boot too—quite often. He fought in different armies for money as also for his own sadistic pleasure, visiting his home rarely, maybe whenever he felt like beating his wife and children. Well, young Kepler's mother often followed her husband abandoning her children, maybe for her own masochistic pleasure of getting beaten up. Hush, listen but not a word spoken loud, the mother had a reputation of being a wicked witch. Maybe it matters not, since she is long dead and gone. But you never know, since vindictive ghosts may exist after all. Kepler's mother's aunt was also believed to be a witch and was burnt by people out of fear—looks like witchcraft was a hereditary hobby. To cut the story short—ah, I love shortness naturally—Kepler was brought up in a poor, chaotic household. All his life he suffered from a variety of ailments, some real but most of them imaginary. Yes, sir, he has written them down too. At the age of four I nearly died of smallpox, he says, my hands were badly crippled; later on, he goes on, I suffered continually from skin ailments, severe sores, scabs, a worm in the finger, whatever that means, so on and on and on. Let us forget all these complaints. One more thing about Kepler, which I almost forgot. In all his life he took a bath only once, that too at the continuous nagging of his wife, and almost died as a result. What did you say, your honour, he is a *stinker*? No, no, no, he is a *thinker*, not a stinker.' Jeppe guffawed loudly and then added with unusual seriousness, 'Yes, he is a brilliant thinker, one of the greatest, who has transformed astronomy. He wanted to be a priest, but had to teach astronomy and mathematics to make a living. How fortunate for those two subjects! Tycho Brahe is the King of Observers and Johannes Kepler is a King among Thinkers.'

Jeppe seemed to grow old. He mumbled to himself, forgetting our presence. 'Ah, playing the fool to make others laugh renders you sad, does it not, my dear Jeppe? Well, sadness and joy cancel each other out leaving nothing behind, I suppose. Maybe they are the two faces of the same coin. Maybe, maybe, there is something for you to think about, my dear Jeppe.'

Jeppe shook himself out of his reverie. His face had assumed a serene expression. He spoke now in a serious tone. 'There they are, two erudite men discussing the heavens. Listen to them and you shall reap the treasures of vast learning.'

With that Jeppe withdrew to the background, only temporarily as it turned out, leaving the two astronomers in focus

'Observing the *nova*, the new or the guest star as we know it, was the greatest event of my life, which led me to dedicate myself to the divine pursuit of observing the heavens and recording the positions of the celestial objects,' Tycho was saying.

'I still remember vividly the momentous occurrence on the eleventh day of November in the year fifteen hundred and seventy-two,' Tycho continued his reminiscence. 'It was evening, after sunset, when according to my habit, I was contemplating the stars in the clear sky, I noticed that a new and unusual star, surpassing all the other stars in brilliancy, was shining almost directly above my head; and since I had almost from boyhood known all the stars of the heavens perfectly, there is no great difficulty in attaining that knowledge, it was quite evident to me that there had never been any star in that place in the sky, namely in the constellation Cassiopeia, even the smallest, to say nothing of a star conspicuously bright as this.'

'I was so astonished at this sight that I was not ashamed to doubt the trustworthiness of my own eyes,' continued Tycho. 'But when I observed that others too, on having the place pointed out to them, could see that there really was a star there, I had no further doubts. A miracle indeed, perhaps the greatest of all that have occurred in the whole range of nature since the beginning of the world!'

'So far so good,' I was startled a bit, when suddenly Jeppe intervened. 'My master has been describing his spotting a new star, a revelation that truly belongs to the realm of astronomy. But humans, in their folly, think that there is more to it than meets the eye, you know. They believe that stars and planets control their lives, their destiny, and foretell things to come, events to befall. This so-called science, known as *Astrology*, is as bogus as my so-called prescience. Hush, not a word more about this! Your honour should not talk lightly of these high matters. Like most astronomers of his time, my master believed in astrology and so did Kepler, to some extent, or maybe they pretended to believe, who knows?'

Casanova, who was listening to Jeppe's monologue with an indulgent smile, commented, 'Well, let us give credit where it is due. Tycho happened to say: *Astrologers do not bind the will of man to the stars but grant that there is something in man that has been raised above the stars.* As for Kepler, he clearly claimed: *Astrology is the step-daughter of astronomy, a dreadful superstition, sortilegous monkey-play.* Still, even in your modern age, *signore*, there are some astronomers, I am sure, who harbour such superstitions. Well, so be it. But let us listen to what the new star was supposed to have prophesied.'

Tycho went on to describe the astrological implications of his *nova*. 'The star was at first like Venus and Jupiter, giving pleasing effects; but as it then became like Mars, there will next come a period of wars, seditions, captivity and death of princes, and destruction of cities, together with dryness and fiery meteors in the air, pestilence, and venomous snakes. Lastly, the star became like Saturn, and there will finally come a time of want, death, imprisonment and all sorts of things.'

'None of which came to pass of course,' sneered Jeppe. 'But that is not the end of the story—as we are about to learn.'

'I assume that you are aware of all the inane speculations and the absurd furore that followed in the wake of the *nova*,' continued Tycho.

'As I was only a year old at the time, I am afraid I cannot be expected to have learnt about it first hand,' smiled Kepler. 'Nevertheless, I heard about it later on.'

'But the crucial question was whether the *nova* belonged to the realm of the stars. This would have contradicted Aristotle's idea that the celestial region is eternal and immutable. I am sure you know about his dictum.'

'Naturally,' answered Kepler. 'Aristotle held that only the region below the Moon was subject to changes as it is composed of inferior matter obeying different laws of nature as opposed to the superior quality of the material that makes up the region above it, which is regulated by totally different laws.'

'Yes, indeed,' concurred Tycho. 'Did the *nova* then belong to the sub-lunar world of changeable earthly elements? Was it perhaps a comet condensed from ethereal vapours or from fumes of human sin as some imagined? I put an end to all those foolish conjectures, did I not?' Tycho paused to savour the memory of his triumph and continued. 'I had to make repeated measurements of the position of the *nova*, its distance from the celestial pole, to show that it was indeed a star and not an inhabitant of the sub-lunar region, thereby contradicting the cherished ideas of Aristotle. But what was this strange new star composed of? I believe it was formed out of celestial matter, but of less perfect variety than that of normal stars and thus gradually dissolved away.'

Tycho stroked his beard as he mused about the days of his glory.

'I wrote down all my findings in my book *De Stella Nova*, or *On the New Star*,' recounted Tycho. 'A masterpiece many said afterwards. Deciding whether to write that book or not was a struggle for me, for many a nobleman of my acquaintance felt strongly that writing a book was an occupation beneath my rank. Nobility has its own norms and manners, does it not?'

A serious look came over Kepler's countenance as he remarked, 'Am I not aware of that fact! I know of the occasion on which you accompanied one of the Royal Councillors to supper. Although you drank a bit over-generously and felt pressure on your bladder, you had less concern for your health than for your manners, and remained seated at the table. By the time you returned home, you could no longer pass water.'

Jeppe suddenly jumped up and screeched, 'From Urania to uremia, from Urania to uremia!' and cackled loudly. Kepler, wanting desperately to get out of this awkward situation, hastened to change the subject.

'Three years after you left for your Heavenly Abode, I too was fortunate enough to discover a *nova*. Its location was in the constellation *Ophiuchus* who holds in his arms the serpent, depicted by the constellation *Serpens*.'

I remembered how Aristophanes had concluded his dramatic presentation of the Orion story with Ophiuchus healing both Orion and the scorpion as also the fact that the serpent was supposed to have been the mentor of Ophiuchus.

'I was also able to demonstrate that the *nova* belonged to the region of the stars and not to the vicinity of the Earth,' Kepler went on to say. 'Again, one more blow to Aristotle's philosophy.'

'That is not all,' Tycho said excitedly. 'Five years after the appearance of the *nova* in 1572, I observed a comet and proved that it was far above the Moon and had started its journey way beyond the planet Venus. Aristotle, like many before him, had claimed that each planet was fixed in a rotating crystal sphere, had he not? If so, the comet in its journey ought to have shattered the crystal spheres. Did we here the tinkling sound of the falling shards of crystal?' Tycho laughed at his own comment. 'Oh no, and why not? Because the crystal spheres do not exist at all. Often, those who watch the sky never think of the implications of what they observe. That is why I rebuked them in my book on the new star with the words, *O crassa indgenia, O caecos coeli spectores—O thick wits, O blind watchers of the sky!*'

'Aristotle thought that the Sun and the planets circled around the Earth imbedded in their crystal spheres, and Ptolemy gave a fine but complicated model of this Earth-centred Universe. But we know that the planets go around the Sun, do we not?' added Kepler.

'Yes, what about the Earth?' asked Tycho. 'According to your Copernicus, the Earth goes around the Sun too. I think not, for the Earth is a hulking, lazy body unfit for motion. Look at my beautiful model of the Universe in which the planets move around the Sun, but the Sun, along with the planets, revolves around the Earth.'

Kepler remained silent. He knew that Copernicus was right. Yet, one had to admit that Tycho's model was the same as that of Copernicus but as viewed by an Earth-bound observer. It just complicated matters unnecessarily.

'I wanted you to prove my model,' continued Tycho. 'That is why I set you on to my observations on Mars from which I expected you to deduce its orbit.'

Kepler's eyes became remote and thoughtful as he recalled the past. His voice was tinged with gratitude as he spoke.

'I consider it an effect of Divine Providence that I arrived in Prague when the time was ripe for studying Mars and you asked me to take up the problem. Because for us to discover the secret knowledge of astronomy, it is absolutely necessary to use Mars. Otherwise, that knowledge would remain eternally hidden.'

'You anticipated that you would complete the calculation of the orbit of Mars in eight days,' Tycho pointed out. 'How long did it take in the end?'

'More like eight years,' Kepler smiled. 'Vanquishing Mars, the god of war no less, was an immense task. But what a revelation to discover that the orbit was an ellipse and not a circle as had been believed since antiquity! And Mars did not move with uniform velocity either. He speeded up near the Sun and ambled slothfully far away from the Sun when the latter's influence on him diminished.'

'Did you ever think of the physical causes that control the planetary motion?' enquired Tycho.

'Indeed I did,' Kepler responded. 'My goal after all was to show that the celestial machine is not so much a divine organism but rather a clockwork as much as all the variety of motions are carried out by means of a single, very simple magnetic force of the body, just as in a clock all the motions arise from a very simple weight. The Sun is the fountain of strength, a great magnetic body, whose rotation also turns the magnetic emanations that propel the planets in space.'

'On the other hand, some believed that the angels, who serve God, push the planets around,' Tycho pointed out.

'I feel that the subtle reflections of some people in regard to the blessed angels do not concern us. We are discussing natural matters of much lower rank.'

'Well, I wish I had lived long enough to see the fruits of your labour, which was, after all, based on my observations.'

'So do I, for the reason that, a few years after you were freed from your earthly shackles, the Italian astronomer and philosopher Galileo Galilei introduced his glazed optical tube, or telescope as it has come to be known, to astronomy. He made astounding discoveries with its aid. Incidentally, he too sighted the two different guest stars we had observed. I cannot imagine what marvels you might have revealed with the use of the optical tube.'

'Did you ever employ that instrument for your own observations of the heavens, Johannes?'

'I wrote to Galilei entreating him to send me an optical tube, but he sent me only his book. The Duke of Bavaria lent me one for just five weeks and took it back promptly thereafter.' Kepler's regret of having missed the use of the telescope was evident.

'I understand that Galilei, because of his observations with his optical tube and moreover because of his staunch adherence to the Sun-centred Universe of Copernicus, ran into conflict with the Church, and had to recant his beliefs under the threat of torture. Most unfortunate indeed! Of course, all this must have happened after my time.'

Kepler was passionate: 'That was truly tragic. While in theology it is authority that carries the most weight, in philosophy it is reason. The Holy Office, namely the Church, nowadays is holy which, though allowing the Earth's smallness, denies its motion. To me, however, the truth is more holy still, and with all due respect to the Doctors of the Church, I prove philosophically not only that the Earth is round, not only that it is contemptibly small, but also that it is carried among the stars.'

Tycho closed his eyes in thought. 'We all strive to find the truth underlying what we see. How far we succeed it may not be given to us to decide and often not even to know.'

Kepler fell silent for a moment and recalled with a touch of sadness. 'During the last night of your life, you kept whispering again and again: *Let me not seem to have died in vain.*' And then he added with emphasis, 'There was no need for such thoughts. You were and will always be the King among Astronomers, Master.' The two men fell silent, each engrossed in his own thoughts.

After a short while, Kepler slowly rose to his feet and came around the table at which he was conversing with Tycho Brahe. At the same time, the whole scene, including Tycho and Jeppe, receded back and faded away.

Vanquishing the God of War

Kepler smiled as he approached us. Ah, here is a man who is friendly, kind, and loveable, I thought. He spoke now in a simple style, distinctly different from the one he had been using in his discussions with Tycho Brahe. Obviously, most of the time, the two astronomers had been quoting from their own writings as in the case of Copernicus.

'Yes, it took me a long time in arriving at the true nature of the path taken by the planet Mars around the Sun,' reminisced Kepler. 'I had to match my own calculations to the remarkably accurate observational data of my master, the revered Tycho Brahe. First of all, I firmly believed in the Copernican model in which the planets went around the Sun that stood still at the centre. The first step was to take into account the motion of our own Earth around the Sun before tracing the orbit followed by the planet Mars. It happened to be quite a tricky task indeed!'

'An ingenious piece of work in itself,' Casanova commented.

'Well, my next assumption was that all planets, including our own Earth, described similar curves as they circled the Sun. What geometric figure would that be? As you must have heard time and again, everyone before had believed that this path *had* to be a circle, the perfect, the most beautiful curve of all. And, in addition, the planetary motion was supposed to be uniform with constant velocity. I too tried my hand at a circular path for Mars. Things did not work out if one made it move with unwavering speed in a circle centred round the Sun. So, I moved the Sun away from the centre and made Mars run faster and slower in its course. Ah, it seemed to work! Trial and error, trial and error, I spent five long years on this arduous exercise. Finally, theory seemed to fit the observations, my calculations running to some thousand pages no less. Except for a miniscule discrepancy of eight minutes of arc! As you must know very well, one goes a full circle by going around by three hundred and sixty degrees and a degree is divided into sixty arc minutes. So you can imagine how small the discrepancy happened to be. But could I just ignore this mismatch, when I knew so well how exacting and precise my master's observations were? Heaven forbid! I had no choice but to abandon the sacred circle as the chosen path for the celestial body. And I tried to make it an oval in shape. But then, what kind of oval? Another two years of relentless struggle and I had the trajectory that beautifully agreed with the observation. I could even derive the equation for this curve, but failed to realize what it was. Another year passed. Then it dawned upon me. It was an *ellipse*! I said to myself: *Ah, what a foolish bird I have been!* And I could derive the law that ordained the varying speed of Mars.'

'The god of war marched in an elliptical path around Helios, running fast as he basked in the brilliance of the great god when he was in his vicinity, but sauntered slowly far away from him,' paraphrased Casanova. 'Kepler had discovered his first two laws of planetary motion for Mars in particular, but would be applicable to all the planets.'

Kepler smiled. He had recounted very briefly the heroic struggle he had put up in vanquishing the god of war. But he seemed to take no pride at all in his spectacular achievement. 'The elliptical orbit followed by Mars was almost identical to a circle, the difference being hardly perceptible. That was just a detail. What mattered was the fact that I had abandoned the ideal of uniform circular motion! But then, why should we consider only the circle to be beautiful? Any curve Nature chooses to use is beautiful. A drop of rain is spherical and it is beautiful. So are snowflakes, made up of water as well, that take myriad forms, which I had examined carefully. So is the beehive with its orderly compartments whose shape and structure I had studied in detail. Yes, after eight long years of incessant thought and toil, I had discovered, by appealing to the precise observations of my master Tycho Brahe and carrying out painstaking calculations of my own, that the orbit of Mars—and that of any other planet for that matter—is not a circle, but an *ellipse*! I recorded in great detail my long journey with all its twists and turns, false starts and dead ends, and the final result in my book, *Astronomia Nova* or *New Astronomy*.'

'Another great book to appear after the two volumes by Copernicus and Tycho Brahe,' remarked Casanova.

'All my life I happened to ponder upon the mystery of the Universe,' mused Kepler. Perhaps he had forgotten our presence. 'Why is the Universe constructed the way it is? Why are there only six planets, including our beloved Earth, with five gaps in between? Why are they arranged at particular distances from the Sun? At one time, I had arrived at some conclusions in this respect making use of the five Platonic Solids and even recorded my ideas and findings in my first book, *Mysterium Cosmographicum* or *The Cosmic Mystery*. I imagined that I had at last unravelled the mystery of the Universe. I even wrote in the preface to the reader: *I saw one symmetric solid after the other fit in so precisely between the appropriate orbits that if a peasant were to ask you on what kind of hooks the heavens are fastened so they do not fall down, it will be easy for thee to answer him. Farewell.* But I soon found it otherwise. All this before I found out the two laws that govern the motion of the planets. Perhaps the ultimate mystery of the Universe will never be understood—a thought that never occurred to me, a sentiment that I never expressed. In any event, I did try to find the inherent harmony pervading the cosmos, did I not? The planets moved slower and slower as their distances increased from the Sun. I thought that I could hear the music made by them—fast and slow. In my book *Harmonice Mundi, Harmony of the Worlds*, I even wrote down the musical notes created by each of the planets as I had imagined. Beneath it all is the harmony of the worlds, the music of the spheres, the mysterious melody of the Universe. Yet, it is music that can be heard not by the ear but by the mind and mind alone.'

'And underlying that harmony of the worlds was the third law of planetary motion, which Kepler went on to discover and write down in his book,' added Casanova.

Kepler seemed to be trying to listen to some remote imaginary strains of heavenly harmony. Then he looked up and nodded at us acknowledging our presence.

'I hope—and may I presume—that everything we have discussed is not only clear but interesting as well,' said Kepler eagerly. 'But, fact would be far more interesting when blended with a bit of fantasy, would it not? Indeed I cannot take leave of you without relating to you my *Somnium* or my *Dream*. It is based on true astronomy enshrined within the framework of Copernican cosmos, but presented in a fanciful way.'

'Let us listen with undivided attention, *signore*,' exhorted Casanova. 'The story we are going to hear is the very first example of what you call science fiction nowadays.'

'Our hero of the *Somnium*, Duracotus by name, hails from Iceland, situated in the northern-most part of the Earth,' Kepler related the story. 'His mother, Fioxhilda, is a wise woman tutored by a demon who lives on the Moon, which she calls Levania. After his five-year apprenticeship at Hven under the August tutelage of the celebrated Tycho Brahe, Duracotus is privileged to learn some of his mother's secrets. Demons inhabit both Levania and Volva, namely the Earth, but they cannot cross the gap between the two worlds as light is lethal to them and they would perish in the dazzling brilliance of the Sun. However, during an eclipse the Earth's shadow would make a bridge of darkness between the two celestial objects, along which the demons can travel back and forth. When these facts are revealed to him, our hero wants to make a journey to Levania. His mother summons the demons from that world, who administer to him a dozing draught to make the journey painless and take him up to the point where the pull of the Earth just balances that of the Moon. When the demons let go of him at that point where no external force exists, his limbs curl up like those of a spider since the body itself attracts its minor parts, namely the limbs, because the body itself is the whole. To cut the story short, Duracotus does manage to find his way on to Levania and behold the inhabitants. They are not human: some are serpent-like, while others are equipped with fins so they can swim in the swamps of Levania.'

After just a momentary pause, Kepler announced, 'Well, I awoke suddenly to find my head covered with a cushion and my body tangled in a rug. That was the end of my tale. After all, it was just a *Somnium*, a *Dream*, was it not?'

Kepler shook with laughter, which too was filled with warmth and simple, pure joy. And we laughed with him.

Kepler looked up at the sky. Was he searching for the planets that were forever obeying his laws? Was he contemplating his own life, his own work, and his own imminent exit from this Earth, a planet that followed his dictum like its companions? His countenance was now suffused with absolute serenity tinged with slight sadness. Then he spoke in a barely audible whisper:

I measured the sky, now the shadows I measure
Skybound was the mind, Earthbound the body rests.

Faint sounds of several instruments seemed to be coming from far away, from different directions creating an exquisite, indescribable melody. That music grew louder, played on for a short while, and then faded away. So did Kepler merging into the background of the night sky as was the case with his venerated predecessor Copernicus.

'Kepler's last words were his epitaph, which he wrote himself,' said Casanova. 'He had firmly believed in the Copernican scheme of the cosmos. He had based his work on the heliocentric model for the solar system thereby providing an indirect proof of that framework. Now, *signore*, let us go and meet an absolutely fascinating astronomer, a towering philosopher, who hailed from my own country.'

Chapter 7
The Starry Messenger

New Vision, New Worlds

Galileo

'Ah, welcome, welcome,' greeted the man effusively. 'What took you so long? I have been waiting for you, my friends. Well, I know where you have been. You met Tycho and my friend Johannes, didn't you? What a pair they make, what contrast! The nobleman with the false nose and the commoner assistant with imaginary ailments.' He threw back his head and roared with laughter and then added seriously, 'No, I should not make fun of them. Tycho is a fantastic observer and Johannes is an excellent philosopher. And most of all, before visiting them, you met our great teacher, Copernicus himself. How I wish I had that privilege! I envy you. By the time I was born our teacher was gone for twenty-one years. Oh, well, allow me to introduce myself before I forget my manners, which I often do. I am Galileo Galilei. Son of the accomplished musician, Vincenzio Galilei, I am a philosopher, experimenter, mathematician, and, most of all, a stargazer.'

I was transfixed by the sight of the one and only Galileo, whom I could easily recognize from his portraits. He had a large head with a receding hairline that made for a high forehead. He sported a full beard, which made his head look even larger. He had an intensely expressive face: his eyes, brows, and lips displayed his changing moods in forceful concerted movements. His eyes would glitter mischievously, widen with wonder; his brows would arch questioningly, knit together in a frown; his lips would widen with a smile, curl up in a sneer.

As I looked around, I realized that I was on the top of a tall building overlooking the town below. It was a strange lovely town, covered by a network of canals in place of streets, bridges arching over them. People moved around in gondolas, many of which were quite picturesquely decorated. Some even played guitars and sang as they enjoyed the ride in those gondolas. Far away, there was the sea with ships sailing in and out of the harbour. Of course, this was Venice of bygone days, not the modern version I had visited recently.

C.V. Vishveshwara, *Universe Unveiled*, Astronomers' Universe,
DOI 10.1007/978-3-319-08213-4_7

'Let me show you my glazed optic tube, my spyglass, or telescope as it has come to be commonly called,' said Galileo taking out a long tube. He detached from its ends two lenses. 'Ah, look at these lenses,' pointed out Galileo. 'They are plane on one side. But on the other side, one of them is *convex*, it bulges out. The other one is *concave*, it is hollow—just the opposite of the former. I am sure you know these things. I have shaped these pieces of glass by meticulously polishing them. It takes quite a bit of accurate work, you know. Let us attach them to the tube, one at each end. There, we have our optic tube ready. We look through it with the eye held close to the concave lens—like this. Ah, what wonders we see then! You will discover for yourself since I have arranged a spyglass specially for you.'

Of course, I was familiar with the Galilean telescope. I had used it even as a kid. But, it was such a rare pleasure and privilege to have it demonstrated by Galileo no less.

Galileo turned the telescopes towards the town below and beckoned, inviting me to take a look. As was to be expected, people who were far below now appeared so close, that I felt I could actually touch them.

'I set up my optic tube on the top of the highest building I could locate in this city of Venice, just as I have done right now' related Galileo. 'Many important people mounted up the stairs, some of them huffing and puffing because of their great age or bulk, in order to see the ships that were so far off that it was two hours before they were seen without my spyglass, steering full-sail into the harbour. The effect of my instrument was such that it made an object fifty miles away appear as large as if it were only five.'

Galileo waited until I had fully enjoyed the sight of the town below.

'Looking at people and objects at a distance was only a pastime. But, between you and me, the foolish people—they happened to be important ones as many fools are—were so impressed that they gave me a permanent job carrying a generous remuneration,' laughed Galileo. 'But then, the greatest event of my life happened: I turned my optic tube heavenwards!'

As he spoke his last words in all seriousness, Galileo turned the telescopes slowly upwards. At the same time, night descended, the sky darkened, stars and planets shone brightly.

'You will see for yourself the wonders I observed through my spyglass,' said Galileo. 'I recorded my findings in a short book with the title *Sidereus Nuncius* or *The Starry Messenger*. I wrote my book in Latin as the custom was in those days. In any case, take a look at the sky above through your optic tube.'

Peering through my telescope, I was overwhelmed by the incredible sight of countless stars, although this was to be expected.

'Oh, yes, I too was astonished when I looked at the night sky through my spyglass,' confided Galileo. 'You will see a fantastic number of stars. Most of them cannot be seen without the aid of an optic tube. They are unbelievably numerous, are they not? Some of them are crowded together in an amazing manner.'

'Now look at the Milky Way,' Galileo indicated the faint band of light that festooned the night sky. 'You will hear quaint tales related to the Milky Way later on, my young friend. Its mysterious appearance had tormented philosophers through the ages, but now my spyglass could resolve all wordy disputes by revealing its true nature. When I turned my optic tube towards the Milky Way, it dissolved into a mass of innumerable stars planted together in clusters. For, the Milky Way is nothing but collections of innumerable stars grouped together. Upon whatever part the optic tube is turned, a vast crowd of stars is immediately presented to view. Many of them are rather large and quite bright, but the number of smaller ones is quite beyond determination.'

'Now, how about our nearest neighbour?' asked Galileo pointing to the Moon that shone with a pearly glow in the sky. 'It is most beautiful and it is pleasing to the eye to look upon the lunar body, is it not?'

'Take a closer look at the Moon,' urged Galileo. 'Almost the entire face of the Moon glows bright. But there is an uneven darker region that seems like a cloud covering part of it. This had been noticed for ages.'

Oh yes, I had noticed this since my younger days as I gazed upon the Moon on many occasions.

'Yes, I turned my optic tube towards our heavenly neighbour,' said Galileo, his head held high with pride. 'And what do you think I observed? Observed for the very first time, observed as never before! The entire face of the Moon is covered with spots. They look even more interesting when only a part of the Moon is illuminated by the Sun. I made a drawing of what I saw. Let me show it to you.'

Galileo brought out the sketch of the Moon that he had made after observing it through his telescope, showing all the spots or patches he had mentioned.

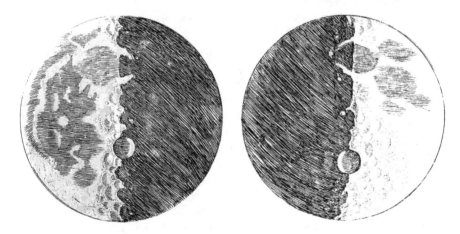

'What a beautiful drawing, *signore*!' exclaimed Casanova. 'Even your modern-day telescopes reveal almost exactly the same picture of the Moon, do they not?'

By now, the real Moon above us resembled exactly Galileo's drawing, partially illuminated by the Sun, with a jagged line separating the bright and the dark sections. As I looked through my telescope, Galileo continued his description.

'Spots, spots everywhere,' exclaimed Galileo. 'They are no ordinary spots. Look, they are bright on the side facing the Sun and, moreover, they cast shadows on the other side, do they not? As I said, those spots are no ordinary ones, my friend. They form, in fact, chains of mountains and depths of valleys just as it happens on the Earth.'

I felt that I could actually see those mountains and valleys. This was a totally different view of the Moon.

'Now, for something beautiful,' announced Galileo. His excitement was infectious. 'The bright and dark parts of the Moon are exactly like day and night on the Earth with a dividing line between them. Look, look, a point lying in the dark area, some distance away from the dividing line, has brightened up. Why? Because, the first rays of the Sun, grazing the surface of the Moon, have lit up a mountain peak while the surrounding region is still in the dark. See, more peaks are brightening up, while the surrounding regions too are getting bathed in sunlight. What a lovely sight! All this happens right here on the Earth too.'

This was indeed amazing. One could feel the thrill of watching something happening far, far away.

'Yes, yes, it was all wonderful to watch,' recounted Galileo. 'But, you know, by measuring the distance of the lighted peak from the dividing line and from the known size of the Moon, I could estimate the height of the mountain. It was quite simple.'

'Simple but ingenious,' added Casanova. 'Like the measurements of Aristarchus, whom you met long ago.'

'Oh, what a surprise! I found out that some mountains on the Moon are loftier than those on the Earth.'

'Not quite I am afraid,' put in Casanova in a low voice. 'That is because, in those days, they did not know the exact heights of the mountains on the Earth as you do now.'

'What does it all mean?' mused Galileo stroking his beard. 'It means that the Moon, with all its mountains and valleys, is just like our Earth. For ages, philosophers had claimed that the Moon, like all heavenly bodies, is perfectly spherical and absolutely smooth. But, I had demonstrated that this was not true. But, foolish people, would they give up their idiotic ideas? Some said the Moon was indeed smooth as it should be. All the mountains, peaks, and valleys were *within* the smooth surface. Like the marbles you saw in the amusement park.'

I remembered the amusement park, where I had spent not only some joyful time, but also had learnt about epicycles that too guided by Ptolemy himself.

'Humans are a strange lot,' reflected Galileo shaking his head. 'They do not give up their prejudices easily. In any case, the important fact is that the Earth and the Moon are alike. Both of them happen to be just two similar celestial bodies.'

Galileo paused to give time for me to digest all the details about the Moon he had demonstrated and the conclusions that had followed.

'Now, my dear friend, let me tell you about another observation I made. I feel it deserves to be considered the most important in my work. I am talking about the discovery of the four Moons of the planet Jupiter,' announced Galileo. 'Ah, those Moons, they looked like little stars through my spyglass. But were they fixed like ordinary stars? Oh, no, they moved along a straight line that passed through Jupiter. There happened to be four of them in all. Three of them were quite bright, but the fourth one was very faint indeed. Different ones appeared sometimes on one side of the planet and sometimes on the other. Then again, sometimes, they disappeared altogether. Do you see what I mean?'

I was delighted to watch the hide-and-seek the little stars, or the Moons, were playing around Jupiter.

'Here, let me show you the diagram I made marking their positions.'

Galileo pulled out a sheet from his pocket and spread it out. It showed a series of very simple sketches of the planet Jupiter drawn as a circle and the positions of what he had taken to be regular stars marked by little crosses.

Obseruationes Jouiales
1610

2.ꝰ. gbris. marꞓ H. 12	O * *	
30. mane	* * O	*
2. ꝛbr:	O * *	*
3. morƞ	O * *	
3. Ho. s.	* O	*
4. mone.	* O	* *
6. mane	* * O	*
8. marꞓ H. 13.	* * * O	
10. mane.	* * * O	*
11.	* * O	*
12. H. 4 uesp:	* O	*
13. mane	* * • O	*
14 Lure.	* * * O	*
15.	* * O	
16. Clariſ.ᵉ	* O * * *	
17. clariƞ.ᵉ	* O * *	
18.	* O * * *	
21. mone	* * O * *	
24.	* * O *	
25.	* * O	*
29. uesp:	* * O	
30. mane	* * O *	
January 4. morƞ	* * O	*
4. uesp.	* * O	*

'From these meagre notes, I concluded that what I had observed were not stars, but Moons circling around Jupiter just as our nearest neighbour goes around our Earth.'

Galileo inclined his head and looked intently at me, holding my attention.

'Needless to add, it was I who discovered those Moons of Jupiter for the first time in all human history,' proclaimed Galileo.

Galileo was proud of his discoveries. Justly so, I felt.

'In essence, just as our Earth circles the Sun, while the Moon does the same as it goes round the Earth, we have exactly the same situation in the case of the planet Jupiter and its four Moons. Again, therefore, there is nothing special about the Earth. It is not the only centre of revolution, it is not the centre of everything, as had been considered all along. Why couldn't the Earth itself go round the Sun then, as Copernicus taught us? He was absolutely right, no doubt about it.'

'To complete the story, those Moons Galileo had discovered are collectively known as the *Galilean Satellites* as they should be,' said Casanova and added, 'They bear the names, *Io, Europa, Callisto*, and *Ganymede*. In mythology, the last one was the cup-bearer of Jupiter, while the other three were his sweethearts.'

'Is the Earth at the centre or the Sun? Is Ptolemy correct or Copernicus? That was the question,' Galileo extended his open palms and raised and lowered them alternately as though he was weighing two imaginary objects held in his hands. 'Why not put this question to a planet and ask whether it went around the Earth or the Sun? How about the brightest of them all, namely Venus, the planet named after the goddess of love? So, I focussed my spyglass on Venus. And what did I discover? Ah, the phases of that lovely lady!'

Galileo nodded his head a few times vigorously and went on.

'I do realize that you know all about the phases of our Moon. Oh, fortunate one, you learnt about that phenomenon from that towering astronomer of ancient Greece Aristarchus himself. He was the first one to think of the Sun being the centre of the Universe, was he not? There is no limit to my astonishment when I reflect that Aristarchus and Copernicus were able to make *reason* so conquer *senses*, that in defiance to the latter, the former became the mistress of their belief.'

'That transition from *appearance* to *reality*, so well articulated by our distinguished host just now, was a giant leap in human thought, *signore*,' said Casanova emphatically. 'All your science has striven since then to establish reality gleaned out of appearance, however elusive, however contrary to common sense it might have looked at first on the surface. But let us listen to my illustrious compatriot's further observations.'

'Suppose the Earth is at the centre and Venus and the Sun go around it,' continued Galileo. 'Then Venus would always be nearer to the Earth than the Sun. One can easily show that the planet would appear to have a crescent shape alone as viewed from the Earth. On the other hand, let us go along with Copernicus and assume that Venus and the Earth, in that order, circle the Sun. So, Venus is sometimes between the Earth and the Sun and sometimes on the farther side of the Sun. In this case, it would exhibit a complete set of phases as observed from the Earth, but quite different from those of the Moon though. Look through your optic tubes and see for yourself which of the two possibilities happen in reality.'

I eagerly glued my eyes to the telescope in front of me. Oh, yes indeed, the bright beautiful planet went through all the phases but unlike what I had seen in the case of the Moon, as Galileo had described. I was not surprised. Galileo seemed to know it all.

Galileo's eyes shone with joy as he declared, 'Yes indeed, Venus displayed the entire range of phases. The love goddess showed her full face in all its glory, whereas sometimes she displayed only part of it tauntingly, and finally hid it shyly in complete darkness. Was this not positive proof of the Copernican model? Yes, indeed! Oh, yes, yes!'

Galileo waited, relishing his past momentous observation and the conclusion that followed, before going on.

'Let me recount two more discoveries I made with my optic tube,' Galileo went on. 'I now turned my most powerful telescope—I had made several of them you see—towards the farthest planet of all, namely Saturn. I could see its disk all right. But, on closer examination, I was startled to see what I saw. And what *did* I see? The planet Saturn was not just one entity, but was composed of three parts, which almost touched one another and never moved nor changed with respect to each other. The middle one was three times the size of the one on each side.'

As Galileo spoke, Jupiter had made way for Saturn and I could make out the central disk as well as the adjoining members of the planet.

Galileo pondered for a moment, running his hand along his telescope, and continued, 'I was puzzled to say the least. The two attendants of Saturn were quite unlike the satellites of Jupiter. They showed no relative motion at all. Two years later, I was even more astonished to discover that the planet was single. The other two parts were gone. What was to be said of this strange transformation? What had happened to the two attendants? Had they suddenly vanished or fled? Had Saturn perhaps devoured his own children? Or were the appearances indeed illusion or fraud, with which the glasses had so long deceived me, as well as many others to whom I had shown them? I did not know what to say in a case so surprising, so unlooked for, and so novel. But, marvel of marvels, the two attendants appeared again now in the guise of arms or handles attached to the main component!'

Galileo shook his head frowning in bafflement.

'Poor Galileo,' laughed Casanova softly. 'What he had observed were the rings of Saturn. His optic tube was not powerful enough to show the details. When viewed edge-on, those rings had become invisible.'

It took a little while for Galileo to regain his composure after experiencing once more his lack of understanding of Saturn's mysterious behaviour. He shook his head again and again before continuing with his exposition.

At that moment, light started seeping into the night sky as though it was dawn now. The Sun seemed to be hastening to appear in the sky, realizing that its presence was needed for Galileo's next demonstration.

'The Sun, the blazing source of light, a perfect heavenly body, venerated by people of all ages as a supreme god!' exclaimed Galileo. 'But is the Sun perfect? Tell me, my friend from the future. No, there are blemishes, spots on that god's face as I discovered. But let us not look upon that face through the spyglass. For, there is no doubt that we shall go blind. Let us, instead, direct the spyglass as if we were going to observe that body. Having focused and steadied the optic tube, expose a flat white sheet of paper placed about a foot from the concave lens close to which you normally place your eye. Now look, a circular image of the Sun's disk has fallen upon the paper, with all the spots showing exactly as they appear on that heavenly body.'

Oh yes, I could see on the sheet of paper the face of the Sun marked with a few dark spots.

'As you can see, these spots are carried across the disk of the Sun from day to day. By following their movement, I could determine that the Sun takes between three to four weeks to spin once on its axis.'

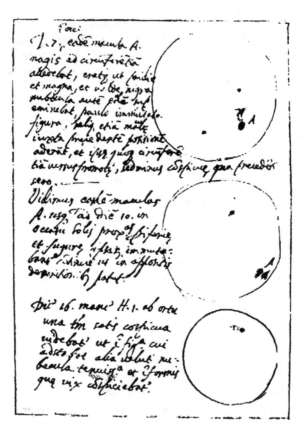

Galileo sat down on a chair next to the telescope, mopped his brow with the back of his hand, and heaved a sigh. The excitement of reliving the moments of his phenomenal discoveries had exhausted him. He closed his eyes and sat motionless, while I watched him with utmost respect. To my immense astonishment, ever so slowly, he seemed to grow older. His hair and beard turned grey, wrinkles appeared on his face that clearly showed signs of extreme weariness.

Crime and Punishment

'I was convinced beyond doubt that Copernicus was right,' Galileo said gathering his strength. His voice was still firm and his mind was clear. 'Yes, the Sun was the centre of the Universe and the Earth went around it. How could anyone doubt this fact when I had proved it by my own observations? I was determined to spread the truth about our Sun and his family of planets and spread it I did.'

Galileo's voice turned sombre as he recalled the past events. 'But the Church did not take this kindly. How could the Earth—and man with it, man who had been created in the image of God—be banished from its central position? More importantly, how would people react if they came to know the falsity of all the teachings of the Holy Scripture that they had come to believe for ages? Let me not go into the details of the painful dealings with the Church I had to carry out in the initial stages of my struggle. I was ordered to abandon the doctrine of Copernicus and, in the words of the Church's decree, *not to hold, teach or defend it in any way, orally or in writing.*' Galileo's eyes gleamed as he announced, 'Yet, I wrote my book discussing the Copernican system as opposed to that of Ptolemy, *Dialogue Concerning the Two Chief World Systems: Ptolemaic and Copernican.* I wrote the book in Italian, our common language, so that everyone, including the vegetable vendors and boatmen, could read it and understand the true nature of our Universe. That was a point against me since all scholarly books were supposed to be written in the ancient language, Latin, which the common man could not read or write so that knowledge remained the property of a chosen few, especially those belonging to the Church. Well, I presented my arguments as a discussion among three people, Sagredo, Salviati, and Simplicio. Of these three, Sagredo is an intelligent questioner eager to learn.' Galileo's face lit up as he described the second member of the trio. 'Salviati, who champions Copernicus, is brilliant, profound, and possesses a sharp wit and tongue, which he uses to advantage in argument. Well, I do not feel ashamed to say that I had my own self in mind when I created this character.' A malicious glee appeared on Galileo's face now. 'And Simplicio, as his very name indicates, is a blundering simpleton, who firmly believes in Aristotle and repeats that ancient scholar's ideas time and again, advocating the Ptolemaic system of the Universe. Salviati, needless to add, demolishes Simplicio and holds up Copernicus for all the eyes of the world to behold.'

Galileo's face darkened as he brooded over the events that had followed his writing the *Dialogue.* 'At first the Church did not mind what I had written. The book was duly published. But then the tide turned. It was thought that I had deliberately lampooned the Church through the character of Simplicio. The Pope, the highest authority in the Church, was infuriated. Then came the shattering blow. I was summoned to Rome, the highest seat of the Church, to be tried. This was the worst thing that could have happened to me or to any other man.'

Galileo bowed his head in silence. By now he had aged completely.

Then the scene shifted.

'We are in Rome, *signore*.' Casanova said in a hushed voice. His expression had turned sad and serious. 'We are standing in the Piazza della Minerva. Behind us stands the Church of Santa Maria built in the thirteenth century on the ruins of a Roman temple dedicated to Minerva. Look over there on the right. That building is the Convent of Santa Maria Sopra Minerva, the seat of the inquisition, its victims relentlessly pursued by the Dominican order. It is here, at the moment, Galileo faces the inquisition, standing trial for his heretical views and their propagation in violation of the order issued by the Church.'

Casanova shook his head. 'Minerva, the goddess of wisdom! What an irony, *signore*, wisdom seems to have been thrown to the winds by those who are in power.'

I was reminded of the last lines of Robert Frost's poem *Stars*:

And yet with neither love nor hate,
*   Those stars like some snow-white*
Minerva's snow-white marble eyes
*   Without the gift of sight.*

Had Minerva, the goddess of wisdom, turned a blind eye on the proceedings taking place within the temple dedicated to her? I felt the atmosphere in the piazza oppressive and threatening.

'Let us go in, *signore*, to witness the end of the tragic event of the past, an event that cannot be reversed,' Casanova spoke his words gently and led me into the Convent.

The interior of the Convent was gloomy. A number of people had thronged around. Seated on high carved chairs were men with stern faces dressed in impressive robes. In front of them stood Galileo. He appeared to be almost on the verge of collapse, far from what he had been—the exuberant stargazer, robust and happy, who had described his observations and explained the system of the Universe.

'The entire Congregation of the Holy Office, the highest authority of the Church, has gathered here,' informed Casanova in a barely audible whisper. 'They have just passed their judgement on Galileo.'

Galileo bowed low before the Congregation and knelt down with great difficulty. I could hear his feeble voice that shook as he made his forced confession and caught a few of his last words.

'I Galileo, son of the late Vincenzo Galilei of Florence, aged seventy, kneeling before you, do swear that I have always believed, and now believe, and with God's help will in the future believe all that is held and taught by the Holy Church. The Holy Office intimated to me that I must altogether abandon the false opinion that the Sun is the centre of the world and is immovable, and that the Earth is not the centre of the world and moves, and that I must not hold, defend or teach in any way whatsoever, either verbally or in writing, the said false doctrine, which is contrary to Holy Scripture...Therefore, with sincere heart and unfeigned faith I do abjure, curse and detest the said errors, which are contrary to Holy Church. I swear that in future I will never again say or assert verbally or in writing, anything, which might again give ground for suspicion against me...'

Galileo's voice trailed off as the scene slowly faded away.

'They say that even then, as he rose from his kneeling position, Galileo had mumbled inaudibly, *Eppur si muove—And yet. . .it moves,*' remarked Casanova. 'That is absurd, to say the least. By the time he faced the Congregation of the Holy Office, Galileo was already a broken man. He had been shown the implements of torture in order to intimidate him. If those hounds of inquisition had even suspected that he had uttered those words repudiating his own abjuration, they would have torn him to pieces. For the rest of his life, Galileo became practically a prisoner confined to his home in Arcetri. Let us move on to that last abode of the unparalleled scientist who had changed Man's view of the Universe as none had done before him.'

Twilight Days

We were inside a room of modest proportions with very few pieces of furniture. Seated at a rough-hewn table, was Galileo. On the table rested his favourite telescope, the very first one he had fabricated, which he had turned heavenwards.

Galileo was silent. He seemed to be far away, somewhere in the past surrounded by his memories. Those memories were probably not happy ones as his frown betrayed. Presently, he started to mumble to himself, his musings often accompanied by exaggerated gestures. Strangely enough, most of his words seemed to have been written by him long before.

With a wistful smile, Galileo picked up his telescope and caressed it lovingly.

'Ah, my spyglass, my glazed optic tube,' he addressed it affectionately. 'My friend Kepler thought that you were more precious than any sceptre. You revealed so many wonders, so many new phenomena. But my colleagues stirred up against me as if I had placed these things in the sky with my own hands in order to upset Nature and overturn the sciences. But you, Kepler, you confirmed everything I wrote, without rejecting an iota of it. I indeed congratulate myself on having an associate in the study of Truth, who is a friend of Truth. For it is a misery that so few exist who pursue the Truth and do not pervert philosophical reason.'

Galileo's lips twisted into a sardonic smile. 'Even after my demonstrating Jupiter's Moons with my optic tube, none of the philosophers and scholars conceded their existence. They claimed that, seven being a sacred number, there could only be seven objects in the sky: Sun, Moon, Mercury, Venus, Mars, Jupiter, and Saturn. Cesare Cremonini, teacher of philosophy at Padua refused even to look through my optic tube. So did my dear colleague Julius Libri!'

Galileo's expression turned into one of intense glee, almost verging on the malicious, as he recalled his own past reaction to this foolishness. 'Soon after, Libri died and I could not help commenting: Libri did not choose to see my celestial trifles while he was on Earth; perhaps he will do so now *en route* to heaven!'

Galileo exploded with uproarious laughter slapping his thigh, but fell quickly into his melancholy mood again.

'Ah, Kepler, most of my colleagues are incapable of identifying either Jupiter or Mars, and hardly even the Moon. What is to be done? Let us laugh at the stupidity of the crowd, my Kepler. Why are you not here? I wish I had more time to laugh with you. How you would shout with laughter, my dearest Kepler, if you were to hear what the chief philosophers of Pisa said against me to the grand duke... But the night has come and I can no longer converse with you.'

Galileo closed his eyes in weary loneliness. 'I believe that good philosophers fly alone, like eagles, and not in flocks like starlings. It is true that because eagles are rare birds: They are little seen and less heard, while birds that fly like starlings fill the sky with shrieks and cries, and wherever they settle befoul the Earth beneath them.' After a pause he added, 'The crowd of fools who know nothing is infinite. Those who know very little of philosophy are numerous. Few indeed are they who really know some part of it. And only One knows it all.'

'Ah, the Universe!' exclaimed Galileo with awe after a momentary pause. 'To investigate the constituents of the Universe is one of the greatest and noblest problems in nature. But has anyone so far proved whether the Universe is finite and has a shape, or whether it is infinite and unbounded? O, foolish man, does your imagination first comprehend some magnitude for the Universe, which you then judge to be too vast? If it does, do you like imagining that your comprehension extends beyond the Divine Power? Would you like to imagine for yourself things greater than God can accomplish? And if it does not comprehend this, then why do you pass judgement upon things you do not understand?'

Galileo wiped away the glistening beads of perspiration that had appeared on his brows with the back of his hand.

'No son of Adam had seen further than I since the beginning of the world,' Galileo closed his eyes and heaved a deep sigh. 'This Universe which I with my astonishing observations and clear demonstrations had enlarged a hundred-, nay, a thousand-fold beyond the limits commonly seen by wise men of all centuries past, is now for me so diminished and reduced, it has shrunk to the meagre confines of my body.'

I was startled as Galileo raised his head, opened his eyes and stared vacantly into space. He had grown very old indeed, his hair and beard had turned white, and his face was covered with a network of wrinkles. 'Alas, Galileo has become infirm with age and his eyes have lost their power of vision,' said Casanova in an almost inaudible whisper. 'He has become totally blind, *signore*.'

Galileo sat motionless as he brooded over his own condition.

'Bereft of my powers by my great age and even more by my unfortunate blindness and the failure of my memory and other senses, I spend my fruitless days which are so long because of my continuous inactivity and yet so brief compared with all the months and years that have passed. And I am left with no other comfort than the memory of the sweet former friendships. I shall therefore remain silent, and so pass what remains to me of my laborious life, satisfying myself in the pleasure I shall feel from the discoveries of other pilgrim minds.'

Galileo's countenance turned serene and peaceful with his acceptance of his fate combined with the conscious realization of his achievements.

'I must not forget my book *Two New Sciences* which I believe is superior to everything else of mine hitherto published, for it contains results which I consider the most important of all my studies.'

'Galileo wrote that book during his last years,' said Casanova. 'It was based on a number of experiments he had performed over many years and the conclusions he had drawn from them. A monumental work indeed!'

After a moment of absolute stillness, Galileo's final words reverberated in majesty.

'There will be opened a gateway and a road to a large and excellent science into which minds more piercing than mine shall penetrate to recesses still deeper.'

Galileo then bowed his head, his clasped hands resting against his forehead. 'I render infinite thanks to you, God, for being so kind as to make me alone the first observer of marvels kept hidden in obscurity for all previous centuries. *Grazie, grazie, Dio...*'

The rest of his prayer was said in silence.

I felt sad, so sad as I had never felt before.

'Yes, *signore*, Galileo had a difficult life. We did not learn about all the tragic twists and turns it took,' said Casanova. 'But let us not forget, he is one of the greatest men of science ever. His book, *The Starry Messenger*, heralded a new age in astronomy. But Galileo himself ought to be hailed as the Starry Messenger. There are many other things he discovered, some of which you will learn about later on. They happen to be so important to the modern science of your own times. Galileo will always be remembered by those who wonder and think about the stars, the planets, and the entire Universe.'

I could not agree more with Casanova.

'You heard what Galileo said at the end,' Casanova recalled. '*There will be opened a gateway and a road to a large and excellent science into which minds more piercing than mine shall penetrate to recesses still deeper.* Yes, a man with a great mind most piercing came into this world the same year Galileo passed on from it. You are fortunate enough to learn about him and his work soon enough.'

Casanova smiled. Not his usual mocking smile, but a smile of warm friendship. The melodious music of Mozart had started playing in the background again.

'As always, it has been an immense pleasure to spend these precious moments with you, *signore*,' said Casanova. 'Once again, let me make it clear that I am not going to bid you *addio*, my final farewell. Please allow me to say just this, *arrivederci per il momento, e che la pace sia con voi*: goodbye for the moment and may peace be with you.'

As Casanova slowly melted away, so did the soft soulful music. At the same time, I was engulfed in thick yellow fumes.

Chapter 8
The Fall and the Firmament

The Golden Goal

When the fumes cleared, I realized that I was in the strangest room I had ever seen. There was a variety of appliances made of glass and metal. A young man was stirring some material in a large vessel that was being heated and from which the clouds of yellow fumes, that had engulfed me, were still rising up. He was pale and seemed to be somewhat weak. Although he was in his shirtsleeves, he was profusely perspiring.

There was a knock on the door and, without waiting for a response from the young man, an older gentleman entered. He was quite handsome and appeared to be aware of the fact. He was splendidly dressed in a maroon velvet coat with silver buttons, embroidered vest, frilled shirt, and a cravat of white lace. Most striking was his auburn coloured hair cascading down to his shoulders in a profusion of curls—obviously an expensive custom-made wig.

© Springer International Publishing Switzerland 2015
C.V. Vishveshwara, *Universe Unveiled*, Astronomers' Universe,
DOI 10.1007/978-3-319-08213-4_8

'What a strange place!' exclaimed the gentleman in his deep resonant voice looking around with curiosity. 'Ah, there you are. I do not know how you can stand all this smoke and smell,' said he spotting the young man at the fire.

The latter extinguished the fire, wiped his hands on a cloth lying nearby, and hastened towards the visitor, while putting on his coarse jacket which he had tossed on a nearby chair.

'Mr Edmond Halley, what a pleasure to meet you again after such a long time, sir,' the young man beamed. 'I hope you remember me. I am Humphrey Newton, assistant to the great man with the same last name, but let me hasten to add, no relation of his as some jump to the mistaken conclusion.'

'Of course, I remember you well, Humphrey,' Halley assured him. 'What is going on here? I could not find you in the library, so I ventured here.'

'You must have heard about Mr Isaac Newton's alchemical experiments in search of the Philosophers Stone that can turn base metals like lead into pure gold,' answered Humphrey. 'As you can see, he has set up several pieces of equipment for this purpose. Please allow me to show them to you. Here are two furnaces made of iron and copper to heat different substances; *alembics*, vessels for mixing, heating, and distilling; *solem*, a distillation tube; and a *tribikos* or funnel.'

'Oh yes, these are used in scientific experiments also,' said Halley. 'The science of chemistry that involves the study of the properties of various substances—solids, liquids, and gases—has gained much from alchemy. But tell me, what are those ingredients you are handling?'

'Seven metals mainly—gold, silver, iron, tin, mercury, lead, and copper. Sulphur is also used for processing, which gives out horrible yellow fumes. Mercury, the *prima material*, the first member of all substances, is considered to be the soul of alchemy. But, it too gives out noxious vapours.' Humphrey motioned towards a tray that contained the ingredients he had mentioned: glittering gold and silver, shining red copper, a dull piece of lead, and a crucible of bright mercury. He added in a whisper, 'Between you and me, Mr Halley, I think this vapour causes some kind of nervous reaction in Mr Isaac Newton leading to irritability, which is not beneficial to him.' He lowered his voice even further and mumbled almost inaudibly, 'Or for that matter to others around him, I am afraid.'

'I understand your concern for Mr Newton, Humphrey. I am sure these experiments are not pleasant for you either,' sympathized Halley. 'But do you have any idea why the eminent thinker is indulging in alchemy? I am sure it is not for material gains, since I know he is not interested in amassing money.'

'No, not for any material benefits is he pursuing the path of alchemy, which many have followed in vain since time immemorial,' agreed Humphrey. 'This is what I understand from the snatches of his conversations with others who are interested in this esoteric pursuit including the well-known scientist Mr Robert Boyle. In order to be a true devotee of this ancient art, one has to find, read, and assimilate all the ancient writings, because the ancients were the wisest people and were the repository of all knowledge and wisdom. The primary prerequisite for practising alchemy is that the alchemist be pure in soul through religion, or metaphorically speaking, he has to transmute himself into pure gold. Mr. Newton thinks, I believe, that he is ideally suited for the role as he satisfies all the requirements including the possession of the highest degree of intellect that enables him to avoid the pitfalls of the others.' Humphrey lowered his voice again to a whisper. 'All alchemists assume pseudonyms or code names in their interactions with one another. Do you know the pseudonym Mr Newton has chosen for himself? It is *Jeova Sanctus Unus*, One Holy God, based upon an anagram of the Latinised version of his own name *Isaacus Neutonus!*'

I knew very well that Isaac Newton, one of the greatest scientists of all time in fact, was a very complicated person. And I was now glimpsing one of his unusual sides. Yes, this was altogether a different world I had stumbled into.

'Let me show you something most interesting that Mr Newton has produced during the course of his experimentations,' said Humphrey, opening one of the boxes that had been stacked in a corner. Inside, on dark blue velvet lining, lay a beautiful crystal harbouring within it a scintillating star-like structure.

'What is it?' enquired Halley with astonishment.

'A *Star Regulus* or *Regulus of Mars*,' exulted Humphrey. 'It is the crystalline configuration of the metal antimony produced by processing it with iron, copper, tin, and lead. Please look carefully at its structure, which may be visualised as rays of light radiating outwards. But I overheard Mr Newton thinking aloud: *The components seem like lines of force acting inwards towards the centre like gravity in fact*. Those were his very words, Mr Halley.'

Halley shook his head in amazement. 'Humphrey, mysterious are the ways in which genius works! Imagine how the profound intellect of Mr Newton might have linked the *Star Regulus* to the idea of a *central force* that gravity seems to be. And it is precisely in connection with that force, which Mr Newton has so magnificently explained and employed in describing the motion of the heavenly bodies, have I come to see him as you might very well have surmised. Let us adjourn to the library, Humphrey, if you please.'

The Magnum Opus

Slowly, the scene shifted to Isaac Newton's library.

The furnishings indicated a preference for the red colour; even the books on the shelves had been bound in red leather.

'Look at the red colour everywhere,' Humphrey made a sweeping gesture with his hand. 'I am told that the red colour too contributes to the nervous irritability of my master just as the mercury vapour does. Returning to alchemy, no less than one hundred and thirty-nine books among Mr Newton's collection are on that subject, while only thirty-one pertain to the science of chemistry.'

'I can see that,' commented Halley who was casually turning the pages of some of the books. 'You showed me the lovely *Star Regulus*, which seems to symbolise the *central force* or a force that acts towards a centre. Well, the most important visit I made to Mr Newton's lodgings here in Cambridge some time ago was to explore the role of such a force in the context of planetary motion. When I asked him about it, Newton readily said, why, of course it is an ellipse. How do you know that, I enquired. I have proved it and I shall show it to you says he rummaging through his papers, but could not find it. I was astounded by the piles of paper containing an enormous amount of work he had done. With great difficulty, I persuaded him to write up his findings so the world could learn the secrets of Nature, which he had revealed for himself alone till then. The rest is history as the saying goes. He has now thrown himself—body, mind, and soul—into the glorious task of composing his masterpiece, a book of science such as the one that has never been written before nor, perhaps, will ever be written in the future. The monumental volume will bear the title, *Philosophiae Naturalis Principia Mathematica—Mathematical Principles of Natural Philosophy*—or *Principia* for short. I have come to see how his work is progressing and discuss the book's publication, the responsibility of which I have shouldered in its entirety.' Halley added with a smile, 'That includes bearing the costs of publishing as well.'

'You will have to wait a little, I am afraid, Mr Halley,' said Humphrey tilting his head towards a room at the back of the library. 'Mr Newton is in his study, which he hardly leaves, thinking all hours.'

'Is he taking care of himself properly?' asked Halley anxiously.

'So intent, so serious upon his studies is he, that he eats very sparingly, nay, oft times he has forgot to eat at all, so that, going into his chamber, I have found his meal untouched, of which, when I have reminded him, he would reply—*Have I!* and then making to the table would eat a bit or two standing. As you know, he keeps neither dog nor cat in his chamber, which makes well for the old woman his bed maker, she faring much the better for it, for in the morning she has found both dinner and supper scarcely tasted of, which the old woman has very pleasantly and mumpingly gone away with.' Humphrey permitted himself a subdued laughter taking care not to disturb Isaac Newton in his study.

'Does he sleep well at least?' enquired Halley.

'He very rarely goes to bed till two or three of the clock, sometimes not until five or six, lying about four or five hours.'

'With all this strain, I hope his health has not suffered.'

'Fortunately, his health has remained satisfactory. I once expressed my concern and can you imagine what he says to me—*Do not trouble yourself, Humphrey. If I die, rest assured that I shall leave you a legacy.* What an extraordinary notion!' Humphrey shook his head.

Halley suppressed his mirth at this remark and decided to wait for Newton to take a break from his work.

All of a sudden, the door to the study flew open and Newton emerged from his study. His countenance indicated a vast penetrating mind. His wig with flowing hair, which he must have worn earlier to receive some visitor and had forgotten to take off, was askew displaying here and there his real hair that had become grey as a result of his severe studies. Though not tall in stature, yet he seemed to be strong, sinewy, and well built. His eyes were very full and protuberant, which had rendered him near-sighted in youth, but was the reason of his seeing so well in age.

Newton, absorbed as he was in his own thoughts, seemed to be totally unaware of our presence in his library. He strode towards one of the book cases, briefly consulted a volume, and jotted down a few lines in a thick note book he was carrying. Then, as abruptly as he had emerged from his study, he hastened back and disappeared.

'Well, Humphrey, while Newton is busy in his study, let me attend to the pleasant task of discussing Mr Newton's prodigious work with our honoured visitor here who has been following our conversation keenly and watching with great interest all that you have shown me. It is best I undertake the job in the vast expanse of space inhabited by stars and planets. I shall return soon, Humphrey. Farewell till then.'

Halley smiled at me and inclined his head in silent invitation to follow him.

Newton's lodgings slowly dissolved giving way to the night sky.

A Matter of Gravity

I was happy to return to the infinite space sprinkled with distant stars, the Earth glowing far below. I was floating now in the august company of Edmund Halley.

'Wonderful feeling, is it not?' remarked Halley. 'Ah, the exhilaration of weight-lessness with no gravity pulling you down! On the other hand, as you very well know, sir, Newton's greatest discovery was related to the nature of gravity, which is so important a force in Nature.'

Halley relaxed for a while enjoying this feeling of total freedom only the far reaches of space could offer.

'I know that you have learnt all about Newtonian physics in the course of your deep discussions with a most learned friend, philosopher, and guide of yours,' began Halley. 'Nevertheless, please allow me the privilege of discussing some important aspects of Newton's monumental work, his book of books, *Principia*, a modern copy of which you have acquainted yourself with.'

Halley reflected for a moment drumming his finger on his chin. 'Before we come to *Principia*, let us begin our discussions with Isaac Newton at Cambridge, shall we?' Halley resumed. 'Newton, owing to the lack of sufficient funds when he joined Trinity College at Cambridge, was obliged to become a Sizar, a student who paid his way through his studies by doing odd jobs by waiting on his tutor. What a lowly position for one who was destined to be a towering genius, my dear sir! Well, by great good fortune, he became a Fellow when vacancies opened up. Do you know how such an opportunity arose? A Senior Fellow had been sent away from the college on account of insanity and two others because they had unceremoniously fallen down the stairs after a long bout with the bottle! Grape and gravity make a dangerous combination, do they not?'

Halley threw back his head and laughed heartily. I could not help but join in.

Halley pulled out a pure white lace handkerchief from his jacket pocket and wiped away the tears caused by his mirth. He would repeat this performance in a rehearsed manner whenever he laughed.

'In his youth, Newton did his best work not within the confines of the University, but outside in the free atmosphere of his home, the Woolsthorpe Manor,' Halley went on. 'That was during the two years around sixteen hundred and sixty-six that have come to be known as *Anni mirabilis*, the Miracle Years of Isaac Newton, when Cambridge was closed because of the plague. Perhaps it is a splendid idea to shut down universities for a while from time to time to allow the scholars to think in peace so that new ideas may blossom forth, maybe some of them permanently!'

Halley permitted himself another blast of laughter before continuing, 'Newton was to reflect upon the work he did during those years. Listen, dear sir, to what he had to say:

All this was in the two plague years of 1665 and 1666, for in those days I was in the prime of my age for invention, and minded mathematics and philosophy more than at any time since.

'*All this*! Those two simple words represent a staggering amount of work. In those two years, mere two years, Newton had laid the foundations of his mechanics, optics, fluxions, and the supreme pinnacle—gravitation.'

Ah, the genesis of Newton's universal law of gravitation inspired by the sight of a falling apple! I remembered what I had read about that event as described and commented upon by a host of people like Voltaire, De Morgan, Brewster, and Newton's young friend and his first biographer, Dr. Stukeley, to name a few.

'Yes indeed, the falling apple has become a fine symbol of earthly gravity, has it not?' Halley read my thought. 'But, let us not imagine that the law of gravitation flashed in the mind of Newton as a result of divine inspiration. On the other hand, the discovery was initiated by comparing the gravitational pull on all objects on the Earth like the apple with the gravitational attraction exerted on the Moon by its parent planet, the Earth. Once again, here is what Newton had to say in a note he wrote years later:

I thereby compared the force required to keep the Moon in her orb with the force of gravity at the surface of the Earth, and found them answer pretty nearly.'

Halley held one of his hands up, the forefinger pointing heavenwards and the other down with the forefinger indicating the Earth. 'Yes sir, the Moon in the sky conspired with the apple on the Earth to lead Newton to the law of gravitation. He realized that if gravity decreased inversely as the square of the distance, then he could explain how he *found them answer pretty nearly*. Again, it took him nearly twenty years to prove that a spherical body like our Earth acts as though all its mass is concentrated at its centre only under the influence of an inverse square law. Ah, then came Newton's stroke of genius—the Universal Law of Gravitation as we know it today, a sweeping generalization of a particular observation.

Halley held out his two hands. A magnificent copy of the first edition of *Principia*, bound in thick leather, appeared on his open palms turned upwards. He opened the volume respectfully and extended it for my inspection.

I quickly leafed through the book, although I was quite familiar with it. Once again, I was overwhelmed by the contents. It was a veritable forest of definitions, propositions, theorems, corollaries, problems, and scholiums. Page after page was covered with complicated geometric figures—lines, triangles, rectangles, polygons, circles, ellipses, and other curves. I was reminded of Galileo's famous statement:

Philosophy is written in that great book which ever lies before our eyes—I mean the Universe—but we cannot understand it if we do not first learn the language and grasp the symbols in which it is written. The book is written in the mathematical language, and the symbols are triangles, circles and other geometrical figures, without whose help it is impossible to comprehend a single word of it; without which one wanders in vain through dark labyrinth.

Halley, who was watching me, remarked, 'Rather humbling, is it not?'

I nodded quietly. I felt that it must be an immensely difficult book to read, let alone understand.

'Indeed, it is not at all an easy task to navigate through that tome,' agreed Halley. 'Newton himself told a friend: *to avoid being bated by little smatterers in mathematics he designedly made his Principia abstruse, but yet so as to be understood by able mathematicians.* But then, even the ablest mathematicians found the *Principia* difficult to understand.'

Halley paused and offered, 'Let us take a closer look at this living scientific monument. We can only have a glimpse of the treasures it contains.'

Halley opened the book and revealed the title page. The full title of the work stood out boldly at the top: *Philosophiae Naturalis Principia Mathematica*, Mathematical Principles of Natural Philosophy.

As a prologue to the book, there was something I could never have expected.

Halley smiled, a smile that exuded his joy, his pride, and his contentment.

'Yes, sir, I had the honour of contributing that *Ode to Isaac Newton*, which stands at the very beginning of the volume. After extolling Newton's achievements in my own poetic terms, I concluded my Ode with my tribute to that man of unique genius:

Nearer the gods no mortal may approach.

That in fact sums up my humble feeling about one whose contribution to human knowledge happens to be literally beyond measure.'

Halley fell silent. Perhaps he was thinking of how to proceed further in his description of *Principia*. It was a great pleasure and privilege just to watch a brilliant man like him immersed in silent contemplation.

'I fully realize that you have already learnt about the main results contained in the book, sir,' Halley went on. 'I shall just read out some salient parts of the book that should interest you in a broad sense. I shall complement those quotations, if I may, with my own observations, which, I piously hope, will not do injustice to the thoughts and ideas of the venerable author.'

Halley turned the pages of the book before him, selecting sentences, and reading them in his sonorous voice. He interspersed his recitation with his comments and reflections. The whole experience proved to be a most inspiring one, to say the least.

'Right at the beginning, Newton puts forward an idea, which is extremely important. It is an idea that is taken for granted and used routinely. Here it is in Newton's own words:

Wherefore the reader is not to imagine . . . that I attribute forces, in a true and physical sense, to certain centres, which are only mathematical points, when at any time I happen to speak of centres as attracting, or as endowed with attractive powers.

What does the sentence have to say? Forget how huge the celestial bodies like the Sun and the Moon are. Forget their structures. Assume they are just points. Assume they are characterized by, but not physically endowed with, mass and power of attraction. Then Newton will show you how those objects move. He will do it entirely from his laws of motion and his law of gravitation. This is exactly what people do when they study the motion of any object, from the minute spec to the largest celestial body. And Newton started it all. Simple but profound, sir.' Halley shook his head in admiration. He seemed to be carried away by these basic ideas that had shaped science for centuries since Newton.

'Here is another idea, a rather an important one that again occurs right at the beginning. Newton discusses motion and rest, the two states of all bodies in nature.

But motion and rest, as commonly conceived, are only relatively distinguished; nor are those bodies always truly at rest, which commonly are taken to be so.

In other words, states of rest and uniform motion are relative. This is Newtonian relativity initiated by his celebrated predecessor, Galileo Galilei. They both knew that one cannot tell who is at rest and who is moving with constant velocity by performing any mechanical experiments.'

I knew that those ideas of Galileo and Newton formed the starting point for Einstein's revolutionary theory of relativity, which was to emerge far in the future. It was astonishing to listen to Newton's very words in this matter.

'The *Principia* is divided into three Books, as you are aware. And we are now starting on the First Book,' Halley announced ceremonially. 'You already know the astonishing results Newton derived from his three laws of motion and a single law of gravitation, especially in relation to our planetary system. In essence, Newton now held the entire Universe in his palm. Did he stop there? No, sir, no!'

What else did he do? Like Halley, I was getting excited too.

'We know that the Sun is very heavy compared to the planets. We can therefore assume that the Sun is at rest while each planet goes around it. This has come to be known as the *one-body problem*. Now, if you consider two objects of comparable masses, say the Earth and the Moon, you then have to take into account the motion of both objects. This is the *two-body problem*. Newton solves this too. Did he stop there?'

No! I answered Halley's rhetorical question under my breath.

'Let us march on. Take then the Earth-Moon system with the Sun interfering in their affairs,' smiled Halley. 'This is the famous—or infamous, according to your personal perception—the *three-body problem*, as it has come to be known.'

Sounds like a human situation to me, I thought.

'Equally or still more complicated than that,' Halley's smile widened. 'At least the human problem can be solved once in a while. The physical problem cannot be solved exactly at all. The influence of the Earth and the Sun on the Moon is so complicated that Newton complained that his head *never ached but with my studies on the Moon*! All the same, Newton made a good beginning in handling the problem with suitable approximations. This is more or less the essence of the First Book. All in all, it is a magnificent treatise on mechanics as applied to the Solar System, each result a pioneering discovery.'

From what I had learnt already, the First Book by itself was sufficient to establish the *Principia* as a great work. But I knew that there was much more to come. Halley paused in preparation for further exposition of the *Principia*.

'Now we come to the Second Book,' Halley began. 'In the First Book, Newton had considered all motion to be taking place in empty space. Now he treats motion in a resisting medium such as air or water. For instance, he considers motion of objects under gravity, including pendulums when there is air resistance. I must add that Newton deals with the flow of fluids as well in his Second Book. In all this he keeps in mind experimental results, his own as well as those of others. All this is typical Newton. But allow me to point out to you two of the astonishing subjects he deals with. First is his mathematical treatment of wave motion.'

Wave motion—I knew very well that it was one of the most important phenomena in all of physics as well as in our daily life.

'You are absolutely right, sir, in your recognizing the inestimable importance of waves! Newton derives certain fundamental results for the first time, thus leading the way for further discoveries.'

What results? I wondered.

'Ah, what he does now is simply amazing. He wants to find out the shape of an object that meets with the least resistance from the medium saying that it *may be of use in the building of ships*. I am sure that this problem has proved to be unavoidable not only in building sailing ships but also in designing their flying counterparts that have become a reality in your own time.'

Which meant that Newton was far ahead of his time!

'Very much so! And he proves that fact again and again. That, dear sir, is the briefest glimpse of the Second Book. We shall desist from going any further into it, not because it is uninteresting or unimportant, but simply because it is not directly relevant to our story of gravitation.'

Halley appeared to be totally relaxed and in a mood of profound anticipation. I could feel that something extraordinary was in the offing. After a while, he opened the *Principia* to reveal the first page of the Third Book.

System of the World, the title of the Book was printed in large letters. Halley read out what Newton had written in the beginning:

> *In the preceding books I have laid down the principles of philosophy; principles not philosophical but mathematical ... It remains that, from the same principles, I now demonstrate the frame of the System of the World.*

That sounds like too fantastic a claim to make—it was the same feeling I had experienced earlier when I had seen the title of the Third Book.

'With another man, in another book, it could very well have been a resounding hollow boast. But not with Newton,' Halley averred echoing George's words. 'Those lines always sound to me like—how shall I put it—like the heraldic proclamation of a triumphal march. And what follows is truly magnificent beyond measure.'

I wondered what exactly Newton meant by the caption the *System of the World*.

'Just that,' Halley's two words were the response to my unspoken query. 'Newton explains practically everything that was known about the heavens at our time. I know that you have learnt this from a distinguished scholar. And much more. On what basis? Just his three laws of motion and one law of gravitation. Incredible!'

The Third Book ran to some two hundred and odd pages that would reveal the clockwork of the cosmos.

'First of all, Newton describes and accounts for the motions of the Moons of Jupiter and Saturn as well as our own constant companion,' continued Halley. 'And, of course, he deals with the orbital motion of the planets around the Sun. All this within the framework of his gravitational theory.'

That was the central theme of his whole work, I supposed.

'To begin with, yes. But, now, Newton performs an incredible feat, invading completely uncharted territories. This is exactly according to the guiding principle Newton sets down right in his *Preface*. Let me show it to you.'

Halley thumbed back to Newton's *Preface*.

For the whole burden of philosophy seems to consist in this—from pheno-mena of motion to investigate the forces of nature, and then from these forces to demonstrate the other phenomena.

This was indeed a revelation. 'That seems to sum up the entire purpose of science!' I could not help exclaiming aloud.

'How right you are, sir, how perceptive!' remarked Halley. 'So, Newton sets out to demonstrate what he calls *the other phenomena* now. For instance, he not only derives the shape of the Earth taking into consideration its self-gravity as well as its rotation, he extends it to other planets too. As a matter of fact, the shape of rotating fluids happens to be a fundamental problem by itself. Many mathematicians must have worked on it since Newton's time, I am sure.'

Halley leafed through *Principia* for a moment and found what he was looking for.

'Ah, here we are. Listen to what he writes later on in the *Principia*:

Bodies projected in our air suffers no resistance but from the air. Withdraw the air and the resistance ceases; for in this void a bit of fine down and a piece of solid gold descend with equal velocity.'

Why, that is exactly what the astronauts verified on the Moon by dropping a feather and a hammer! This was incredible.

'Precisely, as the travellers into the vast outer space of your modern days demonstrated it so dramatically. I wonder what Newton would have said about it. Maybe nothing, since it would have come as no surprise to him.'

I fully agreed with Halley's conjecture.

'Now another phenomenon again of fundamental importance, namely the occur- rence of tides. I certainly realize that you are already quite familiar with how tides are generated as explained by Newton on the basis of the differential gravitational attraction exerted by the Moon on the oceans. Until Newton gave his explanation, the tides had remained a mystery. Even the great Galileo had not been able to resolve it. He had put forward an ingenious but completely erroneous theory in this regard.'

'So far we have been considering Newton's work related to our Earth, have we not?' said Halley. 'Now let us fly back from the Earth to the sky once again.' Halley held out his hands apart to indicate the vast space around us.

'You remember how it all began, sir, I mean the creation of the *Principia*?'

How could I ever forget it? It all began with Halley's meeting with Newton in order to consult him about a *comet*. Did it not?

'I am overjoyed that you recognize that fact, sir. Ah, comets, comets!' exclaimed Halley. 'It was but natural for Newton to study the comets amongst other celestial objects. He worked out their orbits and considered the detailed observational data of some of them. In this context, he took note of the comet I happened to observe with a period of 575 years. It was the same comet that had appeared at the time when Julius Caesar was assassinated!'

I remembered Calpurnia's warning to Caesar in Shakespeare's play:

When beggars die there are no comets seen
The heavens themselves blaze forth the death of princes.'

'Well, that used to be one of the beliefs, apart from all kinds of other superstitions about comets,' Halley remarked.

At that moment a comet came drifting from afar with its tail blazing. It was an extraordinarily beautiful sight indeed.

'Ah, here sails in the comet that happens to be my favourite one,' said Halley joyfully. 'It makes its round once in 76 years. I observed it with great care and interest, did I not?'

Yes indeed, it was none other than the most well-known comet of all—*Comet Halley*!

Halley acknowledged the compliment with a slight bow and remarked, 'You know sir, having a celestial object named after you bestows upon you, in a sense, a sort of immortality. It is nothing to be proud of. As a matter of fact, it makes you feel humble.'

I remembered what Mark Twain had said in 1909:

I came in with Halley's Comet in 1835. It is coming again next year, and I expect to go out with it. It will be the greatest disappointment of my life if I don't go out with Halley's Comet. The Almighty has said, no doubt: 'Now here are these two unaccountable freaks; they came in together, they must go out together'

And his wish had been fulfilled.

Yes, even to this day, we look forward to the appearance of Comet Halley with great anticipation.

'Newton went on to theorize about the nature of the comets and the origin of their tails,' Halley waited for my musings to be over before continuing. 'He had some ideas that you may find a bit outlandish after all the observations and theories that have evolved since our time. For instance, Newton conjectured that some comets may fall into the stars, including the Sun, and fuel them. You have learnt about the *novae* that were observed by Tycho Brahe and Kepler, have you not? Newton thought that comets falling into the stars were the cause of those *novae*. But, stranger still you may find Newton's notion that cometary matter, dissipated and scattered throughout the heavens, rejuvenates the Earth and the planets. Whether correct or not, Newton's writings on comets make quite interesting reading even to you in your modern day, I am sure.'

Halley patted the big book with affection. 'Well, sir, I think we have had a good glimpse of the *Principia*. Do you agree?'

All that I had learnt, was just a glimpse of the book? It was unbelievable.

'Yes, it contains a great deal of other things we have not discussed. After all, the *Principia* provided the foundation for the work of the brilliant mathematicians who came after us.'

As a matter of fact, I knew the tributes that had been paid to the *Principia* by mathematicians like Laplace, Lagrange, Euler, and so on. For instance, Laplace wrote: *This admirable work contains the seeds of all the great discoveries that have since been made about the system of the world.*

'Now, I just want to read to you, sir, parts of the General Scholium, which concludes the Third Book. It is a lofty testament to gravity,' Halley remarked. 'Here are some of Newton's thoughts.

Hitherto we have explained the phenomenon of the heavens and our sea by the power of gravity, but have not yet assigned the cause of this power ... I have not been able to discover the cause of those properties of gravity from phenomena, and I frame no hypotheses.'

Yes, I had read about Newton's famous statement: *Hypotheses non fingo—I frame no hypotheses*!

'I must add that Newton is very emphatic here, as he was in his correspondence, about his firm refusal to hypothesise about final causes. This was indeed a radical departure from the age-old tradition. In essence, Newton was asserting that he demonstrated *how* gravity worked and not *why*.'

Isn't that true of all science? One will never reach the final cause, which keeps receding with every step one takes forward?

'You are perfectly right, sir,' endorsed Halley. 'To continue with the General Scholium, a single sentence summarizes the central theme of Newton's entire work:

It is enough that gravity does really exist, and act according to the laws which we have explained, and abundantly serves to account for all the motions of the celestial bodies, and our sea.'

Well, that was a superb summing up. I was quite impressed by this statement.

Halley joined his hands together as though in prayer, closed his eyes, and pondered for a moment. Opening his eyes slowly, he said, 'There is one topic we must touch upon without which our discussion of the *Principia* would be incomplete.'

What might that be? I waited for Halley to go on.

'God!' announced Halley. 'As you may know, Newton was a very devout Christian and extolled God and his creation in his General Scholium. Here, allow me to show you. After describing the Solar System, Newton writes:

This most beautiful system of the Sun, planets, and comets, could only proceed from the counsel and dominion of an intelligent and powerful Being.'

That sentence seemed to echo in spirit what Copernicus had written about placing the Sun at the centre of the planetary system.

'Yes indeed. Now here is something about God that seems to be an offshoot from his Second Book on motion in a resisting medium. It gives you an impression of almost childlike innocence and simplicity.

He is omnipresent not virtually but substantially; for virtue cannot subsist without substance. In him are all things contained and moved; yet neither affects other: God suffers nothing from the motion of bodies; bodies find no resistance from the omnipresence of God.'

That was absolutely charming. As Halley had pointed out, there was childlike innocence in that statement.

'Finally, Newton concludes his thoughts about God with his view on the purpose of scientific exploration.

And thus much concerning God; to discourse of whom from the appearances of things, does certainly belong to Natural Philosophy.'

Halley closed the *Principia* slowly, gently, with utmost care.

Halley seemed to be in a mood of peaceful contentment, having described some important ideas enshrined in the *Principia*, especially through Newton's own words. He was most likely thinking back on all the interactions he had with that great man. I had read in sufficient detail Halley's own remarkable life and work. He had contributed so much to science in so many different ways, including charting the southern stars, so much so that he had been hailed as *the southern Tycho*. He had mapped the magnetic field of the Earth at various places as an aid to navigation. In fact, Halley had proved to be a skilled navigator and had been made captain of a ship several times. There was even evidence that he had served his country as a discreet spy, but his indiscretion in his amorous adventures had won him a rather dubious reputation.

I looked up and saw Halley watching me from the corner of his eye and smiling at my thoughts.

'Coming back to Newton, one cannot underestimate his marvellous investigations in areas other than those that are covered in the *Principia*,' resumed Halley. 'His experiments and hypotheses concerning the nature of light made new pathways in many directions. He was the first one to reveal the component colours of white light using a prism. He wrote it all in his other magnificent book *Opticks*, that too in English unlike his *Principia* which was in Latin, so that the book was accessible to everyone. The study of spectra, following Newton's discovery, was destined to encompass so many different fields including astronomy. Yes, astronomy which he enriched with his invention of reflecting telescope that employed a concave mirror instead of a lens thereby reducing dispersion of light and increasing the clarity of image. Well, there is no end to the list of Newton's achievements, is there, sir?'

Halley raised his head and looked deep into space as his gaze became distant and dreamy. In his open palm, which he held forth, appeared a glass prism. A single shaft of light shot out from somewhere in the encompassing space, perhaps from a distant star, passed through the prism splitting into brilliant colours. Slowly the glass prism turned into a single drop of water and the multi-coloured light spreading out from it arched into a rainbow around Halley. He bowed deep as he repeated softly as if to himself the last words from his Ode to Isaac Newton:

Nearer the gods no mortal may approach!

Halley then dissolved slowly into space that hosted myriad glittering stars, while the rainbow surrounding him lingered on for a long time.

Ever so slowly, I opened my eyes. My bathwater was still covered with glittering star-like specks surrounded by what looked like miniature planetary systems. They gradually dissolved away, leaving behind just the bathwater.

It took some time for me to come out of the trancelike state I was in. I climbed out of my bathtub like a sleepwalker. Like a sleepwalker, I flopped on my bed and was in no time fast asleep.

Chapter 9
Lives of the Stars

George and I had agreed to meet once again in the University. This time it was not to be in his office, which George calls dusty, musty, and rusty! We decided to get together in the loveliest spot you could think of, namely the University Park. Being located on the outer edge of the sprawling campus, it is far from the madding crowd of stressed out students and frustrated faculty. Beautifully landscaped, yet retaining the natural charm of spontaneous growth, the park is a work of art harmoniously blending a variety of colours through its plants, shrubs, and trees. It was late afternoon and the gentle sunlight seeping through the foliage of the trees added to the beauty of the surroundings.

When I entered the park, I could see George already relaxing on one of the comfortable benches. Now that his students have graduated and his teaching duties are still to begin, he can afford to take life a bit easy. George looked up, smiled, and waved to me.

'Well, Alfie, how have you been?' enquired George. 'Knowing you, you must have been assiduously going through the manuscript. What do you think of it?'

'You want my honest opinion?'

'Yes, free, frank, and fearless.'

'All right, here we go,' I grinned. 'I love it, George. It is so vivid in its details, still so imaginative in recounting the lives of the great scientists. I felt as though I was really meeting all those fantastic people face to face.'

Of course, I didn't divulge the secret source of my experience. But there was no denying the fact that my extraordinary flights of mind were being triggered as much by George's book as by my precious... Well, better leave it at that, I thought.

I could see that George was quite pleased by my opinion, which was indeed absolutely honest.

'All right, you had your private meetings with the pioneers and your flight among the planets,' smiled George. 'The time has come to reach for the stars, then.'

'Yes, indeed,' I agreed. 'We discussed the birth, existence, and death of the stars when you told me all about the black holes a few years ago. You called it *The Stellar BED*. You remember?'

© Springer International Publishing Switzerland 2015
C.V. Vishveshwara, *Universe Unveiled*, Astronomers' Universe,
DOI 10.1007/978-3-319-08213-4_9

'You forget nothing, do you, Alfie?' George remarked. 'That was a nice acronym, wasn't it, BED for Birth, Existence, Death? Well, doesn't hurt to recapitulate our earlier discussion. As Francis Bacon said, repetition maketh a perfect man.'

'George, Francis Bacon never said that,' I laughed. 'His famous saying is, *Reading maketh a full man; conference a ready man; and writing an exact man.*'

'What, Bacon never said anything about repetition and perfect man?' George feigned surprise. He has these impish moments from time to time. 'Well, he should have, don't you think? Even great minds have their blind spots, can't be helped.'

'Some humorist added another line of his own to Bacon's quotation though,' I said. '*Bacon maketh a fat man!*'

'That is a good one,' laughed George. 'Anyway, let us remind ourselves of stellar evolution.'

George leaned back against the bench. He seemed to be scanning the flowers of varied brilliant hues in front of us.

'I wonder how those flowers would look like were they to be transformed into stars. Or if we could see the stars in all their gorgeous colours as we see the flowers. We can only imagine that,' mused George. 'But, before we talk about stellar colours, let us begin with their formation. Let us make the constellation Orion our reference point, shall we? I don't have to ask you whether you know that constellation, do I?'

'Of course, I know Orion,' I nodded. 'Orion is one of the handful of constellations I can identify in the night-sky.'

'Very well, then. Beneath the stars that form Orion's belt, you can see a fuzzy patch. That happens to be *The Great Orion Nebula*. It is a huge cloud of gas.'

I had seen the picture of Orion Nebula, a composite image of photographs taken using different colour filters. It was spectacular and looked like blazing tongues of brilliantly coloured flames leaping out, mostly red tinged with yellows and greens.

'There are lots of such gas clouds you can see in the sky,' continued George. 'And those gas clouds are the breeding grounds of stars or the nurseries of new stars. In time, the gases in the nebulae may clump together here and there. Inevitably gravity pulls the gas within such a clump inwards. What happens then? The clump starts condensing and grows increasingly dense and hot. Finally, when the temperature zooms up to some millions of degrees, nuclear reactions are triggered off that produce immense amounts of energy.' George paused before adding dramatically, 'And, lo and behold, a star is born!'

'You took hardly a few seconds to describe this fantastic process, George. But in reality, how long does such star formation take?'

'You mean the gestation period of stellar pregnancy? Let us see. Well, for the gas cloud to shrink from a diameter of trillions of kilometres to the present size of the Sun, it would take, say, about ten million years. Give another twenty million years and the star will stabilise.'

'George, you talk as if millions of years are mere minutes,' I laughed. 'I think it is all a professional show-off for you guys studying the stars. Am I right?'

'Can't fool you, can we?' George laughed too. 'In any event, I am sure you are familiar with the big numbers we use. A million is one followed by six zeroes, a billion nine zeroes, a trillion twelve zeroes and so on. They used to be called astronomical figures. Now they belong to finance too with all the deficit budgets, national debts, and what not, the what-not includes scams too.'

'Of course, I know all those numbers. And don't tell me what a light-year is. It is the distance travelled by light in one year which amounts to about ten trillion kilometres.'

'My, my, am I impressed, Alfie,' George widened his eyes in mock admiration. 'Yes, the light-year will come in handy as we go on. You reminded me of the Stellar BED, didn't you? Birth B is over. Now we come to E for existence. E also stands for equilibrium the essence of existence. Nice alliteration, if you hadn't noticed.'

'I did notice, George,' I gave my assurance. 'Especially, since you had made exactly the same remark some years ago when we discussed your BED.'

'Not *my* BED, my boy. Any way, we exist because we maintain our equilibrium. And that equilibrium arises only if we keep our balance among opposing forces. Stars are no exception. What are these opposing forces in the case of a star? On the one hand, it is the outward pressure created by the heat, which is in turn generated by the nuclear reactions going on within the core of the star. On the other hand, we have the inward pull of gravity. The two are exactly balanced, so that the star stays in equilibrium shining away. A star like our Sun remains in equilibrium for about ten billion years. Our Sun is a middle-aged guy having spent almost half of his life span.'

'Ah, youth has dreams, age memories, and midlife only crises!' I wanted to be dramatic too.

'That is splendid, Alfie,' George nodded with genuine appreciation. 'I know, no need to ask who said that. You will say that modesty prevents you from telling me. I think I shall include your saying in the manuscript, with proper acknowledgement of course. Now, coming back to the Sun, there will be no middle-age crisis though. A happy stable life for the next five billion years. The crisis will occur only towards the end of those ten billion years since its birth, as we shall see.'

'You mentioned the two opposing forces. The pull of gravity I know. But what kind of nuclear reactions are we talking about?'

'Well, before we go into the question of nuclear reactions, let us take a look at the structure of the Sun and what goes on in the interior,' said George. 'First of all, the Sun is a huge ball of fire, a sphere of burning gas. It is 75 percent hydrogen, about 24 percent helium, and a small quantity of other elements. What we see of the Sun is the light from the surface, which is at a temperature of about 6,000 degrees. Most of the mass of the Sun is concentrated around the centre, nearly 90 percent within the inner half of its radius. The temperature at the centre is some 15 million degrees. Horribly hot, wouldn't you say?'

The hottest temperature I had experienced was around forty degree Celsius when I was travelling in India. But, fifteen million degrees? I couldn't even imagine what that meant.

'The temperature is so high that the electrons are ripped off from the atoms,' continued George. 'They fly around madly leaving the nuclei shamelessly naked. The nuclei themselves streak around at a frenetic pace and come close to one another. The density and pressure are extraordinarily high. But this mixture agitating violently still behaves like a gas, you know. This is the nuclear furnace where the solar energy is generated by nuclear reactions. So, that brings us to the question of nuclear reactions, which you are eager to know about. To begin with we have these hydrogen atoms at the Sun's core stripped of their electrons. This we saw just now. At a temperature of millions of degrees in the stellar core, the hydrogen nuclei—or equivalently the protons—and the free electrons would be flying around at breakneck speeds colliding incessantly. This enables them to interact with one another. Now the stage is set for nuclear transformations. Four hydrogen nuclei undergo a chain of reactions, combining to form the final product, namely a helium nucleus. Now, a helium nucleus is made up of two protons and two neutrons. A neutron, as you know, is an elementary particle just like the proton. Except that it is neutral carrying no charge and it is a wee bit heavier than the proton.'

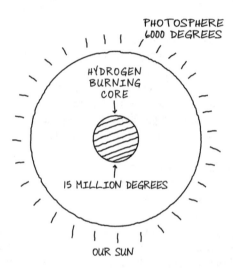

'Yes, I do know something about elements and their composition, as also elementary particles like protons, neutrons, and electrons,' I told George.

'Very well, then. Now here is the extraordinary fact that underlies this nuclear fusion or the thermonuclear reaction in which hydrogen is converted into helium. When you add the individual masses of the four hydrogen nuclei we started with, the sum exceeds the mass of a helium nucleus, the final product, by a minute quantity. This mass difference is converted into energy in the form of heat and radiation during nuclear fusion. The conversion is in accordance with Einstein's famous formula $E = mc^2$, which says that energy released is equal to the mass difference times the square of the speed of light.'

'A formula, which even a child can recite nowadays, like a nursery rhyme,' I said and added, 'But, very few understand it.'

'Let us get an idea of the immense amount of energy generated in the interior of the Sun,' George went on. 'The energy produced in a single thermonuclear reaction, when four hydrogen atoms combine to form a helium atom is minute. But the Sun is huge containing an enormous amount of hydrogen. Each second at the Sun's core, 564 million tons of hydrogen are converted into 560 million tons of helium.'

'That means every second 4 million tons of mass are transformed into energy. And you have to multiply this stupendous amount by the square of the speed of light to get the actual amount of energy released. How much would that be, George?'

'Let us see. You know what a *watt* is?'

'Come on, George, everyone who pays electricity bills knows that. It is the unit of power, the energy generated or consumed per second, named after James Watt, the inventor of steam engine. Maybe a tautology, but a hundred watt light bulb obviously uses up a power of hundred watts.'

'Good. Each second, the Sun generates some 400 trillion-trillion watts of power. Remember a trillion is one followed by 12 zeroes, so the energy released within the solar core measured in watts is 4 followed by 26 zeroes. Now the total amount of electrical power being used in the whole world is about 10 trillion watts. In other words, the energy the Sun produces in one second is equivalent to what mankind would use in 40 trillion seconds or in a million years!'

'Don't tell me that we are bathed every second in this immense amount of energy coming from the Sun.' The magnitude of energy production in the Sun was incredible.

'Wait a minute,' George held up his hand. 'We are talking about what goes on within the core of the Sun. The vast amount of energy we talked about is created in the form of high energy X-rays or *gamma rays* as they are called. This radiation travelling towards the surface is continually scattered by the matter in the Sun's interior. As a result, it keeps losing energy as it travels and becomes more or less the visible radiation streaming out of the Sun.'

'Now, don't forget that this radiation goes out in all directions,' George continued. 'Only a fraction of it falls on our Earth making sure we enjoy a comfortable life. So, to answer your original question, we are not bathed every second in a horrendous deluge of destructive amounts of energy.'

'All right, so the mass of hot gas in the Sun remains a spherical globe of fire instead of simply spreading out as the self-gravity of the gas pulls it towards the centre. There is perfect balance between this inward pull of gravity and the outward pressure of gas. And the Sun continues to exist retaining its size and shape.' I summed up the equilibrium the essence of existence—E of *The Stellar BED*—on behalf of George.

George nodded his approval of my performance several times smiling with satisfaction.

'So much for Birth and Existence of a star like our Sun,' resumed George. 'Now we come to Death. Well, it won't be a sudden death like, say, cardiac arrest or something like that, you know. It will be somewhat long-drawn affair with complications.'

'George, please try not to be so morbid about death,' I protested. 'After all, we are talking about stars that have had their days of glory or should we say nights of grandeur?'

'Alfie, my boy, life and death are the two faces of the same coin. We have to take them philosophically, you know,' said George solemnly.

I thought for a moment. Sometimes memories come up slowly.

'Ah, I remember now, George,' I exclaimed. My data bank had yielded the information I was looking for. 'During my wanderings in Europe, I visited Frederiksborg Castle in Denmark. To my great surprise, I came across the coat of arms Niels Bohr had designed and adopted when he was conferred the prestigious Order of the Elephant, normally given to royalty and heads of states. Those men are supposed to be as wise as elephants, which was of course true in the case of Bohr. The coat of arms featured a *taijitu* the symbol of *yin* and *yang* along with the motto in Latin, *contraria sunt complementa*—opposites are complementary. According to Bohr, *There are trivial truths and the great truths. The opposite of a trivial truth is plainly false. The opposite of a great truth is also true.* Bohr believed that life and death are also complementary like many other things. There is wholeness only when they are put together.'

'There you are, Alfie. And a fascinating facet of Niels Bohr,' George seemed to be moved by what I had recounted. He thought about it before getting back to the final stages of our Sun and any other similar star.

'All right then, let us get back to the Sun,' George said and picked up Sun's evolution again. 'So, in the Sun's core the heat production through nuclear reactions keeps going on. In other words, the hydrogen keeps burning as the process is commonly described. As we have already noted, the nuclear furnace at the heart of the Sun continues for ten billion years. Of which five billion years are already gone. Sun shines on. In another five billion years, the situation takes a rather dramatic turn. It is a complicated story from now on. Let me give you a brief, simplified account of the somewhat complex events that follow.'

George paused, probably trying to compress ten billion years of Sun's life into a few minutes of exposition. I knew that this was not at all difficult for him especially since he had done it already for his book.

'Now, in ten billion years most of the hydrogen at the Sun's core would have been converted into helium,' George went on. 'So what happens? There is no longer enough heat production to balance the gravity of the core mass. The core shrinks under its own weight and gets heated up. And this heat is transmitted to the shell surrounding the core. The shell is of course made up of hydrogen too, which starts burning. The outer part of the Sun in turn gets heated up and bloats up. And as you know an expanding gas cools down. So, the surface temperature drops because of this expansion. The Sun, grown gigantic by now, glows red. In short, the Sun has turned into a *red giant*.'

'Everything in the world of astronomy is gigantic,' I observed. 'But, how gigantic is gigantic in the case of a red giant?'

'Well, the Sun's radius in its reincarnation as a red giant would have shot up to some two hundred times its original value, you know. Big enough, wouldn't you say?'

I thought about it for a moment. 'What happens to the planets then?' I wondered.

'Ah, the Sun, now turned into an immense ball of fire, will gobble up the planets Mercury and Venus. Like a mythological monster swallowing up some unfortunate victims!'

'You know what your remark reminds me of, George?' I said. 'Not a mythological monster, but the story of the so-called Cannibal Count Ugolino of the thirteenth-century Italy. He is said to have eaten his own sons, starved as they were all in prison. Dante writes about it in his *Inferno*. And Rodin sculpted the episode into his *Gates of Hell*. Actually, the whole story is supposed to be apocryphal.'

'Ghastly, Alfie! I am often amazed by the amount of weird information stored in your brain. With all that stuff churning in your head, how can you even sleep?' George shook his head.

'More to the point, think of what Copernicus said,' I ignored George's comment about my brain. 'Didn't he say that the Sun sits upon his royal throne, while his children, the planets, circle him? Once again here is an example of the Sun, transformed into a red giant eating up his own offspring, Mercury and Venus. Well, what about our beloved Earth?'

'Destiny reserves no better fate for the Earth either. If it is not gobbled up, it will still be scorched to a cinder. Obviously, all life would be extinguished. The Earth is now a dead planet enveloped by thick clouds rising out of the boiling oceans.'

'Enough of red giant's wanton havoc! What is going on with the Sun's core in the meantime?'

'Ah, the Sun's core. The nuclear reactions within the Sun's core have come to a halt. But the hydrogen in the outer envelope keeps burning. And the region where this happens keeps growing outwards. Helium, the ash of the hydrogen burning, so to speak, goes on getting deposited on the core. Now, the core, which is under tremendous gravitational compression, heats up enormously from 15 million degrees to some 100 million degrees.'

'The story is getting hotter and hotter,' I remarked.

'You said it. At this temperature even helium is ignited. As a result, some amounts of carbon and oxygen are produced through nuclear reactions.'

'How long does this process take?'

'Oh, in a star like our Sun this *helium flash*, as it is called, lasts just a few seconds. In heavier stars it can be longer.'

'It looks like the star is trying to prolong its own life, undergoing a lot of turmoil.'

'That is what happens with the onset of old age and the approaching death, Alfie. We humans try to postpone the end with medication and what not. But no such luck for the stars. During the death-throws of the star like our Sun, it sheds its outer layers which expand away in glowing colours. The expanding shell is called a *planetary nebula*.'

I had seen photographs of planetary nebulae. They present a lovely picture glowing with variegated colours of reds, yellows, and greens.

'Again my question, what has happened to the inner core in the meanwhile?'

'The core has contracted to the size of the Earth, that is about a hundred times smaller than its original radius or a million times smaller in volume.'

'That means it has become extraordinarily dense.'

'Absolutely. The density is now around a million grams per cubic centimetre. Or tens of tons per cubic inch. Can you imagine it? Do you have any idea how dense that is? Well, a chunk of matter the size of a sugar cube, if brought to the Earth, would be as heavy as an elephant.'

'Amazing! The Sun had once become a red giant. But now, is this a totally new phase of our nearest star?'

'Exactly. The Sun has now become a *white dwarf*. It is white because the surface shines bright as it has been heated up when the core contracts. But it slowly cools down and shines no more. We can call it a *black dwarf*. The Sun has entered its eternal resting place now. That means. . .'

'The letter D in BED that stands for the death of the star has been reached!' I completed the sentence for George.

'Let me get something clear,' I said after a moment. Sun's metamorphoses were so interesting. 'There is no longer any heat production within the Sun's core that has turned into a white dwarf. Am I right?'

'Correct. All nuclear processes have come to an end.'

'Then how does the white dwarf hold up against its own gravity?'

'Aha, if some student of mine had asked such a question in class, I would have said that it was an excellent question. And now I say that unto you,' commended George. Needless to add, I felt most gratified.

'The interior of a white dwarf is where two revolutions join hands, Alfie.' George is fond of making these mysterious remarks. He claims that he is following my example in this regard! Absurd, I would say.

'George, I know of the Copernican revolution. We talked quite a bit about that,' I said. 'What do you mean by *two* revolutions?'

'Oh, the two modern revolutions, Alfie, that happened at the beginning of the twentieth century,' George clarified. 'Quantum physics and relativity! We pause before plunging into the strange world of these two theories.'

George paused for a long while, a satisfied smile on his lips. He needs occasional rest, having used up sufficient energy what with all his detailed explanations. I too needed a bit of time to digest everything I had heard.

'All right, first the quantum, relativity comes later,' began George slowly. 'Quantum physics, as I am sure you know very well, describes the subatomic world populated by elementary particles such as protons, neutrons, electrons, ons and ons, and even the photons that make up electromagnetic radiation. This area of physics was created by brilliant minds—Max Planck, Albert Einstein, and several others. To begin at the beginning, let us talk about the blackbody radiation and Max Planck or the birth of quantum physics. As you will see, Alfie, the spectrum of the blackbody radiation is extremely important for stars and even for cosmology.'

Radiation everyone knows. But what is this *blackbody radiation*, I wondered.

'Let me answer your unasked question,' smiled George. 'Blackbody radiation is nothing but radiation that is in thermodynamic equilibrium. Let me elaborate.

Suppose you consider an enclosure painted pitch black so that the walls absorb and re-emit the radiation that falls on it. Now heat this enclosure to some sufficiently high temperature so that radiation is generated within it. Now after a while, the emission, absorption, and re-emission processes keep going on and the radiation comes to a thermodynamic equilibrium. If this radiation is made to escape through a tiny hole without affecting the equilibrium, then the radiation coming out would be *blackbody radiation*. This is the ideal situation. But in lots of cases in nature, one encounters approximate forms of blackbody radiation, as in the case of stars for instance. Have I made myself clear enough or is it all just a black box?'

'Clear as always,' I responded.

'All right then. Let us see how this radiation's brightness, or intensity, behaves as we change the temperature. First, take the simple case of an iron rod. Keep heating it. As the temperature rises, it will glow with increasing intensity and its colour will go from dull red to blindingly brilliant blue-white. Similar behaviour occurs in the case of blackbody radiation. If you take its spectrum, the peak of intensity moves from longer wavelengths to shorter ones as the temperature is increased. Or equivalently, the intensity maximum shifts from a lower frequency to a higher one when the temperature is raised. Correspondingly, the overall colour changes from, say, red to blue. Let me show you the curves displaying the spectra of blackbody radiation for different temperatures.'

George opened the manuscript of his book, which he had brought with him, and showed me the plots.

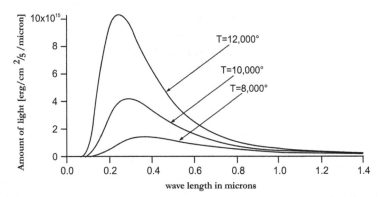

'You see, these curves illustrate the colours of the stars also. Red giants, whose temperatures are lower than 5,000 degrees, appear orange or red, because that is the wavelength where the spectrum peaks. Our own Sun glows yellow at 6,000 degrees and so on. All this is fine. But how do you account for the blackbody spectral curves theoretically? This was a puzzle that classical physics could not solve. Physicists could explain those curves at either high or low wavelengths, but not the full curves. Max Planck did the job with his neat little trick and so was born quantum theory!'

I could see that George was quite excited, as he had thrown open the portals to the strange new world of the quantum.

'What was Planck's revolutionary idea?' George posed his rhetorical question. 'Well, let me give you an analogy. Think of radiation as, say, milk. For centuries it was considered to be like milk, a continuous fluid, which you could extract in any amount you wished, however small. This worked well at large scales of classical physics. But now, Planck said, no, radiation had to be exchanged in discrete units, or packets of energy or the *quanta*, at the subatomic scales. This was like demanding that milk be bought and sold only in small bags of a fixed amount. *Planck's Law* simply states that the amount of energy carried by each *quantum* is proportional to the frequency of the radiation, the constant of proportionality bearing the name, what else, *Planck's constant*. And, lo and behold, he could reproduce the blackbody curves with this extraordinarily simple equation. With this, quantum hypothesis came into its own explaining many other phenomena in quick succession. Niels Bohr gave his atomic model. Erwin Schrödinger derived his famous equation ushering in the probabilistic interpretation of physical phenomena. A system could be in more than one state with different probabilities. I am sure you have heard of the *Schrödinger cat* that could be both dead and alive simultaneously—two probable states.'

'Oh yes, some time ago, I saw a quaint poster that depicted a cat as a bandit with the caption: *Schrödinger's cat, wanted dead AND alive!*'

'I wish you had pinched that poster, Alfie,' laughed George. 'Well, to continue, our hero Einstein came on the scene. Sitting in his patent office, he was to publish three groundbreaking papers one after another: Brownian motion, which demonstrated the real existence of atoms and molecules; photoelectric effect; and of course the theory of relativity. Planck had assumed that energy of radiation could be exchanged only in discrete packets, namely the quanta, as in the case of milk. Now, Einstein had shown that radiation *existed* in those discrete units. It was like showing that milk was not only bought and sold in packets of minimum units, but it existed in those units within the barrels to begin with.'

'That is a nice analogy, George,' I said.

'To give due credit, Alfie, I read it in Einstein's biography written by his Czech colleague Philipp Frank. Yes it is a lovely analogy,' admitted George. 'Now, the quantum story gets weirder and weirder. Werner Heisenberg put forward his *Uncertainty Principle*. You could not pinpoint both the location and the momentum of an elementary particle—they were sort of fuzzily smeared out. Finally, we come to the brilliant Austrian physicist Wolfgang Pauli's *Exclusion Principle*. And it is this Exclusion Principle that is responsible for the pressure that holds up the white dwarf against the inward pull of gravitation.'

George paused before continuing his white-dwarf story.

'All right then,' George nodded and continued. 'As we have seen already, the matter in the white dwarf is in a highly compressed state. The electrons are no longer held by individual nuclei, but move freely and behave like a gas—*electron gas* as it is called. This gas obeys the exclusion principle that generates the required pressure. Let me explain very briefly. A subatomic particle like an electron is supposed to be in a *quantum state* described by a set of *quantum numbers* that indicate its various attributes like energy, angular momentum, and so on.'

'It is like assigning to a person numbers that give his height, weight, age etc. Isn't it?' I pointed out.

'Nice comparison, Alfie, but the quantum numbers can take only certain values unlike the quantities you mentioned,' clarified George and continued. 'The exclusion principle essentially means that no two identical particles, like electrons, can occupy the same state with the same quantum numbers. So, two electrons having opposite spins can occupy the same state. Since their spin quantum numbers are of opposite signs.'

'You mean an electron spins like tops? An atom with electrons both orbiting and spinning would be like a miniature solar system then.' This was interesting.

'No, Alfie, you cannot visualize subatomic particles like macroscopic objects such as the planets,' corrected George. 'We just say electrons are endowed with spin quantum number half. We say plus and minus half correspond to opposite spins. Two electrons with opposite spins can occupy the same quantum state. Any other electron in the electron gas has to occupy another state. This kind of repulsion, so to speak, gives rise to the pressure acting within the electron gas.'

'So, it is like a bus that allows only two passengers, that too of opposite genders, to sit together in each seat,' I visualized the situation. 'If a seat is occupied, another passenger would be kept away.'

'Crazy, this seems to be a special day for analogies,' George shook his head. 'By the way, Pauli's exclusion principle applies only to spin-half particles that are called *fermions* named after the celebrated Italian physicist, Enrico Fermi. Particles like protons and neutrons are also fermions.'

'Are there particles that have spins other than half and are therefore not fermions?'

'Yes, indeed. They have integers for their spin quantum numbers, like photons that have spin one. There are other elementary particles, *mesons*, some of which have integral spin quantum numbers. They are not subject to the exclusion principle. These are called *bosons*, in honour of the distinguished Indian physicist, Satyendra Nath Bose. Any number of bosons can congregate in the same quantum state together.'

'Like a cocktail party you mean. Don't groan, George,' I grinned.

'This is getting out of hand,' said George suppressing his smile. 'In any case, we now know how the pressure in the electron gas is created that can oppose gravity and keep the white dwarf stable.'

'So, all is well with the white dwarfs,' I said. 'I suppose that all stars would rest in peace as white dwarfs once they have used up their nuclear fuel. Am I right?'

'Well, that was the general belief among the astronomers,' answered George. 'But then, the most unexpected event happened. Subrahmanyan Chandrasekhar, the Indian astrophysicist who was to win great fame eventually in his lifetime, showed that only stars up to about 1.4 solar masses could become white dwarfs. This is known as the *Chandrasekhar limit*.'

'Ah, Chandrasekhar!' I exclaimed and went on to convey what I had learnt while studying Indian culture. '*Chandrasekhar* is one of the names of Lord Shiva, the

Indian God. *Chandra* means the Moon and *Sekhar* means the wearer. Shiva wears the crescent Moon above his forehead as an ornament.'

'What a wonderful name for an astronomer, Alfie!' George's eyes gleamed on hearing what I had told him. 'And to think that Chandrasekhar was called Chandra by his friends many of whom probably didn't even know that the word meant the Moon!'

I too felt happy with the piece of information I had passed on to George.

'Well, coming back to the white dwarf, Chandra worked on it during his stay at the University of Cambridge in the 1930s at the rather early age of about nineteen. And it is in his work that the two revolutions, quantum physics and relativity happened to meet. We have already discussed the quantum effect of producing pressure in the electron gas. Let me say a few words about Einstein's *special theory of relativity*, as it is called, without straying too far from the white dwarf.'

I remembered George explaining to me in great detail both the special theory of relativity and the general theory of relativity of Einstein some years ago. But it did no harm to receive a brief recapitulation from George again.

'Long before Einstein's time, Galileo and Newton had realized that observers moving with constant velocities with respect to each other could not decide who was at rest and who was in motion by any mechanical experiment. It is like playing billiards either in the comfort of your basement or in the luxury of an ocean liner. This was known as *Galilean* or *Newtonian* relativity. Here was a most important tacit assumption: both observers measured the same time—time was absolute. Now, with the advent of electromagnetism, this changed drastically. I don't want to go into the complicated story. Let us just say that a new kind of relativity was required and Einstein did the job. Now, even time became relative, its measure changing among the *inertial observers*, observers moving with constant velocities relative to one another. This special theory of relativity introduced strange effects like length contraction, time dilation, mass changing as a function of velocity, and so on. And of course, from his relativity theory Einstein was able to show the equivalence of mass and energy. We talked about all this once upon a time, do you remember?'

'I do, I do,' I said emphatically.

'Relativistic effects become important and perceivable whenever one encounters enormous speeds comparable to that of light. In our ordinary macroscopic world, we do not normally come across such speeds. On the other hand, the electrons within the white dwarf fly around with speeds that are indeed comparable to that of light. So, Chandra realized that relativistic effects have to be taken into account in describing the behaviour of the electron gas mathematically. Essentially, he derived the formula that related the pressure of the electron gas to its density. From this he could study how the radius of a white dwarf varied as a function of its mass. For low masses, the radius increased with mass. On the other hand, it started decreasing with increase of mass eventually reducing to zero. This meant that there could be no white dwarf beyond this limit of about one and a half solar mass, the Chandrasekhar limit. That was indeed a fantastic and most important piece of work.'

'All right, George. Stellar core of mass below the Chandrasekhar limit enjoys eternal life as a white or a black dwarf after the death of the parent star.' I could see how important Chandrasekhar limit must have been for stellar evolution. 'Then what happens to a massive star that may end up with a core heavier than Chandrasekhar limit?' This was a natural question following our discussion so far.

'Fine, let us drop the curtain on the white-dwarf drama for the time being,' said George. 'That means the time has come for us to talk of heavier stars, their evolution, and their ultimate fate.'

'I guess a star of high mass behaves quite differently from the one comparable to our Sun.'

'Exactly, both during its evolution and in its final fate,' affirmed George. 'Because of the higher mass, the gravitational compression acting towards the centre is much more than in a lighter star like our Sun. Consequently, higher temperatures are generated in the core and more power is released. Correspondingly, the gas pressure in the interior of the heavy star is also greater, which counteracts the stronger gravitational force.'

'We know that the Sun's life span is about 10 billion years. Now, the heavier star has a lot more mass to burn than the Sun. So, it must live longer than our Sun. Am I right?'

'Sorry, Alfie, you are wrong there,' George shook his head. 'As I just told you, tremendous temperatures are generated within the heavy stars. That means the energy generation takes place at enormous rates. What happens as a result? These fat cats burn themselves up pretty fast, even though they have a lot of mass to begin with.'

'Ah, just like humans!'

'All right then. Let us compare the life of the Sun with that of a heavier star,' George went on to give an illustration. 'Suppose we scale down our Sun's life-span to one year. On this time scale, a star seven times more massive than the Sun lasts for about a day. And, the one that has eighteen solar masses survives for just nine hours! Well, many of such stars may shed part of their mass and live longer.'

'Aha, there is a case for going on a diet,' I laughed glancing at George's paunch.

'There you go again, Alfie,' George wagged his finger. 'You once quoted Philip James Bailey's saying—*It matters not how long we live but how*. So, eat, drink, make merry, and get fat. Happiness may prolong your life, how do you know?'

'All right, George, you win,' I conceded. 'How about the nuclear reactions taking place within the cores of massive stars? Because of the high temperatures now, they must be quite different from those that occur within the Sun. Say I am right!'

'I say you are right!' confirmed George with a smile. 'Once again, let me give you just a glimpse of what happens within the massive star. As a result of the high temperatures, nuclear burning does not stop at helium produced from hydrogen. Helium, the ash of hydrogen burning as it is called, is further converted into carbon. This way, a whole sequence of elements is cooked in the interiors of heavy stars: neon, oxygen, silicon and so forth and finally ending up with iron. Then there can be no more nuclear reactions. You know why? Because, in iron we have hit the most stable configuration. What is the end result of all this toil and trouble taking place within the stellar cauldron? An onion-like structure with an iron core surrounded by shells containing ashes left behind by the successive thermonuclear transformations.'

— CORE OF A HIGH MASS STAR

'I love French-fried onion rings, especially those cooked inside the massive stars,' I said making George groan as usual. 'What about the core? I suppose it is condensed even more than the white dwarf. If so, I repeat the question I asked in the case of a white dwarf. What counteracts gravity?'

'*Similia similibus curantur, like cures the like,* the homeopathic dictum of Samuel Hahnemann, the originator of that system of medicine. Aha, you thought I didn't know any Latin, didn't you?' chuckled George. 'To paraphrase that dictum, similar questions have similar answers, my dear disciple. Now nuclear reactions have come to an end. There is no longer any heat generation, no gas pressure to balance gravity, just as it happened in the case of the white dwarf. The whole situation is quite ironical you know.'

'What is ironical? Is that a pun on iron, the final product of stellar evolution?' I asked.

'No puns, I mean gravity's role in all this,' answered George. 'It was gravity that gave birth to the star and it was gravity that regulated its life. But look at what happens now. That same force turns into an agent of utter devastation. Iron nuclei are crushed and torn apart by the gravitational compression. Not only that. The electrons are forced back into the nuclei turning all protons into neutrons. The stellar core is now composed of just neutrons. It has turned into a vast sea of neutrons, or the *neutron gas*.'

'In other words, this neutron gas has taken the place of electron gas that inhabited the white dwarf.'

'Yes, the neutron gas now behaves exactly like the electron gas.'

'We are back to quantum effects then?'

'Indeed. Like electrons, neutrons are also fermions, spinning particles that obey Pauli's exclusion principle. And the resultant pressure of the neutron gas due to this quantum effect counteracts gravity. We have now a *neutron star* composed of nothing but the neutron gas.'

I had heard of neutron stars of course. Now, I knew what they were.

'My next question. The white dwarf is the size of the Earth you said. How large is a neutron star, George?'

'How small you mean,' George corrected me. 'The radius of a neutron star is typically around a mere ten kilometres. You could cover the distance from the centre to the surface easily. Of course, if you attempted such an adventure, you would be crushed out of existence by the compression of the matter.'

'Let us see. Matter of the order of a solar mass is now packed into a sphere of ten kilometre radius! That means the density of a neutron star must be awfully high.'

'You bet. Since the stellar core is made up of closely packed neutrons, it is of nuclear density, which is phenomenal. About one hundred trillion grams per cubic centimetre. Can you imagine that! Suppose you manage to bring a spoonful of neutron star matter to the Earth. It would weigh as much as the entire humanity put together. Present population of course.'

Illustrative analogies are meant to help understand unusual phenomena. But, in the present case, it was impossible to imagine such high densities. Well, that is what happens in astrophysics I thought. Having learnt a bit about white dwarfs a little earlier, more questions arose in my mind. I continued my enquiry.

'George, you told me that a white dwarf shines merely because of the heat it has retained from its shrinking. When it cools completely and becomes a black dwarf, it no longer shines and becomes invisible,' I recapitulated. 'What about a neutron star? Does it shine? Does it become visible so that it can be detected observationally?'

'Very relevant question as usual, Alfie. Well, the scenario we have now is rather different from the one we had in the case of the white dwarf,' George went on to explain. 'When the core collapses giving birth to the neutron star, the surface of this monstrously heavy new-born baby is at a temperature of about a million degrees. At such high temperatures, the surface would correspondingly emit high energy radiation like X-rays. Eventually, the neutron star will cool down and cease to glow. Obviously the neutron star would be invisible in this state. Nevertheless, there is something extraordinary that can happen. Neutron stars often possess a pair of beacons of radiation, radiation composed of different wavelengths. This could possibly include visible light too. As the star spins, we would observe a flash of radiation each time we come in the line of the radiation beam. It is like what happens in the case of a lighthouse. So the neutron star keeps sending pulses of radiation at constant intervals. Such neutron stars are called *pulsars*. Neutron stars spin so steadily, that the pulses sent out by pulsars are extremely regular. The regularity of the pulses is comparable to the accuracy of atomic clocks, you know. Because of this, when the pulsars were first detected, astronomers wondered whether some extra terrestrial beings might be sending these signals. So half jokingly the new objects were christened *LGM* or *Little Green Men*.'

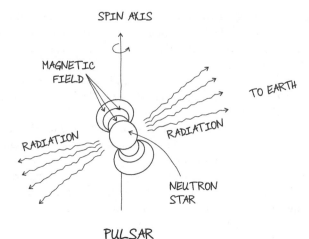

PULSAR

George smiled to himself. He must have remembered something amusing from the past.

'You know, Alfie, I happened to attend this one-day conference convened to discuss the mystery behind the pulsars,' George reminisced. 'One of the experts said that if indeed extra-terrestrials were sending the pulses, it proved that life existed beyond Earth. But it could not be intelligent life, the expert said, if they were wasting so much energy, especially targeting the Earth inhabited by another stupid civilization!' George laughed remembering the past incident.

'That is quaint, George,' I joined George in his mirth. 'At least some scientists do have a sense of humour I must admit. Still, a question remains. You said there is no energy generation going on inside the neutron star.'

'That's right.'

'Then how are these beams of radiation produced?'

'That is a million-dollar question, Alfie,' remarked George. 'Neutron stars happen to come with very strong magnetic fields attached to them. Charged particles moving in these magnetic fields are supposed to produce the radiation beams emanating from the magnetic poles. Nobody knows the exact details of such a mechanism. If I found out the answer, I would be rich and famous. I wouldn't be sitting here doling out my hard earned knowledge, you know.'

'All right, ignorance is bliss as the saying goes,' I said. 'To sum up what you have told me then, stars lighter than the Chandrasekhar limit end up as white dwarfs. And all those that are heavier rest in peace as neutron stars.'

'Patience, Alfie, patience, let us not jump ahead too fast,' cautioned George. 'Even neutron stars, like white dwarfs, have a mass limit. This was demonstrated by Robert Oppenheimer and his co-workers in the thirties. They even thought of what would happen to stellar cores heavier than such a limit.'

'You mean the same Robert Oppenheimer who headed the Manhattan Project that produced the atom bomb?' I was quite surprised.

'Yes, Alfie, the very same Oppenheimer,' confirmed George. 'As you correctly pointed out, he is well known for his leading role in producing the most lethal weapon of mass destruction. Only experts in theoretical physics know and appreciate his work on neutron stars and gravitational collapse.'

George fell silent. After collecting his thoughts, he resumed the discussion. 'Oppenheimer, in collaboration with George Volkoff, was the first one to work out the mass limit for neutron stars. This was a major advance in line with the Chandrasekhar limit to the mass of white dwarfs. Our knowledge of nuclear interactions has steadily increased since those times and a lot has been done in this field. But even then, Alfie, our knowledge is not really complete especially in the case of the immense compressions that exist in the core of a neutron star. In any case, the neutron star mass limit may be taken to be, say, about three solar masses.'

'All this is theoretical. So far, so good,' I commented. 'But, how about observational values for the neutron stars?'

'You are right. One must always keep in mind what is actually observed in nature,' George nodded. 'It so happens that many neutron stars, or pulsars, have been observed with masses of about the Chandrasekhar limit, which is about one and a half solar masses as we have seen. But, pulsars of around two solar masses have also been observed. But we can take as extreme theoretical limit of, say, five solar masses for neutron stars. After all, Alfie, we have to be absolutely certain of the identity in a line-up.'

'All right, what happens to stellar cores that exceed this limit?'

'Aha, we are now getting to the most interesting part of our story, Alfie,' George said jubilantly. 'There is no known force in nature that can withstand the overwhelming gravitational attraction within such a heavy core. Gravitational collapse is now inevitable. The collapse is catastrophic, continuous, and relentless. And, Alfie, what is the final result? It is the formation of a *black hole*. The bizarre end product of a super-massive star that has exhausted its nuclear fuel, the grand finale to the symphony of stellar evolution!'

Ah, black holes! Everyone is enamoured by black holes. After all, gravitation is an irresistible force of attraction, is it not?

'Well, let us get back to Oppenheimer for a moment,' said George. 'In 1939, Oppenheimer, along with H. Snyder, studied the continuous collapse of a heavy star and showed how it ends up as a black hole.'

'This process was obviously a logical sequel to the neutron star mass limit, I suppose. Was there anything fundamentally different from the formalism that had been used for white dwarfs?'

'That was a telepathic question, Alfie.' George regarded me for a moment. 'How did you know that I was going to tell you precisely what you wanted to know? Well, I explained how Chandrasekhar had used Einstein's *special theory of relativity* in his calculation, didn't I? Now, Oppenheimer and his co-workers had to use the *general theory of relativity*.'

Well, I knew it. Something new this way comes!

'In 1905, Einstein gave his revolutionary special theory of relativity that dealt with inertial observers moving with constant velocities with respect to one another,'

George elaborated. 'That theory fused space and time into spacetime among other things. Einstein now wanted to extend the theory and the spacetime formalism to include accelerated observers. Acceleration naturally incorporates gravitation, doesn't it? *If God had been satisfied with inertial systems, he would not have created gravitation*, Einstein declared. For ten years, he grappled with gravity, working incessantly, meandering in unfamiliar mathematical grounds, following false leads and then returning to the right path, on and on. During the course of this strenuous stretch of work, he wrote to one of his colleagues: *I am now working exclusively on the gravitation problem...Compared to this problem, the original theory of relativity is child's play*. And at the end of it all, in 1915, he presented to the world his magnificent general theory of relativity. It was a geometrical theory of gravitation, a total departure from Newtonian gravity, Alfie.'

George paused to catch his breath.

'From the first fundamental ideas to its final completion, the theory was an unprecedented stroke of genius, Alfie,' George said emphatically. 'J. J. Thompson characterized general relativity as, *One of the greatest—perhaps the greatest—achievements in the history of human thought*. It was not an achievement just in physics, Alfie, not even science as a whole, but *in the history of human thought*. According to Dirac, had Einstein not given special theory of relativity, sooner or later, someone or the other would have formulated it. On the other hand, if Einstein had not created general relativity, no one would have accomplished such a feat. In that case, there would have been no black holes, no gravitational waves, no proper description of the Universe as a whole!'

To tell you the truth, I had never seen George so carried away as now when he was talking about the general theory of relativity.

'I can go on with my panegyric to GTR. But, returning to Oppenheimer, his collaborative work on both the neutron-star mass limit and gravitational collapse leading to the formation of a black hole involved general relativity. This was indeed a major step forward in astrophysics. Unfortunately, his work was all but forgotten till the nineteen sixties. Then there was a surge of tremendous activity in the field of gravitational collapse with enormous amount of studies, both theoretical and observational, on neutron stars or pulsars, and black holes.'

'You explained how, when a star collapsed to a white dwarf, the outer shell was thrown out gently as a planetary nebula. What happens to the stellar matter when the core is catastrophically reduced to a neutron star or a black hole?' I enquired.

'I think you know the answer, Alfie, since these things get lot of publicity nowadays,' said George. 'The outer envelope of the star explodes violently. The energy released is enormous, some hundred million times the luminosity of the Sun. Or equivalently, the energy released in a supernova explosion amounts to all the solar energy emitted in a million years! The exploded star shines as bright as the stars of an entire galaxy put together. This is what we call a *supernova*. Remember elements up to iron are cooked within a star? Now heavier elements are formed in the brief, intense heat generated in a supernova. But that is not the end. New stars can be formed out of the debris of a supernova, which now contains all the elements that exist in nature. As a matter of fact, our Sun is made up of such recycled stellar matter and so are the planets. That is how our Earth happens to contain all the elements needed for life. Therefore, Alfie, you and I are parts of the same star that exploded way back in the past. Think about it!'

I didn't have to think about it. It was indeed a spectacular fact.

'My usual question, George, what about the observational situation regarding the supernovae?' I continued my enquiry.

'First of all take our own Milky Way,' George went on to answer my question. 'Way back in the year 1054, Yang Wei-Te, the Chinese astronomer, observed a new *guest star*, as it was called, appearing in the morning sky. It could be seen even in broad daylight for weeks on end. Now, the remnant of that event, which was in fact a supernova explosion, has been identified as the Crab Nebula in the constellation Taurus.'

'I have seen the photograph of the Crab Nebula. A network of glowing multicoloured filaments enveloping a diffuse cloud of gas. It is beautiful.'

'Many celestial objects are absolutely beautiful, Alfie!' acknowledged George. 'The Crab Nebula contains a pulsar, the Crab Pulsar, which is the relic of the star that exploded. We have already talked about how Tycho Brahe sighted a supernova in the constellation Cassiopeia in 1572. Again in 1604, Kepler also observed a supernova in the constellation Ophiuchus. In either case, no neutron star resulted during the stellar explosion. Both of these celestial events were noticed by Galileo also. Since then no supernova has been observed in our Milky Way. There is a saying that a supernova is seen in our galaxy only when there is a great astronomer around. The implication is obvious.' George added in a stage whisper, 'Alfie, never mention this to our astronomer colleagues. They will get terribly upset.'

'My lips are sealed, George,' I laughed. 'But, George, I do read quite often in the newspapers about some supernova or the other being detected. What about them?'

'Ah, several supernovae are observed each year in other galaxies, not our own Milky Way.' George thought for a moment. 'A bit more about supernovae, Alfie. Supernovae resulting from catastrophic gravitational collapse are classified as Type II supernovae. There is another type of stellar explosion that involves the white dwarf. Let me explain. Suppose a white dwarf is paired with another normal star. The two stars form a binary system going around each other. Take for example, Sirius the brightest star in the sky. It is actually a binary star system. The bright star has a faint companion, which is a white dwarf. If some amount of matter is dumped on the white dwarf by its companion, the former can flare up. This is now termed as *nova* in modern astronomy. On the other hand, matter can be accreted in large quantities by a white dwarf from its companion. When the white dwarf's mass reaches the Chandrasekhar limit, it explodes. The end-result is not a *nova*, but actually a *supernova*. This is classified as Type Ia supernova. Incidentally, the *new star* that was observed by Tycho Brahe, or *nova* as he called it, happened to be a type Ia supernova. Such supernovae have the same brightness, since the conditions needed for their occurrence are also the same. As we shall see later on, they come in handy in measuring distances to far off galaxies. And, believe it or not, this has led to one of the most recent and extremely important discoveries in cosmology. The little white dwarf measuring the vast expanses of the Universe as a whole, amazing isn't it? Well, before we come to our own times, there are many aspects of our Universe we have to cover, Alfie.'

George knit his brows as though he had forgotten something he wanted to tell me. And then his face brightened up.

'Alfie, let us get back to the black holes for a moment, shall we?' George resumed. 'Once considered to be invisible and hence undetectable, black holes seem to be everywhere. Many of them of several solar masses have been detected in binaries. But the surprising fact is that most of the galaxies seem to harbour giant black holes at their centres. Our very own Milky Way boasts of a black hole weighing some three million solar masses. There seem to be galaxies loaded with black holes of billions of solar masses.'

'Oh yes, one reads about them in news papers all the time. Do they lead a quiet life like us ordinary citizens or do they get violent like some others we come across once in a while?'

'I thought you led a pretty wild life, not a quiet one, Alfie,' smiled George. 'Fortunately, the one that our Milky Way holds in its bosom is a tame one. But matter can pour in copiously into a black hole from a normal companion star forming an *accretion disk* around it that can lead to spectacular effects. This happens in the case of stellar black holes too. But, the galactic giants! There you can have fireworks of cosmic proportions. Huge amounts of matter can be thrown out from around the black holes with energies comparable to billions of supernova explosions. Centres of such activity are called *Active Galactic Nuclei* or *AGN*s. That is not all. Even objects known as *quasars* are supposed to be energised by such galactic black holes. They are so distant that you do not see the parent galaxy, only the brilliant baby. Obviously, these big black holes can be of prime importance for cosmology.'

Both George and I had been so absorbed in our discussion that we had been oblivious of our own surrounding,

'Good heavens, Alfie, did you notice how long we have been talking? It is getting dark already!' George exclaimed.

Indeed, it was dusk and the stars were appearing here and there hesitatingly to make sure that the Sun was gone and they had the night sky for themselves. As we walked across the campus, George looked up at the sky and pointed at the most familiar constellation that happened to adorn the zenith: Orion the Celestial Hunter!

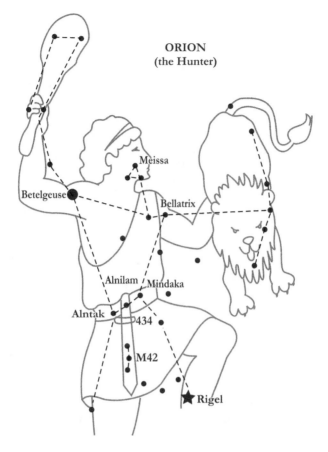

ORION
(the Hunter)

'Beautiful, beautiful,' murmured George. 'Look at the shoulder of the Hunter's arm holding up the club, Alfie. Can you see the bright reddish star there? It is *Betelgeuse* a red super-giant similar to red giant. But it has resulted from the evolution of a star more massive than the Sun. Near his ankle is the hot blue-white star *Rigel*. One can spend hours looking at all those stars that make up the constellation, wondering about them. You know, Alfie, practically every culture has some story or the other associated with Orion. Ancient Egyptians, Romans, Indians, the Chinese, the Bororo Indians of Brazil, the Maori of New Zealand, on and on, they all had their own legends of Orion. It is amazing that one can write the cultural history of the world through constellations. Ah here we are, where our paths diverge, you on your way home and I trudging up to my office. Let us meet soon, Alfie. How about Bruno's for an exotic meal he promised?'

'Great, let us do that, George,' I responded enthusiastically.

'Take care, my boy,' George waved and was on his way towards his physics building.

As usual, I sauntered across the campus enjoying its ambience under the scintillating starry canopy.

Chapter 10
The First Rung of the Cosmic Ladder

I spent the day in total relaxation doing nothing. Doing nothing? Is that possible? Perhaps not. I did go over George's manuscript, studying whatever we had discussed in our recent session—basically about stellar evolution and its three end-points. I found it absolutely fascinating. This was in a way to prepare myself for my next meeting with George. I was eagerly looking forward to the joyful event that would take place in the evening. Well, between you and me, let me confess that there was the added attraction of having a good time at Bruno's, that haven for rare culinary delights.

As I entered Bruno's, the first sight that greeted me was of George carefully reading his manuscript, scribbling some notes on the margins. Being a perfectionist, he would be trying to improve it till the last. Once, he had told me that a talk, like a baby in a mother's belly, keeps developing till the moment of delivery. It is the same with a book, I suppose, it keeps improving till the moment of its delivery to the publisher.

Instinctively, George looked up and broke into his customary spontaneous smile, to which only the most hard-boiled misanthrope would not respond. And, believe me, I am no misanthrope.

At that moment, a lovely piece of music started playing in the background. It was Joaquín Rodrigo's *Concierto de Aranuez*, which is considered to be one of the pinnacles of Spanish music and of the guitar concerto repertoire. Soon enough, Bruno arrived followed by a young man.

'Ah, my two most special guests!' Bruno welcomed us with his customary exuberance. 'Let me introduce our new chef. I told you about him last time, didn't I? He has travelled all over the world and has learnt to cook food of many lands, *squisito*! I shouldn't forget to tell you his name. He is Jose Ramirez, but goes by different names in different places: Yusuf Reza in the Middle East, Jomo Ramyatta in Africa, simple Ram in India and Thailand, so on. For us he is just Joe.'

Joe, who looked more like a Flamenco dancer from Andalusia rather than a chef, bobbed his head up and down with each of his name mentioned, flashing his sparkling white teeth.

© Springer International Publishing Switzerland 2015

C.V. Vishveshwara, *Universe Unveiled*, Astronomers' Universe,
DOI 10.1007/978-3-319-08213-4_10

Bruno extended his hand towards us and continued his introduction, 'Joe, these are our most important guests who come here all the time, George the big Professor at University and our dear friend, Alfie the Wise.'

Joe bowed ceremoniously and said in perfect English, 'I am greatly honoured indeed!'

'Today Joe makes Spanish food for you,' announced Bruno.

'Oh, Spanish cuisine, my favourite!' exclaimed George. 'What is on the menu?'

'What menu?' Bruno raised his eyebrows. 'Look around, you are the only two guests. We make some special dishes for you.'

The young chef Joe took over as his eyes glittered with delight. 'We can start with *Albóndigas*, meatballs in garlic sauce, one of the delicious *tapas*. Then we move on to the main course—*arroz negro*, squid cooked with rice in its own ink, absolutely divine. Finally, we would end the meal with the dessert—*Mantecados*, traditional Spanish crumble cakes served with whipped cream, an unparalleled delicacy. Does such a dinner appeal to you?'

'Sounds great,' I said. 'Except that I happen to be a strict vegetarian.'

'Oh Alfie,' George groaned. 'Heaven help me, as a matter of principle I have to join this cruel man. Yes, make mine a vegetarian meal too.'

'No problem,' Joe assured us. 'We shall replace meatballs with breaded potato balls. The main course will be *pimiento rellenos*, big red bell peppers stuffed with slightly spiced rice, chopped nuts and mushrooms, covered with cheese and baked in tomato sauce. We shall retain the dessert.'

'Sounds great. Vegetarians of the world, Unite. You have nothing to lose but your cholesterols!' George shook his raised fist.

'Food will be wonderful, doubt not,' nodded Bruno. 'How about some wine to go with the meal? It is pure vegetarian, you know,' he grinned.

'What do you have in mind?' George raised an eyebrow.

'Two major types of Spanish wine are Rioja and Ribera del Duero,' answered Bruno. 'People fight which is best, you know. We will serve both, you decide.'

Bruno and Joe withdrew to the kitchen to prepare the special meal for us.

'Knowing you, Alfie, you must be going over the manuscript religiously to absorb all that we have been discussing and more,' began George. I was waiting for this. 'We have talked a lot about planets and stars, especially stars—their types, their evolution, and their end-states. They are the little ingredients that make up the Universe, aren't they? We are now ready to look at the larger picture.'

Our progress towards the large-scale structure of the Universe had been gradual and meticulous. Unless you have the foundations firmly established, how can you build the super-structure? I was sure that the journey forward would retain this character.

'Ideas regarding the large scale structure of the Universe dates way back to antiquity, Alfie,' continued George. 'As a matter of fact, around 300 BC, Zeno of Citium proposed a finite cosmos of stars in an infinite void, the *Stoic model*, although they had no idea of the real nature of the stars at that time. Then came the *Epicurean model* due to Epicurus of Samos, an infinite cosmos consisting of uniform material of unlimited quantity, with worlds scattered and spread all over. All matter was composed of atoms, regulated by natural laws.'

'You mean they were already thinking of atoms, natural laws, and all that?' I queried.

'Not in the modern way, they were all philosophical ideas,' said George. 'Nevertheless, these ideas were revived by Isaac Newton. He first believed in the Stoic model of the cosmos. But then, his young friend, the clergyman Richard Bentley, pointed out that such a model would be unstable since the finite spherical distribution of stars within a void would collapse under its own self-gravity. Newton then switched to the Epicurean model with an infinite number of stars, all attracting one another. This was an equilibrium configuration all right. However, Newton realized that this was like an infinite number of needles standing on their pointed tips, a totally unstable situation. Still, the Universe was very much a reality, wasn't it? So, Newton asserted that this showed the existence and omnipotence of Divine Providence without which the Universe could not exist!'

'God to the rescue, what do you know?' I said. 'George, I have heard that long ago, when physicists were confronted with what looked like innumerable types of elementary particles, they classified them all under the categorical name *Gokons* that stood for *God-Only-Knows-ons*! Is this true?'

'Absolutely! I had completely forgotten about it,' George laughed long and hard.

George leaned back in his chair as he caught sight of the waiter approaching our table. The waiter placed before us a carafe of wine and two glasses.

'Is this Rioja or Ribera?' George enquired.

'I wouldn't know, sir,' the waiter smiled. 'Even if I did know, boss would have asked me not to inform you. He wants you to decide which one is better. Next carafe will contain the other type.' He went back letting us fill our glasses ourselves as we prefer to do.

We poured the wine into our glasses and took a sip slowly. It was simply exquisite, whichever R it might have been.

'Let us get back to the Universe, Alfie,' George resumed. 'The first step that leads to its large-scale structure happens to be our own Milky Way, the mysterious wispy band that stretches from horizon to horizon across the night-sky. As with the constellations, different cultures had different mythological stories associated with it. You can read about them at your leisure as you always do.'

'And Galileo's little telescope resolved the stars and groups of stars in the Milky Way as we saw some time ago,' I said.

'That is right. But then, it was William Herschel who really launched upon its observational study for the first time, you know.'

'Ah yes, you did mention Herschel in passing when we talked about Uranus, didn't you, George?'

'True, his discovery of Uranus was a major event in itself. On the universal scale his findings and ideas regarding the Milky Way are very important, Alfie. Let me briefly tell you about the man first.'

George took a sip of his wine, closed his eyes, and silently enjoyed its taste. I followed his example, but with eyes wide open.

'William Herschel was a talented musician,' began George after his sip of wine.

'Wait a minute,' I interrupted. 'I thought he was an astronomer.'

'Of course, he became a celebrated astronomer in due course. But he began his career as a musician,' explained George. 'Herschel was born in Hanover, Germany, in the year 1738. At the age of fifteen, he joined his father who was a musician, an oboe player, in the Hanoverian Military Band. The young Herschel was an accomplished player of not only the oboe, but also the violin and the organ. Quite remarkable, isn't it? Four years later, he visited England with the Hanoverian Guards regiment. Probably, that country was quite attractive to him. So he chose to settle in England a year later. Herschel became quite well-known for his prodigious musical output, Alfie. He had to his credit no less than twenty-four symphonies, seven violin concertos, and two organ concertos!'

'That is quite impressive, George,' I was indeed astonished at this unexpected side of Herschel.

'Indeed, but his astronomical record was even more striking, you see. Attracted by the stars, Herschel began his celestial journey as an amateur astronomer, constructing his own telescope of twelve-inch aperture. It was one of the finest at that time. Building telescopes became his passion. He fabricated progressively larger ones, completing his big forty-eight inch instrument by the age fifty. Imagine that!'

'I know that many amateur astronomers make their own telescopes as a hobby. But Herschel's case seems quite different,' I observed.

'You are right, Alfie,' George agreed. 'Herschel started making his large telescope for his progressive exploration of the sky, not just as a hobby. Well, let us first consider Herschel's discovery of Uranus. As you know, Uranus was the first planet to be discovered using a telescope. Even with the aid of a telescope, Uranus happened to be a very faint object in the sky. But Herschel's ability to spot a new celestial entity, however dim, was similar to Tycho's. Remember what Tycho wrote in the context of his discovery of a *new star*? Let me read form the manuscript. Here it is, the relevant part of Tycho's statement: *It was evening, after sunset, when according to my habit, I was contemplating the stars in the clear sky, I noticed that a new and unusual star. . .; and since I had almost from boyhood known all the stars of the heavens perfectly, there is no great difficulty in attaining that knowledge, it was quite evident to me that there had never been any star in that place in the sky, namely in the constellation Cassiopeia., even the smallest. . .'*

Ah, I had heard those very words from the mouth of that astronomer himself, hadn't I? I couldn't possibly tell George about it, could I? George went on without noticing my momentary mood of distraction.

'Likewise, Herschel compared the acquisition of his rare gift to his musical training: *Seeing in some respects is an art that must be learnt. To make a person see with such power as I am able to is nearly the same as if I were asked to make him play one of Handel's fugues upon the organ. Many a night have I been to see, and it would be strange if one did not acquire a certain dexterity by such constant practice.*

'Well, Herschel realized with great pleasure that he had discovered a new object, perhaps a comet! This was in the year 1781. But then, to his even greater joy, he could show that it was indeed a planet and not a comet. In his later career, Herschel discovered two Moons of Saturn—Mimas and Enceladus, as well as two Moons of Uranus—Titania and Oberon. Incidentally, Herschel's sister Caroline was also an astronomer who succeeded in discovering her own comets. And William's son John, born in 1792 went on to become a distinguished astronomer in his own right. So you see, Alfie, it was a remarkable Herschel astronomical family!'

'Stars streamed through their veins so to speak,' I commented.

'Most people know only Herschel's discovery of Uranus and none of his other important findings,' pointed out George. 'Let me recount briefly some of Herschel's brilliant contributions to astronomy, Alfie. First let us recognize an exceptional fact, namely William and Caroline Herschel were the first astronomers in history to do a systematic survey of the night sky. William, using the much larger telescopes, discovered that many of the stars that he had previously observed as single stars were in fact double or binary stars. By observing them over a period of time, he realized that they seemed to be moving around each other. This observation was very significant, Alfie. Because it showed that Kepler's laws of planetary motion and Newton's law of universal gravitation not only applied to the planets going round the Sun, but extended even to the stars far away. Those laws were truly universal, you know.'

Herschel's achievements seemed to be growing in number. I was happy that I was learning so much about the great astronomer.

'What next?' I asked.

'What next?' George repeated my question. 'The first of the building blocks that make up the cosmos!'

'What might that be?' The story seemed to be getting more and more interesting.

'The Milky Way!' announced George. 'Herschel had noticed in his survey of the skies that the density of stars varied depending upon what region he was looking at. Two things could account for this varied distribution of the stars. Either the stars were closer together in the regions where they appeared to be more compact. Or the stars were actually spaced apart more or less uniformly. But, the stars seemed to be the denser in some region because that is where the Milky Way extended farther out. It was the latter explanation that William believed to be the case. After further analysis, he concluded that the shape of the Milky Way was that of a disk or a grindstone.'

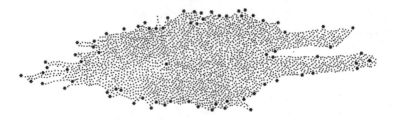

'We now know the correct shape of the Milky Way, which does not really resemble Herschel's model,' said George. 'But what is important, Alfie, is that he was the first one even to come up with the idea. That is what counts. We shall soon see how another remarkable astronomer demonstrated the true shape and size of the Milky Way. For the moment let us join Herschel as he trains his large telescope at the *nebulae*, the diffuse patches that populate the night sky.'

George slowly sipped his wine a couple of times with undiluted enjoyment and, needless to add, so did I.

'Ah, the nebulae! Herschel had received a copy of Charles Messier's catalogue, *Connoissance des temps*, an almanac published in 1781, in which 103 nebulae had been listed. Herschel began observing these objects using his twenty-foot reflector with an aperture of about nineteen inches. Because William was using an instrument larger than anything that Messier could lay his hands upon, he was able to discover new nebulae that Messier had missed. And in the course of the next several years, he had added no less than some 2,500 new nebulae to the list! And his son John would discover more of them eventually.'

'That is an impressively large number, George,' I was genuinely astonished.

'Yes indeed, Alfie. Now, Herschel tried to see whether the nebulae could be resolved into stars. Many could be, but others remained just diffuse clouds. One such example was the case of a planetary nebula around a white dwarf. We talked about that while discussing the end-states of stellar evolution, didn't we? On the other hand, there could be nebulae that could not be resolved because of their great distances. In regard to the nebulae, the German philosopher Immanuel Kant had put forward a rather unusual idea.'

Once upon a time, I had read Immanuel Kant's philosophy. I recalled now his saying that had been etched on my mind.

'George, I remember what Kant wrote: *Two things inspire me to awe: the starry heavens above and the moral Universe within.*'

'That is a deep philosophical statement, Alfie,' George appreciated the quotation. 'Kant had something to say about the starry heavens above. Kant had theorized that nebulae that could not be resolved into clusters of stars had to be *Island Universes*, what we call *galaxies* today. They were being observed at great distances beyond the Milky Way and therefore could not be resolved into stars that made them up. This was a prophetic insight into the nature of the Universe, Alfie. Of course, Kant did not possess exact and extensive observational evidence for his idea in his own time. Kant also believed that the Milky Way was itself an Island Universe, like the distant nebulae. He even speculated that it was shaped like a flat circular disk. As we saw, Herschel did work out from his survey of the stars what he believed to be the shape of our Milky Way, a disk in fact. What is more, many of the nebulae he had observed lay beyond the Milky Way, the denizens of the Universe at large. Isn't that amazing?'

George fell silent for a moment or two. He quickly goes over in his mind whatever he has been relating so far before proceeding further.

'William Herschel, musician of great talent, astronomer of immense ability! He built his telescopes with finesse, used them with skill. Driven by his passion, he explored the Universe with imagination. He had the unique distinction of having set foot on the first rung of the cosmic ladder. In the next century, the realm of the nebulae would be opened up by one of the greatest astronomers of all time. But before that, the Milky Way had to be charted out completing the work Herschel had begun. Enter Harlow Shapley!'

George looked up and exclaimed, 'Ah, I see the much-awaited angelic vision of someone other than Shapley entering the scene, bringing joy to our earthly existence.'

I could not but smile at the last dramatic statement made by George. Of course, I too was happy to see the smiling waiter carrying a tray and making his way towards our table.

We watched as the waiter placed two small plates in front of us. What did Joe say about the starter? It was *Albóndigas*, meatballs in garlic sauce, one of the delicious *tapas*, but the meatballs replaced by breaded potato balls as concession to us vegetarians. I let George try the dish first. As expected, he closed his eyes as he savoured the morsel of food in his mouth slowly, gently.

'Oh, this is out of the world. Try it, Alfie, what are you waiting for?' George nodded his head a couple of times and smiled contentedly at the waiter. The latter returned his smile and went back to the kitchen. I was sure he would report George's reaction, the touchstone of culinary art, to Bruno and Joe. I too joined George in the delightful tasting of the *Albóndigas* and we ate for a while in silence, as one should, helping ourselves to the fast depleting supply of the delicious wine.

'Now for Harlow Shapley. We move in space and time, from England to the United States, and from the nineteenth century to the twentieth,' George took up the story of the Milky Way after we had finished eating the appetizer. 'As usual, let me very briefly outline Shapley's life and work. I shall let you read the details later on when you go home.'

'Fair enough,' I agreed.

Shapley

'All right, Shapely had a rather chequered career before he took up astronomy,' George resumed. 'He was brought up on a farm, worked with a newspaper as a reporter in his teens, wanted to become a journalist, but had no choice but to take up astronomy at the University of Missouri. So Shapley ended up as an astronomer. How do you like that?'

'He was an accidental astronomer then,' I commented.

'You could say that,' said George. 'So was Kepler for instance. He wanted to go into theology but stumbled into astronomy, lucky astronomy! Nowadays, there are hundreds of accidental astronomers, students who register for astronomy since they were not admitted to any other discipline. Alfie, please don't get me started on that dismal subject. Anyway, let us get back to Shapley. After obtaining a doctorate from Princeton University, Shapley joined the Mount Wilson Observatory in the year 1914. He got interested in a class of stars known as the *Cepheid Variables*.'

'What are they?' I interrupted George with my question. Sometimes my curiosity gets better of me.

'I was coming to that,' George said patiently. 'We have talked about Orion, which happens to be a most interesting constellation, haven't we? There is a rather not so interesting constellation, Cepheus the King, which lies opposite the Big Dipper on the other side of Polaris. We shall be speaking of Andromeda, daughter of Cepheus next time, when we discuss the expanse of the Universe. Well, perhaps the only interesting, and most important, aspect of Cepheus is that the constellation is host to a star called Delta Cephei. It is a variable star. Let me explain. Over a period of five days and nine hours, the star passes through a cycle of brightness with clocklike regularity. The difference between its brightest and faintest appearance is so pronounced that you can actually observe the variation with unaided eye.'

'What is the reason for this variability in brightness?'

'It is caused by the periodic change in size, Alfie. The star gets brighter as it swells up, and fainter as it shrinks. Delta Cephei was first observed in 1784. Since then many such stars with both longer and shorter periods have been discovered. These are the Cepheid variables.'

I wondered as to what these variable stars, however interesting, had to do with the Milky Way.

Leavitt

'Before we move on to the inter-relationships among the Cepheid variables, the Milky Way, and Harlow Shapley, let me introduce a remarkable lady,' smiled George. He had obviously read my mind. He added seriously, 'Her name was Miss Henrietta Swan Leavitt.'

George took the last sip of the wine in his glass before continuing.

'After her college education in 1893, Miss Leavitt joined the Harvard College Observatory as one of the *human computers*. That is how the women, who were employed for examining photographic plates in order to measure and catalogue the brightness of the stars, were called. They were also dubbed *Pickering's Harem*, since they had been hired by the astronomer Edward Pickering. The work was gruelling, but the wages were low. Because Leavitt had independent means, Pickering initially did not pay her. Later, she received a royal stipend of thirty cents an hour for her work!'

'Apart from scanning photographic plates, did Leavitt do any observational work?'

'No Alfie, in the early 1900s, women were not allowed to operate telescopes,' George shook his head. 'From the photographic plates, Leavitt could find a number of variables in the *Small Magellanic Cloud*, or *SMC*, an irregular dwarf galaxy which is a next-door neighbour of our Milky Way. Their periods varied from as short as one and a half days to as long as one hundred and twenty-seven days. But, here is the most important thing. The longer the period, the brighter was the star.'

I did not understand why this fact was of utmost importance immediately. I knew fully well that George was going to reveal the secret of those stars.

'Let us now try to understand the concept of a *standard candle* and relate it to the remarkable discovery of Miss Leavitt,' proceeded George as I had anticipated. 'Suppose I have a lamp, Alfie, and move it away from you progressively. If I double the distance, the *apparent brightness* of the lamp, which you actually measure, decreases fourfold. Increase the distance three times, the lamp appears to have become nine-fold fainter and so on.'

'Ah, I know that—the *inverse square law*, just like Newton's law of gravitation!'

'Exactly. If you know the *intrinsic brightness* or *absolute luminosity* of the lamp, you can deduce the distance of the lamp from you by measuring its apparent brightness using the inverse square law. This turned out to be true of the variable stars too. Leavitt's detailed analysis of the photographic plates that had recorded the variable stars in SMC established a direct relationship between the intrinsic brightness of a variable star and its period of variability. So, what do we have in the end? We determine the period of brightening and dimming of a variable star by observation and hence its absolute luminosity. Therefore, we can estimate its distance from its apparent brightness, which we can again find out by direct measurement. In short, the Cepheid variables can serve as *standard candles* for determining distances on cosmic scales. That, in essence Alfie, was the stupendous discovery of Miss Henrietta Swan Leavitt!'

Yes, this was indeed a fabulous discovery, I agreed. I was certain that the detailed measurements of Henrietta Leavitt must have involved a lot of work.

'Miss Leavitt must have been hailed as a great astronomer, an unprecedented accomplishment for a woman at that!' That was but natural I thought.

'Not in her lifetime, Alfie,' George shook his head. 'Leavitt could work only sporadically because of her health problems, including her increasing deafness. In 1921, Harlow Shapley the new director of the Harvard observatory appointed her the head of stellar photometry. But within the year she had succumbed to cancer at the early age of fifty-three.'

George fell silent. I could see that he was truly sad, thinking of the great lady who had been robbed of her deserved place among the stars by her tragic end.

'It is true that she was a member of several scholarly academies during her career. But that is nothing compared to the honours she should have won,' said George. 'As an individual, she must have been a wonderful woman too. Let me quote what one of her colleagues noted in her obituary: *She had the happy faculty of appreciating all that was worthy and lovable in others, and was possessed of a nature so full of sunshine that, to her, all of life became beautiful and full of meaning.*'

George paused and went on.

'Unaware of her death four years earlier, the Swedish mathematician Mittag-Leffler, a member of the Swedish Academy of Sciences, considered nominating her for the 1926 Nobel Prize in Physics. He learnt that it was too late. The Nobel Prize is not awarded posthumously.'

George leafed through his manuscript. 'Let me read the letter from Mittag-Leffler to Miss Leavitt dated February 23, 1925:

Honoured Miss Leavitt, What my friend and colleague Professor von Zeipel of Uppsala has told me about your admirable discovery of the empirical law touching the connection between magnitude and period-length for the S. Cepheid-variables of the Little Magellan's cloud, has impressed me so deeply that I feel seriously inclined to nominate you to the Nobel prize in physics for 1926, although I must confess that my knowledge of the matter is as yet rather incomplete.

'And here are the words of Harlow Shapley, the director of the Harvard Observatory, from his March 1925 letter to Professor Mittag-Leffler:

Miss Leavitt's work on the variable stars in the Magellanic Clouds, which led to the discovery of the relation between period and apparent magnitude, has afforded us a very powerful tool in measuring great stellar distances. To me personally it has also been of highest service, for it was my privilege to interpret the observations of Miss Leavitt, place it on a basis of absolute brightness, and extending it to the variables of the globular clusters, use it in my measures of the Milky Way. Just recently in Hubble's measures of the distances of the spiral nebulae, he has been able to use the period-luminosity curve I founded on Miss Leavitt's work. Much of the time she was engaged at the Harvard Observatory, her efforts had to be devoted to the heavy routine of establishing standard magnitudes upon which later we can base our studies of the galactic system. If she had been free from those necessary chores, I feel sure that Miss Leavitt's scientific contributions would have been even more brilliant than they were.'

George sighed deeply. I could see that he was in a pensive mood having recounted the moving story of Miss Leavitt.

'Much later, one of the asteroids and a crater on the Moon were named after her. But the greatest tribute to the groundbreaking discovery of Miss Leavitt came first through the work of Harlow Shapley who demonstrated the size and the structure of the Milky Way. And then through the exploration of the Universe as a whole by Edwin Hubble. Both of them used Miss Leavitt's Cepheid variables in their measurements.'

After a momentary pause, George added, 'So, Alfie, we are back to Shapley.'

Before George could continue, the waiter appeared again. He cleared our table and placed on it a fresh carafe of wine, the second of the two choices I was sure. Also, pimiento rellenos—the luscious red bell peppers glistened and quivered as steam rose from their hot interior. As before, the waiter looked on until we tasted the dish, while sipping the wine.

'This is indescribable,' George said. 'How can anything taste so good?'

'Aren't you glad that you have become a vegetarian, George?' I said. 'There can be so much variety in vegetarian dishes, often with subtle difference among them.'

The waiter smiled, satisfied that we were satisfied, and returned to the kitchen.

After a few minutes of concentrated epicurean pleasure, George took up his exposition.

'Shapley was now equipped with his standard candles, namely the Cepheid variables,' George said. 'Now, he started his work trying to establish the dimensions of the Milky Way and its shape. For this, he focused on the Cepheid variables couched within *globular clusters*. They are essentially spherical distributions of thousands to hundreds of thousands of stars.'

'I know what they are,' I told George. 'While explaining Newtonian gravity in his famous *Lectures on Physics*, Feynman includes a photograph of a globular cluster and remarks: *This figure shows one of the most beautiful things in the sky—a globular star cluster. If one cannot see gravitation acting here, he has no soul!*'

'Ah, that is Feynman for you, capturing the soul of gravitation,' George was quite happy with what I had told him and so was I. 'Those beautiful things came in handy for Shapley. At the time of his observations, around a hundred globular clusters were known. Shapley concentrated on studying the variable stars within those clusters. He could measure distances to sixty-nine clusters. From these measurements, he could find out their distribution pattern. And in turn, he estimated the dimensions of the Milky Way, its shape, and—what is more—the place of the Sun within this vast system of stars. The Milky Way was found to be roughly like a disk with a central bulge with a diameter of some 100,000 light years. And where is the Sun, where are we? Shapley discovered that we are not at the centre of the Milky Way, not even near enough, but around 30,000 light years from the centre, almost at the edge of our galaxy comprising some 300 to 400 billion stars! Think about it, Alfie.'

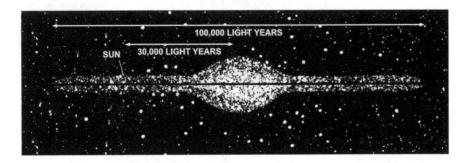

'Copernicus had dethroned man from the centre of the solar system,' I remarked. 'And now, Shapley nudges him away from the centre of the Milky Way. What a pity!'

'Exactly,' nodded George. 'The Milky Way had turned out to be so vast with billions of stars, that Shapley believed that it constituted the entire Universe. As we have already seen, astronomers, starting with William Herschel, had observed diffuse patches in the sky, the nebulae. Did they belong to our own Milky Way, as Shapley believed, or did they lie beyond it? That was the big question.'

'Was the question settled? If so, in whose favour?' This seemed like an astronomical suspense story.

'No happy and easy ending here, Alfie,' answered George. 'In the wake of these two different opinions, a debate was arranged between Shapley and another established astronomer, Heber D. Curtis, who was a proponent of the Island Universe hypothesis. The great debate took place on April 26, 1920, before the National Academy of Sciences in Washington, United States.

'My goodness, thankfully it was not a duel.' I remembered Jeppe's recounting of how Tycho Brahe had lost his nose in a duel. But mum was the word! 'So what happened?'

'Curtis was not convincing,' George replied. 'The judges gave their verdict, point by point of the arguments based on observational data, in favour of Shapley. For the moment, the Milky Way was the entire Universe.'

'We know that it is not true, don't we?'

'Of course, thanks to Edwin Hubble who is about to rise over the horizon in our story,' said George. He added, 'Next time we meet that is. You must wait till then.'

'Let us not forget that since Shapley's time there has been a lot of observations and theoretical modelling of the Milky Way,' said George as a postscript. 'For instance, we now know that the Milky Way turns about its centre. As a result, our solar system is moving round with a speed of about 220 kilometres per second. That seems quite fast, doesn't it, Alfie? However, since our galaxy is so huge, it takes 240 million years for the solar system to go one full round. The last time the Sun was at the position it is now, dinosaurs roamed the Earth! How do you like that?'

We had finished the main course of our dinner. Our waiter, who brought our dessert *Mantecados*, traditional Spanish crumble cakes accompanied by whipped cream, was soon followed by both Bruno and Joe. They were eager to participate in our enjoyment. And they were not disappointment.

'That was a fantastic, fabulous dinner. Let me not even try to guess how you manage to create such a wonder,' George offered his high compliment. That went for me too. 'I have a question and I am sure Alfie has the same query too. The two wines you served were excellent, never tasted better, you know. But which was which, Rioja or Ribera?'

Bruno and Joe exchanged glances and smiled sheepishly. 'We don't know,' volunteered Bruno.

'What, you don't know? How can that be?' George was perplexed and so was I.

'You see it is like this,' Bruno went on to explain. 'When we poured wine into the two carafes, we forgot to keep track. We don't know which one went where.'

George burst out laughing as did the three of us.

'So that was an experiment with negative but very happy result,' concluded George. 'That means we have to repeat it with better control. Bruno, we congratulate you.'

'What for?' It was Bruno's turn to look puzzled.

'For roping in Joe,' replied George. 'We can't thank you enough for the great joy of tasting an exquisite meal.'

As always, Bruno waved away our thanks, while Joe bowed ceremoniously as he had done earlier.

We lingered outside the restaurant for a while. It was a crisp, clear night, the stars shining bright against the dark inverted bowl of the sky. Arched over the sky was the Milky Way that must have been both an intriguing and inspiring sight for eons. George gazed at it silently. It was as if he were looking at it for the first time.

'Isn't it gorgeous, Alfie? It fills you with wonder,' said George. 'As with the constellation Orion, every culture had some story or the other to describe the origin of the Milky Way, *Via Lactea*, in Latin. To the Greeks, it was the starry wake left behind by the chariot of Helios, the Sun God, as the horses drawing it went out of control, when his son Phaethon drove it against the warnings of his father.'

'I know. I have heard the Indian myth depicting the starry path,' I added. 'It marks the course of the sacred river *Ganga*, or Ganges as it is known in the west, as she cascaded down from heaven. Lord Shiva broke her fall, lest the impact should shatter the Earth. He held her captive playfully for a while tied up with the long strands of his hair, and finally freed her so she could descend gently on to the Earth and flow on, bestowing spiritual joy and liberation. The Milky Way is called *Akashaganga*, the River Ganga of the Heavens.'

'That is beautiful, that too with a happy note, Alfie,' George nodded his appreciation. 'It is not at all like the Greek myth that ends in tragedy. Phaethon was struck down by Zeus, since Sun's chariot, out of control, had been wreaking havoc.'

'Alfie, we shall try to understand the expanse of the Universe the next time we get together,' said George. 'Why don't we go to the University Observatory and meet our young astronomer friend Mike Brown? He can explain to us Hubble's monumental work. We can ask Mike to show us some photographs too. What do you say?'

'That would be real great, George.' I was delighted at the prospect of visiting the Observatory as well as meeting Mike.

'All right then. Oh, I almost forgot. Here, I got for you some recordings of the lovely music William Herschel created. I am sure you will enjoy it.' George handed over a compact disc, which he had taken out of the bag he carries nowadays to hold his manuscript.

'Take care, Alfie. We shall meet soon.'

I wished George good night and walked briskly. I was eager to read George's book as usual. Tonight I had the added bonus of Herschel's music.

Chapter 11
The Map of the Milky Way

As soon as I arrived home, I started reading about William Herschel and Harlow Shapley in George's book, while playing the recording of Herschel's music George had given me. The majestic sounds of the orchestra playing one of his symphonies filled my room. As I read the manuscript, I was greatly impressed by the dedication and zeal of the two astronomers in charting out the Milky Way. They had thereby initiated the exploration of the large-scale structure of the Universe. All this brought to my mind two quotations from Plato. I repeated those quotations to myself softly:

Star gazer, my star, if only I were the sky,
I would watch you with ten thousand eyes.

And then again,

Alive, you shone among us like the morning star;
Now, like the evening star, your mantle lights the dead.

After my reading was finished, I carried the record-player to my bathroom, set it on a stool, and started playing one of Herschel's violin concertos. Again, the soft strains of that soothing music spread around my little bath room.

I opened the faucets to fill my bathtub. My bathtub started vibrating gently. I found that the two streams of water from the faucets were rhythmically changing their rates of flow thereby modulating the vibrations and creating sounds that were in total harmony with the music playing in the background.

'So beautiful, boss, the music you are playing,' the bathtub spoke in an almost whisper. 'I hope my own humming does not disturb you.'

Of course, it did not. As a matter of fact, it added to the charm of Herschel's music.

'Thank you, boss,' the bathtub sounded genuinely happy. 'You are so fortunate, boss, but I don't envy you. You read so much, think so much, and you can recall so many things you have learnt like the two quotations you were just reciting to yourself. Those lines are so apt when you think of the astronomers who have illuminated this world with the knowledge of the stars.'

© Springer International Publishing Switzerland 2015
C.V. Vishveshwara, *Universe Unveiled*, Astronomers' Universe,
DOI 10.1007/978-3-319-08213-4_11

I could not agree more with the bathtub's last statement. But, what I had just heard was not its usual style of speech, casual and tinged with playful banter. Moreover, the statement had been crafted beautifully well.

'Like you, I am fortunate too,' continued the bathtub. I could sense its unexpressed appreciation for my unexpressed thought. 'You know why? Long ago my Master used to convey to me so much of his knowledge, so much of his thoughts. Now, you do the same in a different way. If only I could...'

The bathtub hesitated. I wanted to know what it had in its mind. Yes, *its mind*, I was convinced that it had its own mind, its own thought process, however different it might have been from ours, perhaps better than ours!

'Thanks again immensely, boss,' said the bathtub with utmost sincerity. 'What I was going to say is this. I wish I could go out with you and look at the stars, gaze upon them, decipher the multitude of their messages. But then I thought perhaps it is better the way it is. You are my eyes, boss. And I can imagine what you see from your thoughts. After all what did the Master say? *Imagination is more important than knowledge. For knowledge is limited to all we now know and understand, while imagination embraces the entire world, and all there ever will be to know and understand.*'

I was moved by what the bathtub had said. I was indeed fortunate to have such a bathtub for a companion.

'All right, boss, let us not get sentimental about our roles to play,' said the bathtub calmly. 'Let us not postpone the pleasure of having a bubble bath. Yes, it is a pleasure for me too. I share your thoughts, your dreams, your fantasies, don't I? Pour in the mixture, let the bubbles rise and whirl, let all the imaginings in the Universe stir up your mind.'

Well, I followed the bathtub's excellent suggestion. And the brilliant specks that the bubbles had brought with them swirled around me. They glistened in varying brightness and colour. They came singly and in clusters and to my great surprise some of the clusters, containing a large number of the sparkling specks, were shaped like little globes. I closed my eyes and surrendered to the exquisite pleasure of being surrounded by not only the bubbles and miniature stars, but the lovely music that added immeasurably to my experience.

Stargazing Siblings

I thought I knew where I was. During my recent wanderings, I had in fact visited this lovely city of Bath in England. The city was first established as a spa with the Latin name, *Aquae Sulis*, the Waters of Sulis by the Romans sometime in the first century about 20 years after they had arrived in Britain. They built a temple and baths around hot springs on the hills surrounding Bath in the valley of the River Avon. I could go on describing the city, which has become a great tourist attraction. But now, the vision of the town before me looked quite different from what I had seen recently. I felt that I had been transported a couple of centuries back in time.

I could hear beautiful music emanating from the massive, ornate building in front of me. I hesitantly tiptoed inside and stood at the back of the spacious hall. There was a large gathering of people inside listening to the music in absolute silence. On an elevated platform sat the musicians of a full orchestra playing different instruments. Facing them, back to the audience, stood a tall formally dressed figure with silvery grey hair, conducting the orchestra. Waves after waves of harmonious melodies issuing forth from the various musical instruments swept over the listeners. I was sorry that I had missed the beginning of the concert. Soon the concert ended. The conductor turned round and bowed repeatedly to the thunderous applause of the audience. Then the listeners trooped out exchanging admiring comments about the music and the conductor. I too moved on in silence still feeling the power of the music I had heard.

All at once, I realized who the conductor happened to be. It was none other than William Herschel, musician, composer, and conductor. The piece the orchestra had been playing was one of the symphonies he had composed. Probably, this was a rare occasion on which he had found time to conduct an orchestra amidst his busy schedule of observing the heavens. I followed Herschel as he came out of the concert hall and headed towards his home.

The house before me was like any others I had seen on my way. But what was remarkable was the large, seven-foot long telescope that had been installed in the garden supported by sturdy scaffolding. The telescope pointed towards the sky ready to capture light from the celestial bodies. It was getting dark now, night was about to descend. Herschel came out of the house. A handsome and dignified lady accompanied him, carrying a large book apparently to keep record of their observations.

'Caroline, it has been a long time since the new planet was first spotted. It is a pity that you were not here that night when I took a look at that planet for the first time,' said Herschel addressing his sister.

'I know, it was the week of my thirty-first birthday and I was away,' recalled Caroline wistfully.

'First I thought it was a comet,' recalled Herschel. 'I am so proud that you yourself have discovered no less than eight comets, Caroline. But what a strange comet this one was! An imposter with no tail trailing behind,' Herschel laughed.

'Oh, yes, and we traced its motion night after night and tried to fit its path into a normal orbit a comet usually takes, didn't we, Wilhelm?' said Caroline and added with a smile, 'I am sorry, I forgot that I must address you by your British name William. Anyway, the crafty celestial object was determined to move along a circle.'

'At last, we knew what it was a planet no less!' Herschel was elated at the recollection of the past revelation.

'A planet begging for a name,' laughed Caroline. 'Many astronomers wanted to name it after you, remember? You wanted to name it after King George, your benefactor. Finally, following the tradition of selecting names from mythology, the planet was christened Uranus, after the ancient Greek deity of the sky, father of Cronus or Saturn and grandfather of Zeus or Jupiter.'

'The right choice if you ask me,' commented Herschel modestly.

'But, William, everyone recognizes the fact that you are the first person in human history to have discovered a new planet. Not only that. The distance to Uranus from the Sun happens to be twice that of Saturn. Every one hailed that single-handed you had doubled the size of the solar system. What an extraordinary feat!'

Herschel accepted Caroline's praise hesitantly, embarrassed as he was by her admiration.

'You discovered two Moons of Uranus and two years later another two that circled Saturn. But, the most intriguing fact has been the orbit of Uranus. It is not exactly like what one would have expected it to be, assuming that it is going round the Sun under the latter's gravitational pull. No one has been able to explain its odd behaviour. So strange!'

'Well, it is an unsolved mystery indeed, and no one has a clue as to why the planet behaves thus,' Herschel shook his head. 'But, I am sure sooner or later the problem will be solved.'

The scene faded as did the two astronomers, brother and sister, along with their magnificent instrument that had revealed so much of the night sky. Of course, I knew what that wonderful pair did not, namely why their planet behaved so oddly: it was due to the gravitational pull of another planet, Neptune, to be discovered much later. Another coincidence, I had earlier learnt about, struck me. It takes Uranus a little more than 83 years to complete one revolution round the Sun. And strangely enough, William Herschel, its discoverer, lived exactly that long!

Stars on Stage

While I was still musing about the scene that I had witnessed, I was abruptly woken up from my reverie by a deep voice.

'The stars seem to go on and on forever. Is there no limit to their realm?' the resonant voice reverberated. Now fully alert, I looked up at the extraordinary spectacle in front of me. On an elevated platform, illuminated by gaslights, stood a handsome man. He was elegantly dressed in a long velvet coat and a white frilled shirt. He wore a wig of curly locks of hair and held in his hand a silver-tipped cane of dark mahogany. Surprisingly, he bore a striking resemblance to William Herschel.

Then realization dawned on me. Why, I had been instantaneously transported to a theatre in London. And the man on the stage was an actor made up to look like the celebrated astronomer. I was in a large hall filled with people, who were relaxing comfortably on cushioned seats. Thick curtains hung from ornately decorated walls. The hall was dark except for the faint light issuing forth from the distant gaslights propped up on the stage. The single actor on the stage was acting the role of William Herschel describing the astronomical findings of the latter. People loved these shows that were at once both instructive and entertaining.

The actor made his presentation with great fervour, varying his voice, accent, and intonation according to the subject matter he was reciting and the persons he was depicting. He gestured with his hands skilfully to suit his recitation as well as to indicate the astronomical phenomena he was describing.

'Let us begin with the renowned philosopher Galileo, an Italian who was an extraordinary observer of the heavens. Besides, he was a man given to dramatics,' said the actor and made a slight bow as he added, 'Not unlike your humble servant.'

'Galileo's spyglass, though small in size, revealed the immense Universe as never before.' With his silver-tipped cane, the actor pointed to a model of Galileo's telescope on which a hidden gaslight was turned on. 'Recall, if you will, ladies and gentlemen, how he described his vision of the hazy band of light adorning the night sky: *The galaxy is, in fact, nothing but congeries of innumerable stars grouped together in clusters. Upon whatever part of it the telescope is directed, a vast crowd of stars is immediately presented to view. Many of them are rather large and quite bright, while the number of smaller ones is quite beyond calculations.* Ah, what a glorious vision derived from observation! But observation leads to imagination, intuition, and speculation. Indeed, bold and far-reaching were the speculations of the renowned German metaphysician, namely Immanuel Kant.'

The actor swept his hand gracefully towards the background as a portrait of Kant on a large canvas suspended by wires descended from above, stayed on for while, and disappeared as it was hauled up.

'Pray allow me to convey to you, although I am aware of your erudition, ladies and gentlemen, and heaven forbid that I insult it, the gist of what he wrote in his early work *Cosmogony*. I quote: *The Universe by its immeasurable greatness and the infinite variety and the beauty that shine from it on all sides, fills us with silent wonder.* Forgive me, if I fail to be as silent as the great Kant wishes me to be, as I cannot contain my awe at the Universe he depicts. Well, what more did Kant offer that emerged from his unbounded imagination? He went beyond the Milky Way, yet not straying too far from it in his generalisation. He imagined systems of innumerable stars each system gathered together in the shape of a thin disk, similar to the Milky Way, but so far removed from our earthly moorings that individual stars within those systems are indistinguishable even with a telescope. Rather, each system would appear as a dim little patch, circular if its plane is perpendicular to the line of sight, elliptical if viewed obliquely. Such is the true aspect of the faint *nebulae* one perceives in the sky, each made up of countless stars so distant in space that they seem to cluster together indistinguishably, so limited in light that they are the most feeble candles in the heavens, yet sister galaxies to our own. Thus was born the idea that each one of them was an *Island Universe*.'

The actor paused as an enormous frame descended behind him carrying a beautiful picture of the star-studded night sky showing the Milky Way and a few faint patches depicting the nebulae.

'Then came William Herschel,' announced the actor with a flourish and held himself in a dramatic pose before continuing with his oration.

'Yes, ladies and gentlemen, new vistas were destined to be opened up by none other than William Herschel who is hailed as the Prince of Astronomy and

considered by many as one of the greatest celestial observers of all time. But he
made his entrance on the cosmic stage only after following his destiny along a
tortuous path in this world, a story which I have the honour to relate.'

The actor proceeded to recount Herschel's varied career and his discovery of
Uranus. A large-scale model of his telescope was pushed onto the stage for effect.
An actress dressed as William Herschel's sister Caroline and a young boy
portraying his son John joined the main actor. They would go through various
movements supporting the actor's recitation.

'Not being content with all his achievements thus far, Herschel decided to
address himself to some of the most difficult questions of astronomy. Night after
night, his telescope swept the heaven's vault as the Earth turned, while Herschel
worked relentlessly driven by his unquenchable thirst for knowledge. But think of
the treasures he *unearthed* in the night sky, if you would be kind enough to permit
such a paradoxical expression, ladies and gentlemen. He discovered more than
some eight hundred double stars paired in conjugal bliss. He added no less than two
thousand and four hundred nebulae to the mere hundred or so known previously. He
was now confronted by the profound questions that dogged many: What shape is the
Milky Way and where are we in relation to the Universe as a whole? In answer to
these queries, Herschel unveiled his great celestial star map.'

The actor paused as Herschel's depiction of the Milky Way descended from
above.

'In consonance with the ideas of Immanuel Kant, Herschel conjectured that the
Milky Way in fact possesses the form of a disk—or a grindstone, if you please,
which has a central bulge very much like a lens. And, pray, where are we stationed
in the immensity of this stellar realm? Herschel proclaimed that the Sun and his
family of planets are located somewhere in the middle of the celestial grindstone.
Had not Copernicus banished man from the centre of the solar system? Now,
Herschel restored man's lost honour and dignity by placing him again at the centre
of our galaxy.'

The actor lowered his head as if in deep contemplation and fell silent for the
moment to allow his listeners to absorb what he had recounted so far and think
about the implications of the facts he had conveyed to them.

'Now, observe how the celebrated astronomer turns his attention to the empire of
the nebulae, the almost imperceptible patches of illumination scattered across the
sky. He is no longer stargazing, but star gauging. He notes down the time at which
each nebula passes his telescope and its elevation thereby pinning down the position
of the dim denizen of the night sky. His gifted sister Caroline takes down copious
notes as her brother dictates the data, while his son looks on with wonder, inspired
to aspire, determined to follow in the footsteps of his illustrious father. Herschel
ponders and lets his imagination take wing. Perhaps each nebula, like the Milky
Way was an immense aggregate of countless stars. Let us believe for a moment in
Galileo's idea that faintness measures distance. Then these nebulae must be far, far
away lying outside our own galaxy.'

The actor lifted his eyes heavenwards as if he were indeed gazing at one of those
nebulae.

'Let us dream with Herschel then and echo his words: *Those nebulae may well outvie our Milky Way in grandeur.*'

The actor slowly shifted his gaze towards his audience who watched him transfixed with wonder and admiration.

'Let us return to Galileo and hark back to his words of conviction: *With the aid of the spyglass, the Milky Way has been scrutinized so directly and with such ocular certainty that all the disputes that have vexed philosophers through so many ages have been resolved, and we are at last freed from wordy debates about it.*'

The actor gave a theatrical laugh and continued.

'Did the observations of that unparalleled stargazer Galileo and those of the Prince of Astronomy put to rest the disputes about the Milky Way, let alone the nebulae? Did they free the philosophers from wordy disputes? Time, only time, the impartial arbiter of debate and dissent, will tell. For the moment, I offer you my profound gratitude for your attention, patience, and encouragement and bid you farewell.'

The actor bowed deeply to the standing ovation of the audience. He beckoned to his two assistants who joined him and, holding hands, all the three on the stage bowed repeatedly as the curtain descended slowly.

Measurer of the Milky Way

I was quite impressed by the theatrical performance, especially the actor who had presented the facts about the Milky Way and the nebulae in such a picturesque manner. While still savouring the novelty of it all, I could instinctively feel that I was travelling both in space and time as had happened to me time and again. Once, this used to be a rather disconcerting and often disorientating experience. Not any more, in fact I had started enjoying this strange passage and looked forward to it.

I was surprised to see a man on all fours looking intently behind some bushes. I had expected to meet someone gazing up at the sky, not peering down at the ground.

The man looked up. He had a round face with chubby cheeks, above which his eyes glittered with good humour. He gestured towards the spot he had been observing with keen concentration.

'Ants, you know, ants,' he explained as he straightened up. 'I have been studying how they behave, especially how they move around at different temperatures. See, I have equipped myself with excellent instruments. Here is a thermometer to measure the temperature and this is a barometer that finds out the pressure. I have discovered that the ants' movements have nothing to do with pressure. But they do run faster in the heat. Anyway, I have studied their motion day and night very carefully. I even wrote an article entitled *On the Thermokinetics of Dolchoderine Ants.* How do you like that? But my astronomer colleagues thought it was all a joke, although those who study insects took it seriously. Oh, by the way, I do study the stars too you know— that is my main business. It's terribly stressful sometimes, if you have to watch the stars all night and the ants all day. Can't get enough sleep. Look over there. That is the observatory where I work. We are on the top of a mountain called Mount Wilson, you know. And the observatory is obviously called Mount Wilson Observatory, what else?'

In the background, I could see the observatory, a big white building with a shiny metallic dome.

'Yes, ants are fascinating creatures, as interesting as stars,' said this strange man. 'Dear me, I haven't introduced myself, I am so sorry,' he added apologetically. 'My name is Shapley, Harlow Shapley. Let me give you my background so you may know me better.'

Shapley went on to describe his childhood, his education, and his career as an astronomer, all of which I found incredibly fascinating. As the story unfolded, Shapley receded into the background, only his voice audible. Scenes and events from his life appeared in succession illustrating his biographical narration.

'I was born on a farm near the little town of Nashville in the state of Missouri, United States of America. My father Willis Shapley was a farmer and a county schoolteacher too, you know. I loved the animals we had on our farm—hens with little chicks following them, a couple of horses, and cows. My twin brother Horace and I went to this one-room school on the border of our farm. Well, I studied up to just fifth grade. That is not much, is it? Anyway, I studied at a business school for a while and worked as a crime reporter for a miserable little newspaper called *Daily Sun*. Humans commit all sorts of crime like stealing, killing, so on and so forth, don't they? Animals on the other hand, like the ones we had on our farm, don't do such bad things. Reporting crimes is no fun at all. But then I discovered this treasure house of a library where I could read all sorts of things—history, literature, poetry and what not. I loved poetry, didn't I? Reciting verses helped me milk the cows rhythmically!' chuckled Shapley and continued. 'I wrote a nice essay, in fact, on Elizabethan poetry. Because of that, I was admitted to the University of Missouri where I could learn a lot of things. From then on, I never stopped.'

Shapley paused, before continuing, to think fondly of the happy days of learning. 'Did I want to study the stars at the university? No, sir, I never paid attention to them while I was on the farm, nice to sleep after the day's hard work rather than stay awake looking at the night sky, wouldn't you agree? I wanted to become a newspaperman. But, horror of horrors, they were not teaching journalism that year at all—the subject I needed for working for a newspaper. So I had to choose something else. Let me look at the list of things they are teaching this year, I said to myself, and fix something. First to appear on the list was *archaeology*. Archaeology indeed, I couldn't even pronounce that word! So I decided to take up the next subject—*astronomy*! That was it. So, you see, I became an astronomer by pure accident.'

'You may call me an accidental astronomer if you like,' said Shapley with a knowing smile.

Ah, here is another one of them, I thought.

Shapley laughed heartily at his recollection, his eyes buried in his round cheeks, before continuing.

'Well, I know you came to learn more about the Milky Way and not about me. I will be glad to tell you all about it. As a matter of fact, I spent a lot of time making its map,' said Shapley. And with a broad smile he added, 'I succeeded too. The important thing is to measure distances to the stars of the Milky Way in all directions. But how do you measure distances to the stars? You have learnt about Herschel's work, haven't you? He did his star gauging using the fact that light becomes dimmer with increasing distance. I know, I know. You are quite familiar with how exactly that happens, I mean the inverse square law and all that. Furthermore, as you have already learnt, we need *standard candles* whose intrinsic brightness or *absolute luminosity* is known. Then, by measuring their *apparent luminosity*, we can determine distances to the stars. Let me demonstrate it, quite entertaining you know.'

Shapley pulled out of his jacket pocket a whole bunch of brand new candles all of which looked exactly alike.

'Yes, these candles are identical to one another. They give out light that is of absolutely fixed brightness, which has been measured accurately. In other words, they are our standard candles.' Shapley lighted the candles one by one with a matchbox that he had produced out of his pocket too. Shapley snapped his fingers and we were surrounded by thousands of lighted candles. Distant ones shone fainter than the closer ones.

'Isn't that nice? If we measure distances to the candles in different directions, we can know how they are distributed in space. In other words, we can find out the size and the shape of the distribution of the candles. Not only that. We can also find out where we are among the lighted candles. If the farthest candles are more or less at the same distance in all directions, then we must be at the centre. On the other hand, if there are more of them on one side than the other, then we know that we are nearer to the latter edge. This is precisely what I did for the Milky Way.'

He smiled to himself as he recalled his thrilling—although painstaking—explorations of the past.

'But how do you do this in the case of the stars?' Shapley paused to let the question sink in. 'Well, you need bright enough sources of light spread all over the Milky Way for a starter. For that you make use of *globular clusters*.'

All of a sudden a huge collection of stars appeared before my eyes. These stars were distributed in a spherical shape, packed densely around the centre and thinning out towards the outer periphery. They were going in orbits around their common centre at extremely slow rates. They looked like beautiful iridescent bees flying around in orderly patterns to make up the globular cluster. This was incredibly more beautiful than looking at a globular cluster in a photograph or perhaps even through a telescope.

'Ah, there is a globular cluster for you,' said Shapley. 'As you know, these clusters contain anywhere from thousands to hundreds of thousands of stars—even a million—so that they glow quite bright. They are distributed all over our Milky Way.'

The candles were now replaced by globular clusters—nearly a hundred of them—and they shone with varying brightness depending upon their distances. In one direction, there were many of them, more densely packed than in the others.

'All right, we have our globular clusters in place, the distant ones glowing fainter than the closer ones. But we need a standard candle within each, whose actual brightness is known, so we can measure their distances. Astronomers are quite clever you know, even if I say so myself. They found a standard candle that shines not steadily but changes its brightness regularly—a *variable star* as it is known. Again, I am quite aware that you know all about *Cepheid variables*. Look over there, in the nearest cluster, you can see one.'

Oh yes, I could see the star clearly, its brightness varying rhythmically. It was amazing.

'All right, let us recapitulate the most important aspect of the Cepheid variables,' continued Shapley. 'The period of the variable star is directly and exactly related to its brightness. The longer the period, the brighter is the variable star. So, if we just measure the period of a variable star within any globular cluster, we would know the true brightness of that star. If we determine the star's brightness as observed here on the Earth, or its apparent brightness, we can estimate its distance—or equivalently that of the globular cluster from us. Simple enough, right?'

Of course, it was quite simple, I thought.

'Actually, in practice it is quite a bit of hard work,' admitted Shapley. 'In any event, I was able to determine the distances to the globular clusters around us and thereby measured and mapped the Milky Way. And what did I find?'

Shapley gazed at me savouring the suspense his pause had created.

'Let me tell you very briefly,' continued Shapley. 'Our galaxy, the Milky Way I mean, is like a thin disk with a central bulge where lots of stars can be found. And how big is the Milky Way? What I found out was a big surprise to my friends. It is some 100,000 light years across! Pretty large, isn't it? And, let us not forget it contains billions of stars too.'

Yes, the Milky Way is large enough. In all probability, in Shapley's days it must have seemed unimaginably big.

'And where are we? I mean, where is our solar system located within this huge system of stars?' Shapely went on. 'Are we at the centre of our galaxy? No, sir! We are at a distance of around two thirds the radius of the Milky Way from the centre! That means about 30,000 light years away from the centre of our galaxy. Copernicus, as you very well know, had shown that man is not at the centre of the solar system. And now we find that he is not at the centre of the Milky Way either. What does that mean?'

Shapley paused dramatically before answering his own question. 'It means, my friend, that man is not a big chicken after all!' He burst out laughing and continued to laugh for quite a while. Perhaps he was thinking of all the chickens he used to feed on his farm long ago.

Having recovered from his bout of mirth, Shapely said, 'One last thing. The galaxy is so immense that I was convinced that it was everything there is. It was the entire Universe. Those faint luminous patches one saw in the night sky were nothing more than gas clouds belonging to our own Milky Way. Were they Island Universes beyond our own galaxy? I could never, ever accept such a preposterous idea. But there were those who didn't agree with me. We had lot of disputes and debates over that matter. Anyway, at least we had come to know how big our galaxy was.'

Shapley seemed to be quite content with his remarkable discovery. 'I had come a long way from milking the cows to measuring the Milky Way, my friend. Ah, that was indeed the milky way to the Milky Way!' Shapley again laughed long and loud.

'Oh well, let us forget the stars for the moment. They shine on any way, don't they? Maybe it is time for me to get back to my ants.'

Shapley went down on all fours once again and seemed to forget my presence altogether as he peered at the ants running around on the ground.

I opened my eyes reluctantly. Awareness returned hesitantly. Herschel's melodious music was still playing on. How long had I been out of this world, so to speak? Dreams may seem to last for a long time even though they occur for a moment or two. Does it matter? Time is after all relative, isn't it? I stayed on happily in the comfortable warmth of the bathwater until the music came to an end.

Yes, it is wonderful to dream. Undoubtedly it is also wonderful to sleep without dreaming. And that is what happened to me when I went to bed after having enjoyed my bath.

Chapter 12
Through the Glass Brightly

George drove his old car along the tree-lined winding road up the hill. Shafts of mild sunlight of the approaching evening shone intermittently through the foliage overhead. Happy memories of our previous visit to the University Observatory flooded my mind. Soon, the memories were replaced by a sense of anticipation as I saw the tall, white, cylindrical building capped with the gleaming metallic dome looming large in front of us. I was thrilled at the sight of that remarkable place. I am always filled with amazement at the very thought of telescopes. Centuries ago, Galileo had fixed a couple of lenses, mere pieces of glass, to the two ends of a tube and turned it into the most powerful aid in his quest to discover the secrets of the night sky. Then, Newton had replaced one of those lenses by a mirror heralding a new age in astronomy. Then again, Newton's discovery that a triangular piece of glass could split white light into its component colours had revolutionized the study of the stars and galaxies. Since their days, so much progress has taken place in celestial observation, and in theoretical interpretation, that there seems to be no limit to the possible exploration of the Universe.

George and I were surprised to see Mike walking around in circles outside the observatory.

'Hey, Mike,' hailed George. 'What are you up to?'

'*Solvitur ambulando*!' Mike broke into his boyish grin.

'You crazy man, what language is it? And what does that mean?' George asked pretending to be annoyed. I know he just likes everything Mike does.

'It is a Greek proverb, George, meaning *the problem is solved by walking*,' replied Mike.

'For heaven's sake, why didn't you say so in plain English?'

'Because saying in Greek makes every statement seem profound,' laughed Mike. 'Then again, it makes for brevity since the two words of Greek have to be translated into six words of English. Furthermore, we must honour the ancient Greeks who originated the proverb. Also...'

'I have heard enough, Mike. But what is the problem you are trying to solve by ridiculously walking around in circles?'

© Springer International Publishing Switzerland 2015
C.V. Vishveshwara, *Universe Unveiled*, Astronomers' Universe,
DOI 10.1007/978-3-319-08213-4_12

'Oh, that? Nothing to it, some little irritating thing about a disintegrating inte-grated circuit, that is all,' Mike brushed away his problem which, I am sure, was more serious than that. 'Welcome back, George and Alfie, you two beautiful people.'

We entered the observatory, as Mike held open the door.

'Would you like to pay your respects to the Queen of the Firmament first?' asked Mike.

I knew that was the name given to the large, main telescope of the observatory. My memory went back to a huge old cannon I had seen in India. It had been named *Malik-e-Maidan*, the Master of the Battlefield. A telescope resembles a cannon, doesn't it? But of course, the night sky is not a battle field, far from it.

'It will be an honour,' responded George. I made a slight bow to indicate my own desire to pay the Queen my respects as Mike had put it.

There she was in all her majesty, the great telescope, turned heavenwards presumably surveying the sky through the open hatch of the observatory dome.

Mike held up his forefinger against his lips and whispered, 'We cannot disturb her. She sleeps during the day and keeps awake all through the night. We shall come back to her when it gets dark. Let us go up to our observers' lounge where we can relax and talk.'

We followed Mike to the lounge, a cosy compact place. Everything in the obser-vatory had to be compact and efficiently organized because of the limited available space. Arranged around a low circular table were a few comfortable chairs with a small blackboard standing close by. This is where the astronomers met to discuss their results as well as their strategies for further observations. I could see that some writing pads, astronomical photographs, and other material had been placed tidily on the table.

'Well, Alfie, I understand that you want to fly away from the Milky Way,' Mike opened our discussion. 'George tells me that you have learnt all about our home galaxy—Herschel's preliminary studies of the Milky Way and the nebulae, and Shapley's measurements using Miss Leavitt's Cepheid variables. Shapley discov-ered the shape and the size of our Milky Way as also our own place in it, didn't he? No wonder he thought that was it, the Milky Way was the entire Universe. Soon enough, the picture of the Universe was going to change.'

'Yes, lo and behold, Edwin Hubble raises his head, I mean his telescope,' George interposed.

'Oh man, you took away my words,' Mike gave George a big smile. 'Yes, we are going to talk about one of the greatest observational astronomers of all time, Edwin Powell Hubble. Let me give a thumbnail sketch of his life quickly and then outline his work. I know, Alfie, you will learn all this in more detail when you go home and read George's marvellous manuscript. Won't you?'

'George gave you a copy too?' I asked. I was not really surprised. After all, Mike was a professional astronomer and was one of the closest colleagues of George.

'Oh dear, the cat is out of the bag, two bags actually,' George wrung his hands in mock dismay.

'I know that you are supposed to find a title for the book, Alfie. Let me know when you do that,' Mike gave me a wide smile. He seemed to be always happy.

Mike took out a couple of pages on which he had jotted down some notes. I assumed correctly that it was about Hubble. Mike is quite systematic about everything he does. I am sure it is essential in his work, which requires keeping accurate record of your observations.

'All right, back to Hubble,' continued Mike. 'Like Shapley, he too had quite an unusual career, but in a totally different way. He was born in 1889, four years later than Shapley. His father, John Powell Hubble, worked for an insurance company. As a young man, Hubble was quite well-known as an athlete rather than a scholar. Of course, he did earn good grades in all the subjects he studied except in spelling.'

'Ah, I am better than Hubble at something at least,' I commented with satisfaction.

'We are sure you are better at some other things as well, Alfie,' assured Mike and continued. 'Actually, Hubble's athletic record is quite impressive, you know. He excelled at baseball, football, basketball, and he ran track in both high school and college. In fact Hubble even led the basketball team of the University of Chicago to a major title in 1907. That is not all, he happened to be an amateur boxer too.'

'Thank heavens, he did not choose to become a professional athlete,' remarked George. 'Otherwise, what would have happened to the Universe?'

'Nothing would have happened to the Universe, George, as you very well know. It would have merrily continued its life,' laughed Mike. 'In any event, Hubble did choose a path that would lead him eventually to becoming an astronomer. He studied mathematics and astronomy at the University of Chicago, obtaining a bachelor's degree in science. That was in 1910.'

'Then on he must have pursued astronomy as his goal steadfastly,' I guessed.

'Hold it, Alfie, that is not what happened,' Mike shook his head. 'Hubble won the prestigious Rhodes Scholarship that enabled him to go to Oxford. That scholarship does not allow you to study the stars, you have to study law. So, Hubble went on to earn a master's degree in law. He included literature and Spanish too in his curriculum. All this was done as a devoted son fulfilling the promise made to his father. Hubble must have known that his father was dying, because his father passed away in 1913, while the young Edwin was still in England. He returned home soon to take care of his family—mother, two sisters, and his younger brother.'

'That is truly sad, Mike,' George said.

'Indeed it is. Hubble must have been a most dutiful son with strong family bonds,' I added. 'What happened next?'

'I shall be brief with the story,' Mike said. 'Hubble was not keen on practicing law on his return to the United States. He taught physics, mathematics, as well as Spanish at a high school for about a year. I shouldn't forget that he also coached the basketball team! Finally it happened. At the age of twenty-five, Hubble decided to become a professional astronomer.'

'Hooray for astronomy!' cheered George shaking his fist.

'Cheers,' Mike too smiled holding up an imaginary wineglass in celebration. 'Through the good offices of one of his former professors, Hubble joined the graduate school at the University of Chicago. He worked at the Yerkes Observatory and received his doctorate after submitting his thesis.'

'Ah, the Yerkes Observatory, how well I remember my visit to that venerable and beautiful institution,' George reminisced happily. 'I saw a group photograph featuring Hubble along with the faculty, students, and the staff. There was this

Edwin Frost too in the photograph, who served as the director of Yerkes, in spite of his blindness. Also, Chandrasekhar worked at the observatory for a number of years. Great place, Yerkes!'

'So, with his doctoral degree in hand, there must have been nothing to stop Hubble from reaching his cherished goal of studying the stars fulltime,' I concluded.

'Not so fast, Alfie,' Mike held up his hand. 'In the year 1917, the United States found itself in war with Germany, didn't it? Hubble submitted his PhD dissertation in a hurry and joined the army. He was sent overseas holding the rank of Major. When the war ended, he went back to the study of astronomy at Cambridge before returning home for good.'

'What a tortuous journey!' I exclaimed. It was a most unusual life indeed.

'The year 1919 dawned,' Mike announced dramatically. 'The year in which the magnificent career of Edwin Hubble as an astronomer finally began. George Ellery Hale, the founder-director of the Mount Wilson Observatory in California offered him a position on the staff of the observatory. Hubble carried out his great work there for the rest of his life. As he did during the First World War, Hubble again served the army at the Aberdeen Proving Ground during the turbulent years of the Second World War. In recognition of his contribution to research in ballistics, he was honoured with the award of the Legion of Merit. In 1949, the giant two hundred-inch Hale Telescope of Mount Palomar became operative. Edwin Hubble was the first astronomer to use it, taking the very first photographic exposure. He continued to work at both Mount Wilson and Mount Palomar. But his days were drawing to a close. In July of the same year, he suffered a heart attack while vacationing in Colorado. His wife, Grace Hubble, tended to him with great care for the next four years. In 1953, Hubble passed away due to a blood clot in his brain. No funeral was held for him, and his wife Grace never revealed his burial site.'

We sat in silence, immersed in our own thoughts about Hubble. I looked out of the window. It was dark now, myriad glittering stars sprinkled all over the night sky. Edwin Hubble did not need a monument. Instead of a lone sentinel standing at his burial spot, there were billions of stars scattered all over the cosmos shining in his memory.

At last, Mike broke the silence.

'On the night of October 4, 1923, the measure of our Universe changed totally, irreversibly,' announced Mike. 'On that night Edwin Hubble turned the giant Mount Wilson telescope towards the nebula in the constellation Andromeda.'

Mike paused savouring the suspense he was creating with his silence. Like George, he has a flair for the dramatic. And every good actor knows the power of a pause separating spoken words.

'Ah, Princess Andromeda, the damsel in distress chained to a rock by her own parents, Queen Cassiopeia and King Cepheus of Ethiopia, as sacrifice to appease the sea-monster Cetus that was devastating their coastal regions, which Prince Perseus slew in time and rescued the Princess whom he married forthwith thereby bestowing a happy ending on the story. Whew!' Mike drew his breath heavily after the long sentence he had spoken at one stretch. 'All the participants in the story have been immortalized as constellations.'

Mike opened a book lying on the table and showed me the constellations he had mentioned. 'You may borrow this book and a couple of others about Hubble, his work, and his collaborators, Alfie. It will make for wonderful reading, complementing George's manuscript.'

'Now, back to Hubble and the nebula Andromeda,' Mike returned to the great discovery. 'Hubble exposed a photographic plate for half an hour or so, the telescope trained towards the Andromeda nebula getting a smudgy image harbouring a star within it. Comparison with photographs taken earlier revealed that the star was a Cepheid variable. Fortune favours a prepared mind, as the saying goes. As Shapley had done before him, Hubble estimated the distance to the star, hence to the nebula, using the knowledge of the star's absolute luminosity and the inverse-square law. That cosmic measurement owed its success to Miss Henrietta Leavitt. What distance did he measure between us and the nebula?'

'Tell us, tell us,' urged George feigning ignorance.

'I shall tell you, fear not,' Mike said solemnly. 'Hubble's estimate was some 900,000 light years. What did that mean? Shapley had shown that our own Milky Way stretched over a distance of 100,000 light years, hadn't he? So then, Andromeda lay far beyond the Milky Way. That meant that our Milky Way did not constitute the entire Universe. Hubble had once for all demonstrated that the Universe was vaster than ever imagined before.'

I could feel how thrilling such a revelation must have been. It was indeed a unique turning point in the history of cosmology.

'Well, Hubble continued his observations of the nebula in Andromeda as well as the one in the nearby constellation Triangulum, detecting more Cepheid variables in them. There was no turning back from his conclusion regarding the measure of the Universe.'

'Mike, why do you keep on using the term *nebula*? We know they are *galaxies* as the common terminology goes,' pointed out George.

'Aha, that is because in his entire lifetime Hubble refused to use the term galaxy, and referred to those luminous patches in the sky as nebulae and nothing but nebulae. I am following in his footsteps with the hope that someday I shall make some discovery of my own similar to his. By the way, the modern value for the distance to the Andromeda nebula is about two and a half million light years.'

'What did Hubble do with his revelation? Did he send it immediately to some scientific journal for publication?' That would have been the normal course of action for any scientific discovery I thought.

'No sir, he sent it off to New York Times on November 23, 1924,' Mike answered.

'Ah, that is the current trend too, to get your name in the press,' added George.

'To be fair, Hubble cabled his results in the form of a paper to his astronomer colleague Henry Norris Russell on December 31, 1924. Next day, the first day of the New Year, Russell read the paper to his fellow astronomers gathered together during the joint meeting of the American Astronomical Society and the American Association for the Advancement of Science in Washington, D. C. On that day, the astronomers came to know that the extent of the Universe had radically changed thanks to the explorations of one man. A new era in cosmology had arrived.'

Mike sat back in his chair for a while. With his account of Hubble's monumental discovery, the atmosphere in the Observatory seemed to have changed, charged with a new spirit. Perhaps it was just my imagination.

From the material stacked on the table in front of us, Mike took out an old photographic plate and opened one of the books.

'Here is a photographic plate that was used in Hubble's time taken out of our archival collection,' Mike held up the plate and then pointed at the picture appearing in the book. 'And look at the picture in this book. It is a reproduction of the photograph of Andromeda Hubble took.'

The photographic plate looked so primitive, I wondered how those astronomers of the past could extract so much groundbreaking information from it. Then there was Hubble's photograph of nebula Andromeda, no more than a faint oblong smudge that looked more like an enlarged version of a careless thumbprint. At the top right-hand corner of the smudge, two little lines had been drawn enclosing a dot presumably to indicating the variable star. Beneath it was written, in Hubble's own hand, the date of his marking the photograph, *6—Oct 1923*, a small arrow pointing towards the variable star. I looked at it with awe, the date and the dot that had transformed the face of the Universe. It reminded me of the meagre crosses drawn by Galileo showing the positions of the satellites of Jupiter that had centuries ago changed the face of the solar system.

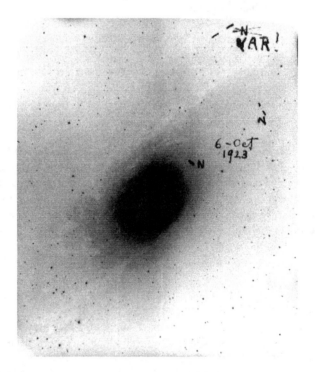

Silently, Mike placed next to the open book a glossy print showing the galaxy in Andromeda that had been taken through a modern telescope. It was spectacular. An ethereal glow emanated from its bright central region like the radiance of the Princess from whom it had inherited its name. Wisps of spiral arms curled around it as if protecting that comely maiden. Here was our neighbour that had paved the way to the realization of the awesome expanse of our Universe!

Having waited for me to take a good look at the Andromeda galaxy, Mike spread out more prints of several other galaxies. With their brilliant colours and detailed structures, they looked gorgeous.

'Look at them, aren't they a feast for the eyes?' Mike exulted. 'They have names based on the constellations they are located in, like Hercules, Ursa Minor, and so on. Or they could be descriptive of their appearance—Whirlpool, Cartwheel, Cigar, Sombrero, Tadpole etc. One of them has the nickname—Black Eye or Evil Eye, how do you like that? Then again, some of them have been named after their discoverers like Hoag's Object, Mayall's Object, and so forth. Alfie, notice they have different shapes too—*spirals, barred spirals, elliptical,* and the *irregular ones* like the Small Magellanic Cloud, our nearest neighbour in which Miss Leavitt had found her Cepheid variables. In fact, those names represent the classification proposed by Hubble himself after observing countless galaxies and determining distances to many of them.'

Hubble seemed to have made a thorough job with his nebulae, it was truly amazing. George sat silent and smiling, enjoying my initiation to this immense wonderland. Mike too remained silent for a while. He must have been reliving his own excitement of the past when he had come upon, for the first time, all the work he was describing.

'So much for the expanse of our Universe,' resumed Mike. 'What comes after expanse? It is expansion, the most unusual aspect of the cosmos! Let us go back to the early twentieth century, a decade before Hubble looked at Andromeda. At a meeting of the American Astronomical Association, which Hubble happened to attend as a doctoral student, the astronomer Vesto Slipher presented his findings suggesting that several spiral nebulae were speeding away from us at enormous speeds of almost up to 1,000 kilometres per second. Compare this with the speed of the Earth going around the Sun, which is merely about 30 kilometres per second. This was incredible and in fact few believed Slipher's observations. And none could make any sense of it. We do not know what sort of influence Slipher's findings had on Hubble. In any event, it would take another fifteen years before Hubble started his systematic study of the flight of the galaxies.'

'Well, often what you have seen or heard sometime or the other can stick in your subconscious and influence you later on,' commented George. 'It happens to all of us and could have happened to Hubble too. Sorry for interrupting you, Mike, please go on.'

'How did Slipher measure the speed with which the nebulae were zooming away from our Milky Way?' Mike posed the question and answered it. 'It was by using the well-known Doppler Effect. I am sure you know what that is, Alfie.'

'Yes, I do,' I nodded. 'We experience it all the time when a fast-moving automobile honks its horn. The pitch goes up as it comes towards us and goes down as it speeds away. I have read the amusing account of how Doppler Effect was verified by having musicians blow their trumpets while travelling on a train.'

'Right you are,' said Mike. 'The ratio of the change in the received frequency, or the pitch, and the emitted frequency of the sound wave is proportional to the ratio of

Doppler

the speed of the source and the speed of sound. If the speed of the source is comparable to the speed of sound, then the change in the frequency would be detectable. And by measuring that change in frequency, one can determine the speed of the sound emitter. I hope I have made myself clear.'

'Perfectly clear, Mike,' George answered for both of us.

'Now, this is true for light or electromagnetic waves also,' continued Mike. 'In place of pitch we have colour now. The spectrum of a source is shifted towards red if it is moving away from us and towards blue if it is coming towards us. Naturally these are called *redshift* and *blueshift* respectively. Then again, as in the case of sound, the shifts would be appreciable if the speed of the source is comparable to that of light.'

'I have heard an anecdote involving the renowned physicist Albert Michelson of the Michelson-Morley Experiment fame,' I related. 'It seems he was hauled before a magistrate by a policeman for jumping a traffic signal that had turned red. Michelson pleaded innocent, claiming that the light had appeared green due to Doppler Effect as he was driving towards it. The magistrate, who knew enough physics, ruled that Michelson was indeed innocent of the charge of jumping the signal, but fined him anyway for speeding!'

'I must remember to tell this story to my class, even if it has been made up,' George clapped his hands laughing. And, of course, Mike was laughing too.

'Well coming back to cosmology, we know how Slipher had measured the velocities of a few spiral nebulae, by the redshifts of their spectra,' Mike continued. 'And a couple of other astronomers reported such recession velocities for other galaxies too. But the exploration of the enormous speeds of galaxies speeding away from us in relation to their distances was to be pursued systematically by Hubble.'

After a studied pause, Mike announced, 'At this point we introduce Milton L. Humason!'

Humason

George's face lit up in anticipation of what Mike was going to say.

'Using the Mount Wilson telescope, the finest one available at the time, Hubble was to collaborate with Humason on his redshift studies of the galaxies,' Mike went on. 'Before we go into that, let us ask the question: What was the path that Humason had taken before attaining the coveted position of being Hubble's collaborator? Well, it was the winding path that led up to the top of the mountain where the observatory was being built, as Humason led his pack of mules carrying construction material! Yes sir, Humason happened to be a high-school dropout, who made a living as a mule driver. He joined the Mount Wilson observatory as a janitor, learnt the trade on the sly with the help of some professional astronomers, and came to be known as a magician of measurements. He was picked by Hubble to determine the redshifts of galaxies. Their observations together led to one of the most astounding discoveries of all time, namely the expanding Universe.'

'Alfie, we learnt about Copernicus the Canon, Kepler who wanted to be a theologian, Herschel the musician, Shapley who had hoped to become a journalist, and now we have Humason the mule driver,' George pointed out. 'The ultimate accidental astronomer indeed!'

'Ah, the accidental astronomer, I like the term,' smiled Mike. 'You must read all about Humason in detail, Alfie. Although he had not been trained in astronomy and had no scent of mathematics, Humason could make the most sensitive determination of redshifts of a number of galaxies, while Hubble measured their distances. From these findings emerged the *Hubble Law*, which showed that the velocities of the galaxies were linearly proportional to their distances from our Milky Way.'

Mike opened one of the books lying on the table and showed me a picture that showed the redshifts of a few galaxies, their velocities, and their distances.

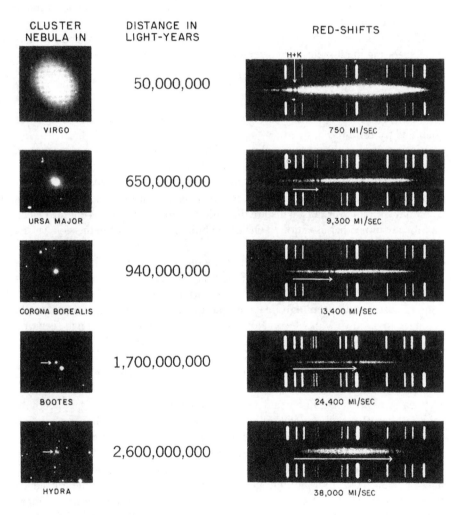

'Take a look, Alfie,' Mike pointed at the photograph. 'You can see the sharp spectral lines of calcium as observed in the laboratory. Compare them with the galactic spectrum of the same element. See how fuzzy the latter is. Still, you can discern the redshift that increases with the distance. It is this kind of photographs that led to the Hubble Law. And that Law heralded a whole new chapter in cosmology.'

I remembered what Winston Churchill had said in the totally different context of the Second World War: *This is not the end. It is not even the beginning of the end. But, it is, perhaps, the end of the beginning.* I was sure what I had heard was, perhaps, not even the end of the beginning, but just the beginning of the beginning. There was going to be a lot more to the cosmic drama.

'The cosmic journey has just begun. Alfie,' Mike echoed my thought. 'I guess you will read more about the Universe from books. And George would be telling you all about the theoretical side of cosmology and the latest developments in the field. How exciting!'

Mike looked out of the window and so did I. It was dark outside and the stars were sparkling in all their glory. I remembered the lines from Shakespeare:

The skies are painted with unnumber'd sparks
They are all afire and everyone doth shine.

'Ah, the time has come to wake up the Queen. Let us go.' Mike rose from his chair and we followed him.

We were back in the telescope chamber. I was again filled with awe, even though this was the second time I was looking at the telescope within a short period of time.

'Yes Alfie, I know that feeling,' said Mike quietly, reading my mind perhaps through the expression on my face. 'I see the magnificent instrument many times every day. But it never fails to inspire me with mysterious wonder. Well, the Queen has already turned her attention to one of the billions of galaxies in the sky. Take a look.'

I peered through the eyepiece of the telescope. After a little bit of adjustment I saw it—a spiral galaxy in all its vivid beauty. The central region glowed bright sending out the curling spiral arms, decked with countless brilliant specks and flaming dust clouds, seemingly reaching out to catch the stray stars that actually belonged to our own Milky Way. No doubt, the many photographs I had seen of a variety of galaxies were fantastic, but paled in comparison to the real one I was looking at now.

Reluctantly I tore myself away from the telescope. It was George's turn to take a look at the galaxy.

'I can't believe it, Mike,' said George with fascination. 'How far are those celestial marvels? Millions and billions of light years? And light has travelled all that distance, for all that time to reach us. What we are seeing right now is how those galaxies were so long ago. Incredible, incredible!'

Mike too glanced through the telescope before escorting us out of the building into the night. Free from the light pollution of the sprawling city quite a distance from the observatory, we could see the night sky filled with immeasurably more stars than we normally did. I knew Mike had his work waiting for him. We had to take leave of him.

'Well, Mike, that was fabulous. Thanks from both of us,' said George. 'Are you going back to walking around to solve the problem as dictated by your Greek proverb?'

'No, George, while we were talking, my brain went on auto-drive. And the problem got solved,' laughed Mike. 'Now, my wild nightlife begins.'

'Take care, Mike, and do visit us soon. Let us live it up,' said George holding up his thumb.

'That goes for me too, Mike,' I added. 'Come down from your mountain of meditation.'

'*Thus Spake Zarathustra*,' Mike smiled. 'Will do.'

He waved goodbye and strode back to the observatory.

We drove down the hill. After the inspiring, miniature cosmic journey, what was there to say?

I got off near George's physics building. I loved to walk home.

'Our next meeting at my office, Alfie,' said George. 'As Mike said, we have to discuss the theoretical basis of cosmology and a bit of modern stuff. All my best to you.'

Every time I look up at the sky, it appears to be new. Among the resplendent stars above, I could see in my imagination diffuse nebulae sprinkled all over. If I focused on them long enough, they resolved into their individual structures from elliptic to spiral. Then they slowly faded out. I had reached home.

Chapter 13
The Universe: Expanse and Expansion

I wish I could describe my innermost thoughts and feelings as I contemplated all that I had heard and all that I had seen at the University Observatory. Is there any limit to the incredible riches the night sky can yield? I remembered and repeated quietly what Edmund Burke, statesman, author, orator, political theorist and philosopher all rolled into one, had written: *Darkness is more productive of sublime ideas than light.* Nothing could be truer than that statement in the case of astronomy: darkness without which the stars had no visible existence and the stars without which some of the most sublime ideas could never have been inspired.

I am lucky to be a fast and voracious reader. With the greatest speed I could muster, but without making any compromise on comprehension, I went over the books in front of me—George's manuscript as well as those Mike had leant me. By the time I finished, *night's candles were about to be burnt out, and jocund day was preparing to stand tiptoe on the misty mountain tops*, to paraphrase Romeo.

The soothing warmth of my bathwater penetrated every part of my body. The magical bubble-bath mixture had created an enchanting surrounding. The water had shed its natural colour and appeared dark like the night sky. There were diffuse luminous patches interspersed among the brilliant little star-like spots that shone against the dark background. On closer look, I could see that they were not just patches, but collections of countless miniscule, glittering specks. In short, I was not just relaxing in my bathwater, but was immersed in my own miniature cosmos. It was me and my Universe! Surrounding me was absolute silence for a while.

'Forgive me for interrupting your reverie, boss,' whispered the bathtub as if waking me up gently. 'I know, silence is golden. But words are diamonds, what boss? Did you say that *darkness is more productive of sublime ideas than light*? I hope I quoted your quotation correctly. But then, the Master could come up with all his great ideas, most of them sublime too, whether it was dark or light. Hmm, maybe those thoughts came to him in the night after all—while reading, writing, eating, drinking, sleeping, dreaming—all in the night time, when it was dark. Who knows? Maybe he happened to announce them in daylight. Maybe, maybe!'

© Springer International Publishing Switzerland 2015
C.V. Vishveshwara, *Universe Unveiled*, Astronomers' Universe,
DOI 10.1007/978-3-319-08213-4_13

'Be that as it may—ah, I like that expression—be that as it may, what about you?' the bathtub continued after a momentary pause. 'I know you had a fantastic evening at the observatory, learning about great men, their great discoveries, looking at the great starry night directly, having a great time in short. You know something, boss? I can see through your eyes, hear through your ears, read your thoughts. I used to think I could do this only when you were near enough. But this evening, I realized I could do it even when you were far away in the observatory. My range has increased, just like the modern telescopes. What do you call it, boss—distance learning, telepathy? Matters not, as long as it works. My Universe is expanding too!'

Listening to my bathtub was as soothing as soaking in the bathwater. I did not have to think at all.

'Aha, I talk too much and you listen too much, is that so?' teased the bathtub. 'But what about the thought process? Am I thinking on my own? Am I reflecting your thoughts like a mirror? Or are you thinking through me? Whatever it is, it is nice. All right, boss, let me not stand—how can I, a bathtub, stand? I mean come—between you and your sublime pleasure, cocooned as you are in your own private cosmos. Otherwise, the headlines will read: *Bathtub betrays boss*! All articles left out for the sake of alliteration. Drift away, boss, drift away!'

My mind was becoming numb. Transition to trance transcended terrestrial techniques... All articles left out for the sake of allitera... My consciousness was dissolving away...

Beauty and the Beast

I was ecstatic to say the least. What a surprise to be in the presence of someone I had already met with immense pleasure. It was the great Greek playwright Aristophanes, standing in front of me with a big smile lighting up his face. And of course, the chorus of frogs were there too grinning widely, some of them winking slyly at me.

'Welcome back, our honoured guest and dear friend, to grace our celestial theatre,' greeted Aristophanes bowing formally. The chorus of frogs imitated him and giggled.

'Ah, last time, we enacted for your pleasure and edification, the story of Orion, the Great Hunter of the Skies, did we not? Now, we proceed without undue delay to present our theatrical performance of the curious chronicle of Princess Andromeda, a one-line gist of which you have already heard.'

As Aristophanes narrated the story of Andromeda, the drama was played out by the starry characters in the background as had happened with the Orion tale.

'Once upon a time, ages ago, there lived in Ethiopia King Cepheus and his Queen Cassiopeia, she of immense beauty, immeasurable by the common standards of man or beast. The King and Queen begot, as it normally happens, a daughter, Andromeda, whose beauty rivalled that of her mother. Beauty, alas, often breeds vanity and vanity acquires a loquacious tongue. Cassiopeia, in her foolish abandon, boasted that her and her daughter's beauty surpassed that of the Nereids, the sea-nymphs. Ah, what, may I ask, did the Nereids do? Insulted as they were, they complained bitterly to Poseidon, god of the seas. Poseidon, infuriated by the impropriety of a mere mortal, namely Queen Cassiopeia, decided to inflict dire punishment on her. Woe betide the proud, what terrible retribution awaited the vain woman? I shudder to think.'

Chorus
Brekekekex ko-ax ko-ax
Brekekekex ko-ax ko-ax
Cassiopeia, O, Cornucopia
Of lovely features that lure all creatures
Big and small, one and all
Dimples rubbed and pimples scrubbed
Washed in beer to make skin clear
Queenie, O, Queenie, why be a meanie?
Why blow your own trumpet?
Annoying the sea-nymphet?
Why slice her ego like a sickle?
And land yourself in a pickle?
Brekekekex ko ax ko-ax
Brekekekex ko-ax ko-ax

'Poseidon seething with rage summoned the terrible sea-monster Cetus and ordered him to devastate Cassiopeia's kingdom. Cetus, assuming the form of a monstrous whale, swam up and down the coast, leaving behind in his wake death and destruction. Alas, alack, what could King Cepheus do now but wring his hands in helpless despair? Well, in addition to wringing hands, he consulted an oracle who claimed, just claimed mind you, to be able to communicate with the gods themselves, extract a miniscule portion of their infinite wisdom, and convey it to the mortals. Offer your daughter as a sacrifice, the oracle advised the King, and the slaughter will cease. Torn upon the horns—ask me not the horns of what animal— of the dilemma between his love for his daughter and the welfare of the people, Cepheus magnanimously chose the latter and had Andromeda chained to the rocks in the sea, the innocent young lady mercilessly forsaken and waiting for the inevitable death in the form of Cetus to approach her.'

Chorus
Brekekekex ko-ax ko-ax
Brekekekex ko-ax ko-ax
Andy girl, O, Andy girl,
To your parents a priceless pearl
We are awfully pained to see you chained
Waves leaping, people weeping
We cannot bear it, nor repair it
Ye gods, swims in Cetus, vicious since a foetus
A fearsome whale bellowing like a gale

See him spread destruction and dread
O, maiden fair, dire despair!
Brekekekex ko-ax ko-ax
Brekekekex ko-ax ko-ax

'I tremble to proceed with my narration as I see in my mind's eye the monster closing in on the doomed damsel,' continued Aristophanes. 'When Cetus saw the comely creature chained to the rocks, he forgot the people and the cattle he was slaughtering, and swam towards his precious prize of a free meal.'

'Oh, what happens now?' Aristophanes pressed the back of his hand to his forehead in despair. 'Ye gods, would you not cast your eyes down on this calamity, on the impending wanton destruction of all the loveliness in the world, on the fearful fate of the miserable maiden? You would? I heave a sigh of relief as strong as the west wind. Lo and behold, the distant vision of Prince Perseus.' Aristophanes paused for effect.

'Take heart, my honoured guest, here arrives hope in the form of our Prince Valiant,' assured Aristophanes holding out his open palms towards the single honoured guest of his, namely me. 'What has been Perseus up to? Well, let me tell you his story so far briefly. Perseus was now returning from the expedition he had undertaken to eliminate Medusa. I am sorry, my distinguished guest, everyone seems to be bent on eliminating everyone else in our tale. But then, who was this Medusa? She was once a beauty with a face that shone like the full Moon, framed by the lustrous streaks of black clouds, namely her dark flowing hair. Nice description, eh? Again, alas, in her overweening pride, she boasted that her beauty was comparable, nay even greater, than that of Athena the Virgin Goddess of Arts, Crafts, and, oh no, of War. The last bit is incongruous with the former ones, would you not agree? Well, be that as it may, Athena, in her untrammelled indignation, transformed Medusa into the most terrible monster. Her face changed from the Moon into a bunch of its craters. Her hair deteriorated into a tangle of hissing serpents. Well, maybe she had forgotten to shampoo her hair, who knows? The end result was that so abominable was her face that anyone who was unfortunate enough to have a glimpse of it would be turned instantaneously into stone.'

The storyline seemed to be getting scarier and scarier, gorier and gorier. In spite of that, or maybe because of that, it was absolutely fascinating to listen and watch.

'Now, our Perseus was persuaded to go and put an end to Medusa's monstrous mayhem. To aid him in this noble endeavour, Hermes, herald and messenger to the gods, lent him his winged shoes and Athena offered him her brilliantly burnished shield. Flying merrily, Perseus sought Medusa and found her fast asleep having left no stone unturned, I mean having turned many into stones. Gazing at Medusa's reflection in Athena's shield, not directly at her, and edging backwards gingerly, Perseus gave a mighty whack and decapitated Medusa, stuffed the still bleeding head, hissing serpents and all, into a sack, and proceeded to fly back to Athena in order to make a pleasing present of the body-less top of the slain subject. That is when he saw what was happening down below, the horrible monster Cetus approaching the hapless Andromeda.'

Chorus
Brekekekex ko-ax ko-ax
Brekekekex ko-ax ko-ax
There is hope, there is hope
Like a mountaineer's rope
Yes, hopes are raised, Heaven be praised
Comes Perseus the Brave the situation to save
Slayer of Medusa the hideous sedusah
None would dare at her face to stare
To be turned into stone without a moan
Brekekekex ko-ax ko-ax
Brekekekex ko-ax ko-ax

'Have you been with me all along our adventure, your eminence?' asked Aristophanes.

I nodded vigorously and leaned forward to catch every word of the narration.

'To return to our original story then, Perseus was distressed beyond measure, enraged beyond words to see the gorgeous girl straining against the chains, her diaphanous dress soaking wet in the seething sea, and worst of all, the inhuman whale—whales are by definition inhuman, are they not?—the menacing monstrosity swimming swiftly towards her, smacking its lips in anticipation of a sumptuous meal. Oh, maiden of incomparable loveliness, why are you in this woeful state, asked our inquisitive Prince. Well, well, well, curiosity in the face of calamity! Be that as it may, to repeat the oft-repeated phrase, our informative Princess recounted her tale of woe in a few chosen words mindful of the fast approaching doom in the form of Cetus. Ha, says our courageous youth, ha again, says he and lunges towards the advancing abomination and plunges his sword through a soft spot between the impregnable scales covering its body. Cetus, roaring in agony thrashes for a while, sinks to the bottom of the sea dead as a doornail, if not a Dodo.'

Chorus
Brekekekex ko-ax ko-ax
Brekekekex ko-ax ko-ax
Hie, Princey, hie; fly high in the sky
Hear the maiden cry? By your hand the beast must die
Ah, the Prince dives down, brows knit in a frown
Fight, oh, fight with all your might
Let the sword flash while the sharp fangs gnash
Bash up the monster's brain
Slash up his body in twain
At last the horror lies slain
Brekekekex ko-ax ko-ax
Brekekekex ko-ax ko-ax

'Seeing with their own eyes the Prince kill Cetus and realizing that all was safe at last, the throng that had gathered on the shore to watch this free show cheered and jumped with joy. The tearful parents freed their daughter from her chains after taking time to hug and kiss her demonstrating their unbounded regal love for her. What about our Prince the saviour? As all such stories end, he and Andromeda were joined in holy matrimony and lived happily forever. Not really forever, but they were given honoured places in the heavenly vault as constellations, to remain so as long as the stars in those celestial patterns endured.'

Chorus
Brekekekex ko-ax ko-ax
Brekekekex ko-ax ko-ax
Enter the Queen and enter the King
While all the town bells loudly ring
Kiss the Princess with immense joy
And hug to their bosoms the heroic boy
See the young ones holding hands
Watch them exchange the wedding bands
All, all, all—man, woman, and beast
Happily partake the wedding feast
In universal joy the story ends
As the curtain slowly descends
Brekekekex ko-ax ko-ax
Brekekekex ko-ax ko-ax

The frogs of the chorus were quite excited and wagged their palms up and down gesturing to me to join them in the refrain. I caught myself repeating under my breath *Bre ke kex ko-ax ko-ax* and looked self-consciously at Aristophanes. He seemed to be quite happy and beamed at me.

'That brings us to the conclusion of the story of Andromeda who shines as a star-studded constellation in the night-sky,' announced Aristophanes. I applauded heartily as Aristophanes and his chorus took a deep bow and melted away.

Beyond the Milky Way

The transition from one scene to another was sudden, as it often happened, leaving me breathless. When I regained my bearings, I realized that I was back at the Mount Wilson Observatory where Shapley had explained to me how he had mapped the Milky Way. My attention was drawn to the commanding presence of a man at the window—tall, handsome, and strong with powerful broad shoulders. He was studying a glass plate, covered with what looked like darkish smudges, with absolute, serene concentration. He seemed to be striking a heroic pose, even though there was no one to watch him. Let alone admire him.

From the photographs I had seen I could immediately recognize him. It was none other than Edwin Hubble himself! It seems that some people thought he looked like a Greek god. At least his wife Grace Hubble thought so, especially when she first met him.

Hubble looked up and took out the pipe he was smoking. He gave me a warm, welcoming smile.

'Ah, there you are, I have been waiting for you, you know' said the famous astronomer. 'I know you have learnt a lot about the Milky Way from my colleague Harlow Shapley. Shapley, shapely, out-of-shapely, especially in his ideas!' he laughed. 'Hush, not a word, old chap, that was strictly for your ears only, I would never say such things in public. Dear me, as Harlow said to you when you met him, I haven't introduced myself to you, I am so sorry.'

'My name is Hubble, Edwin Powell Hubble, pleased to meet you I am sure,' he made a slight bow a bit too dramatically and put away carefully the glass plate with smudges on it. 'That is a photographic plate on which we capture the image we see through the telescope, you know. We shall take a look at it later on, precious it is, quite. Now where were we? Yes, I told you my name, didn't I, old fellah?'

Hubble sounded strange when he kept addressing me as *old chap* and *old fellah*. After all, I happened to be younger than he by more than a century.

'Ah, that is a manner of speaking I picked up at Oxford, old sport,' chuckled Hubble. 'I use it once in a while. We shall talk about Oxford when I relate briefly how I came to be an astronomer. Of course, I am sure that you know much more of my life and work than I know of yours. But one derives great pleasure by talking about oneself, don't you agree?'

Hubble told me the story of his life, which I found most engrossing. It was no twice told-tale vexing the dull ear of a drowsy man, contrary to what Shakespeare had written. As it happened in the case of Shapley, scenes from Hubble's past appeared and faded out from time to time, while his voice could be heard in the background narrating the events.

'Let me begin with my first encounter with the stars,' began Hubble. 'It was in the small village of Marshfield in the state of Missouri, USA, which was my place of birth. Aha, that is the same state in which our friend Harlow Shapley had been born four years earlier some seventy miles to the west of my own place of birth. No streetlights, no pollution, clear dark skies always, mind you. So, one could see thousands of stars above you glittering away in all their grandeur. How could you not watch them and wonder at them? Ah, but it was a great pleasure to view the stars and the planets through a telescope, wasn't it? My grandfather, Martin Jones Hubble—a hefty guy he was—had built a telescope with his own beefy hands. On my eighth birthday, I was supposed to have a party. Instead, I asked my parents for permission to stay up all night and watch the stars. My granddad and I had a whopping good time indeed. When I was a little older, we did it again and talked a lot about astronomy. We both had a great fascination for the planet Mars, you know.'

Hubble paused remembering those exciting moments of his childhood days and continued. 'I joined Wheaton high school in this big city of Chicago. I grew up fast all right and stood six feet and two inches tall—but one full inch shorter than my father, what a pity. Apart from my studies, which included Latin and German, I was fond of sports too. Ah, let me not boast, but in fact I happened to excel in so many of them—pole vault, shot put, standing high jump, running high jump, discus and hammer throw—you name it, I did it. All my life I have loved athletics.'

Oh, yes, I had learnt that his name had appeared in news papers praising him for setting new state records.

'Then on to the great University of Chicago,' continued Hubble. 'I wanted to become an astronomer, you know. But my father, John Powell Hubble, said he wouldn't let me go through school if I was going to become *a thing so outlandish*, as he put it. So, I studied law as per his wish. But I learnt a lot of physics and mathematics also that would help me with astronomy later on. Well, my immediate ambition was to get the highly coveted Rhodes scholarship.'

Rhodes scholarship, it is considered to be quite prestigious even to this day.

'Rhodes scholarship is hard to get,' continued Hubble. 'I had to prepare for it a lot. I knew I would get it and get it I did. I went to Oxford, as I told you earlier, to the great University. I was expected to study law at Oxford as per the terms of the Rhodes. I joined Queen's College, one of the twenty colleges of the University which was considered to be the most beautiful of them all. And I had a ripping good time there, really capital, old chap.'

I could not help but notice that Hubble's accent and his manner of speaking had radically changed all of a sudden.

'Aha, you noticed how my speech changed when I entered Oxford,' smiled Hubble. 'Be a Roman when you are in Rome they say. Oxford, Rome, it is all the same. So, I cultivated the accent and jargon characteristic of my new University. After all, I had to jaw the hours away—I mean talk the hours away—with the old chaps I came to know, right?'

I was sure that many people, especially Hubble's American colleagues, must have thought his accent was quite artificial.

'It was at that time that I took up my father's habit of smoking and built up a nice collection of pipes. Watch this,' said Hubble and lighted his pipe that had gone out. He then flipped the match into the air in such a way that it described a circle and he caught it, still burning as it came down, without burning his fingers. 'Jolly little trick, eh?' laughed Hubble. 'Never failed to impress my friends whenever I did that after scoring a point in a debate.'

Hubble gazed far into space and spoke to himself what appeared to be words he had written long before. '*Ah, I am still dreaming away and awaiting perhaps a rude awaking, wonderful realization, or more dreams upon dreams until the end of all dreaming. The nearest approach to happiness I know is in dreaming, having nought to do with external things.*' After a pause he added, '*Labour which is labour and nothing else becomes an aversion. Work, to be pleasant, must be towards some great end; an end so great that dream of it, anticipation of it overcomes all aversion to labour. So until one has an end which he identifies with his whole life, work is hardly satisfactory.*'

I knew and knew it well that Hubble's dreams had come true beyond his expectations.

Hubble who had become silent for a brief moment continued with his recollections. 'No, my heart was not in the study of law. Some of it was awfully dry. But it had to be done and I decided to face it like a man. However, I did get around to be in touch with astronomy, my great love, through a friendly professor who was the director of the University Observatory. Of course, I took part in all sorts of athletic activities which brought me many prizes. And I was the captain of the baseball team, which was a novelty to the cricket-loving people of Oxford. I read books voraciously, some three hundred of them in my first year, five books per week that is. Impressive, eh? To jump ahead—I love to jump as you know—I completed my studies and got my degree in jurisprudence, the science or philosophy of law. Never mind all that, old chum. I must add that my one or two excursions into astronomy astonished some good science men of the University. I as a law man was supposed to know nothing about the stars up there. In any case, it was time to return home.'

A serious expression came upon Hubble's countenance. 'Many people think that I practiced law in the state of Kentucky after passing the required examination,' he said as though he was making a confession. 'Let me admit it, it is not true. To be honest with you, I might have created that impression deliberately. What I did was to translate some legal documents from Spanish, a language I knew well. That is all. Later on, I taught at a high school—Spanish, physics, and mathematics. Most of all, I was a successful basketball coach you know, quite. But, of course, my heart was set on astronomy and an astronomer I was determined to become.'

Hubble paused. I could sense that a new chapter was beginning in his life.

'Where should I go to pursue my dream? That was the question,' Hubble went on with his narration. 'The obvious choice was my old University of Chicago. So, I wrote to my professor there. I was fortunate to get attached to Yerkes Observatory of the University. I could now directly study the skies and get a good degree, Doctor of Philosophy or PhD as you know. I was most interested in those mysterious patches of light scattered all over the night sky, the *nebulae*. Did all of them belong to our own Milky Way as Shapley believed? Or did at least some of them lay outside our own galaxy as conjectured by those who talked about the Island Universes? Well, the answer to that question was to come sometime in the future. But for now, I studied the nebulae carefully and wrote up my findings under the title, *Photographic Investigations of Faint Nebulae*. I had to submit it to the University in order to get my degree, you see. I wanted to get out of astronomy for a while in a hurry. You know why? There was this war going on in Europe, came to be known as the First World War, didn't it? Along with the English and French people, our country was fighting the Germans. I wanted to fight too.'

I had read all about World War I. Every war brought with it its ravages, the death toll, and the untold misery as its aftermath. The two World Wars were the extreme examples.

'At first, I went for training in preparation for fighting in the war,' continued Hubble. 'It was a tough job all right. My knowledge of astronomy came in handy. You know how? I was able to instruct others in marching in the dark guided by the stars! Eventually, I was made a Major. We sailed to Europe finally to fight in France. The hardest thing was to see wounded men fall, and go forward without stopping to help them. In any event, the war was ending, as the Germans were about to be defeated. I did manage to sustain some injuries. I still carry the scars. But to tell you the truth, I barely got under fire and altogether was disappointed in the matter of the war. When it was all over, I stayed on and visited Cambridge University where once Isaac Newton, the greatest of all scientists, had worked. At last I was ready to return home and join this wonderful place, the Mount Wilson Observatory, where we happen to be right now.'

What a remarkable life, I thought once again, as I had done when I had heard it for the first time. I knew that Hubble's real achievements were yet to come. Hubble watched me, striking his heroic pose. Now he had found someone to admire him, hadn't he? He must have been quite accustomed to the ardent adulation of others!

'Let me show you the telescope through which I have been observing the nebulae, the same one Shapley used for his work on the Milky Way,' said Hubble and walked briskly towards a nearby door, which he opened and waited for me to enter. I gasped, although I had seen a large modern telescope before. This one was not only large, but the history behind it was incredible. I was inside an immense dome at the centre of which stood the huge telescope. It did not have a closed tube. Instead, sturdy girders made up a cylindrical structure. At its lower end was the most important element of the telescope, namely the mirror polished to extraordinary precision. It was one hundred inches in width, I had learnt. That meant it could gather and reflect large amounts of light so that very faint objects in the sky could be seen clearly. The reflected light was brought to focus by a convex lens. Overhead, the dome could be opened whenever observations were to be made, but was kept closed at other times. Hubble led me up the long flight of steps to the observing platform.

'From here, one can directly observe the object of one's interest,' explained Hubble. 'You can also photograph it. Or you can attach a *spectrograph*, which splits the light from, say, a star into individual colours the same way Newton split white light with his prism. All this helps us in finding out a lot about the celestial object of interest. What *is* the object of our interest then? Well, it is a nebula located in the constellation Andromeda. Ah, Andromeda! There is a nice story for you, isn't there? I know you have already had the great good fortune of enjoying the myth enacted specially for you. So we can proceed with our scientific exploration of the nebula Andromeda.'

The Realm of the Nebulae

'Andromeda along with her parents and her prince charming, Perseus, have been immortalised as constellations, as you know,' said Hubble. 'I realize that Shapley has told you about Cepheid variables. The name derives from the constellation Cepheus named after Andromeda's father. Now, here comes the moment when you can see the skies through our great telescope. It is trained towards the constellation Andromeda.'

I was highly excited to look through the large telescope. I could see again the faint nebula with the wispy spiral arms emanating from the luminous main body. It was a unique opportunity to gaze upon the vision that had radically changed the measure of the Universe.

Hubble brought out with utmost care the photographic plate he had been holding when I first saw him.

'Take a look at this,' offered Hubble.

At the centre was the largest of the grey smudges having a granular structure, but distinctly bore the image of the nebula I had seen through the telescope. I was thrilled to look at the original photograph directly, not just a reproduction.

'Well, it was the night of October fourth in the year nineteen hundred and twenty-three. How can I ever forget that date?' recalled Hubble. 'Despite poor visibility, I took a photograph of one of the spiral arms of the nebula Andromeda. I thought I could detect the presence of a nova. So, next night, which was clearer, I took another photograph and marked the plate as H335H, the one you see here.'

Yes, H335H was destined to become the most celebrated photographic plate ever!

'I compared my photograph with those of the same nebula that had been taken earlier,' Hubble went on. I could feel his mounting excitement. 'It was clear that the brightness of one of the spots on the photograph had been undergoing periodic variations. It was indeed a Cepheid variable. I took out my pen and wrote in capital letters 'VAR!' to identify the star.'

I gazed at the star so marked. It seemed to me for a moment that it was actually going through its periodic variations in brightness. An illusion, I was aware.

'The next step was simple,' continued Hubble. 'The same way Shapley had done with his Cepheid variables, I could estimate the distance to the Andromeda nebula. You know what I got? Some nine-hundred thousand—well round it off to a million—light years. How big is the Milky Way as determined by Shapley? A mere one hundred thousand light years across, was it not? Then clearly, the nebula was way outside the Milky Way.' After a momentary pause, striking his theatrical pose again, Hubble announced, 'One little Cepheid variable had proclaimed that our Milky Way was not the entire Universe!'

I gazed at the photographic plate with absolute fascination just as I had done with its reproduction. Once again, the image of the Andromeda nebula reminded me of Galileo's meagre jottings of Jupiter's Moons that had supported Copernicus and opened up new avenues in astronomy and human thought.

Hubble stroked his chin thinking back about his momentous discovery. 'Well, I wrote a letter to Shapley informing him of my findings,' said Hubble. 'Poor chap, on reading my letter, it seems he said: *Here is the letter that has destroyed my Universe!*'

I too thought about the impact of Hubble's discovery. Astronomers had hailed William Herschel on his discovery of Uranus, saying that he had doubled the size of the solar system. Now Hubble had enlarged the entire Universe ten times. He did not stop there. He went on to discover that the Universe made up of galaxies was inconceivably larger than that.

'Take a look through the telescope and see for yourself how the Universe looks like,' invited Hubble.

I looked through the telescope again. I could see everywhere, interspersed among the stars, fuzzy patches. Some were bright enough, but in addition there were innumerable faint ones, some of which were so faint that I had to strain my eye to see them. These patches differed in shape to some extent from one another. Apart from individual galaxies, there seemed to be groups of them here and there.

Hubble waited for me to have my fill of the galactic panorama and continued his narration.

'Every one of those nebulae is like our own Milky Way consisting of billions of stars,' informed Hubble. 'The farther they are from us, the fainter they appear. You noticed that those nebulae com in different shapes. There are spiral-shaped ones like our Milky Way, while some have a bar at the centre of the spiral structure. And there are those that are elliptical.'

Once again, I had learnt that it was Hubble himself who had classified his nebulae into different categories.

'As you saw, many nebulae are grouped together into *clusters*,' continued Hubble. 'Our Milky Way and the nebula in Andromeda belong to the *Local Group* of thirty or so nebulae. In the constellation Virgo, there is a rich cluster made up of some thousand or even two thousand nebulae. Yes, sir, the Universe is quite something.'

The picture of the Universe at large was getting more and more amazing to say the least.

'Well, I realized that it would take thousands of years to photograph the entire sky with our telescope and count all the nebulae that could be observed. I didn't have that long a time, did I?' laughed Hubble. 'So what to do, chum? I did the next best thing I could. I photographed one thousand two-hundred and eighty-three—oh yes, I kept count for sure—sample regions of the sky. My survey showed the total of around forty thousand nebulae. From that I could estimate that no less than a hundred million nebulae must exist within the range of our one-hundred inch giant of a telescope. And there is a two-hundred inch supergiant of a telescope sitting on Mount Palomar. And that one can see up to a billion nebulae. Yes sir, again the Universe is really something. Quite so, old chap.'

Of course, astronomers following Hubble were to show how immense the Universe actually is using more powerful telescopes and more sophisticated methods. But, Hubble had taken the first and the most important decisive step in that direction. A historic event that will never be forgotten!

'Yes, the future was to tell us how big the Universe happens to be in reality,' Hubble echoed my thoughts. 'But the strangest fact related to the nebulae and therefore to the Universe was yet to emerge. Before we go into that part of our story, I must introduce to you my dearest friend and my most esteemed colleague.' Hubble held out both his hands in a joyous welcoming gesture, 'Ah, here comes Milt, Milton Lasalle Humason that is! Everyone loves Milt.'

The Magnificent Mule Driver

I stared down the mountain path in surprise. I could see a man on horseback winding his way up. His coarse clothing was covered in dust and his dishevelled hair blew in the wind. A big dog rode behind him, standing on its hind legs with its forepaws placed firmly on the man's shoulders. Strangest of all, the man was driving a pack of mules carrying a heavy load of lumber.

When he reached the top of the mountain where I was standing, he dismounted, as did the dog. The latter came close to me, sniffed at me, wagged its tail in friendship, yawned wide and then fell asleep curled up on the ground. The horse too nodded vigorously several times at me and bared its teeth in what appeared to be a smile as it neighed its greetings. The man called out to the mules to stop as they tried to wander off and cursed under his breath. Then he spat out tobacco juice from his mouth in a neat jet.

'Pardon me, pardon me,' he said. 'I shouldn't spit tobacco juice, let alone curse in front of visitors, isn't it so? Well, I am Milt as you were told. Mighty glad to meet you, buddy.'

Milton Humason, Milt to all his friends, was stocky in build, with a round face, twinkling eyes, and easy manner that pleased everyone. I liked him instantly.

'Ah, it is hard work, driving those mules bringing up all that timber and other material. They are needed for building telescope's supporting structure, local cottages, scientists' quarters, and all that jazz you know,' said Milt looking over my shoulders.

I looked back and I was startled to see that the fully built observatory was no longer standing there. Instead, a number of hardy men were busy on the construction work of the half-built observatory.

'Oh, yes, we have gone back in time,' chuckled Milt. 'The building work is going on full speed. It will all be done soon. Oh, you are confused about me, this place, and whatnot. I can see that. Let me tell you my story, so you will get a better idea of things around here.'

Milt took a deep breath and went on to give a brief account of his life. As always, this was often accompanied by shifting scenes tracing the passage of time.

'Well, to begin at the beginning, I was born. Don't we all!' guffawed Milt. 'That happened to be at Dodge Centre in the state of Minnesota. When I was fourteen— ah, to be young again, what wouldn't I give—I came here to Mount Wilson, a summer camp it was. What an experience! After that, who wanted to go back to school? So, I dropped out. My parents, bless their souls, allowed me to do whatever I liked. I became a bellboy at the brand new Mount Wilson Hotel. What did I do? Carry guests' luggage, make them comfy, tend to their pack animals, wash dishes and so on, that's what. I dunno if I liked my job, but I sure loved the mountain. So what did I wanna do? Settle down here. Yes sir, and that's what I did.'

Milt closed his eyes with a big smile playing on his face recalling those bygone days before going on with his story.

'Well, as you can see, I became a mule driver,' said Milt waving his hand around to indicate his present occupation. 'Let us fast-forward a little, shall we? I loved the observatory, naturally.' In a stage whisper he added, 'This is strictly between you, me and the doorpost over there. More than anything, I loved my Helen, daughter of the engineer Mr Merritt C. Dowd, didn't I? Any better reason than that to become part of the observatory? Women have a greater pull than all the stars put together, you know.' He winked and gave a long, belly laugh.

Milt paused again to indicate that the story was moving into another phase.

'We got married, Helen and I, and had our Baby Bunting, named William after his grandpa,' Milt went on. 'So, being the man of the family, I was obliged to support it. I had to take up some regular work. So, I got a janitor's job at the observatory. But, lucky me, one of the students taught me how to take pictures of the stars on plates, bless him. You met Mr Shapley, didn't ya? Yes, he noticed me all right. It was all so jolly good—me and my heady Panther Juice, which some people sipped on the sly, and my passing around a deck of cards on cloudy nights when no one had noth'n to do. Wish I could give you a taste of my Panther Juice, but you look underage to me, what to do?' Milt winked again and laughed boisterously. 'Ah, well, I began taking photos for Mr Shapley. And more than that, I enjoyed telling visitors about Mr Shapley's ants.'

'Shapley acknowledged that Milt was one of the best observers we ever had,' put in Hubble who was listening to Humason with pleasure. 'That is one point on which I fully agreed with Shapley. Let me complete the story for Milt. From Janitor, his position was elevated to night assistant. When his abilities were recognized, Milt was appointed to the regular scientific staff. Then on, there was no stopping. He went on to become successively Assistant Astronomer, Astronomer at not only Mount Wilson, but also at Palomar Observatory that was to house the two-hundred inch telescope, the world's largest. And the high-school-dropout-mule-driver was to receive the honorary degree of Doctor of Philosophy from the University of Lund in Sweden, no less!'

Humason, both happy and embarrassed, was pretending to be busy dusting some old photographic plates. All three of us had now moved once again into the Observatory.

'The most enthralling part of the story begins now,' said Hubble. 'The strangest and the most spectacular discovery Milt and I made together about our Universe.'

Galloping Galaxies

'Before we discuss our discovery, I must remind you of a phenomenon you know very well, namely the Doppler Effect,' informed Hubble. 'In the case of sound, its pitch goes up, if the source is moving towards you. And it goes down if the emitter of the sound is travelling away from you, right old chap? For the effect to be appreciable, the source must be moving with a speed comparable to that of sound. Then again, this can happen in the case of light too. Light, or equivalently its spectrum undergoes blueshift if the source is rushing towards you, and shows redshift if the light emitter is zooming away from you. Yes sir, the speed of the source in this case has to be a good fraction of light's phenomenal velocity. Nothing on Earth can move that fast. Then what's the point of all this blueshift, redshift, greenshift, and all that, eh?'

I shrugged as if the Doppler shift of light was of just academic interest and was of no use for practical application, although I knew what was coming. Of course, it was wonderful to listen to Hubble's recounting of the past.

'Aha, I said nothing on Earth moves that fast. Did I say anything about the objects in the sky?' Hubble wagged his finger. 'Let us find out what is happening up there. We go back to the telescope.'

I was again overwhelmed by the gigantic telescope just as I had been when I had first seen it. I know, I had no control over this feeling.

'Hey, Milt, come down for a minute, our friend here wants to see the spectra you have photographed,' called out Hubble cupping his mouth with his two palms.

Humason came down panting as he descended down the ladder adjoining the telescope. He beamed at me in spite of his fatigue clearly written on his face.

'Whew, what a life,' groaned Humason. 'Night after night I am perched like a monkey on the tiny platform five stories above the observatory floor. I wonder how I look like with my face illuminated by the red dark-vision lamps. More grotesque than I am, I am afraid. God, some nights it is freezing cold up there and the merciless winds make it worse.'

Involuntarily Humason cursed under his breath.

'Sorry, sorry, I should watch my language especially in front of our young visitor here,' Humason was apologetic. 'Look at the telescope. It is a monster. I have to make sure it points exactly at the guide star. Sometimes the whole system stops and I have to put all my strength into my shoulder and hold the thing in place.

Sometimes, I have to get on to the iron frame and do some contortionist acts so as to make sure the photographic plate is in place. I have to go through this circus night after night so the plate soaks in enough light. The nebulae out there are awfully faint, you know. Heaven help me.'

Humason took out of his hip pocket a small flask and took a swig of his famous Panther Juice from it to compensate for his plight.

'Milt, why don't you let me have a sip too,' said Hubble taking the flask from Humason. He tilted it up and helped himself to what seemed like a gulp rather than a sip.

'All right, Milt, we do realize the great trouble you take in photographing the nebulae. You are doing a wonderful job, old boy,' said Hubble with genuine appreciation. 'I would like to show our visitor here some of your spectra displaying the redshifts of the nebulae.'

Humason took out a photographic plate from the drawer of a nearby desk and held it up to light. I had forgotten all about the faint appearance of the galactic spectra I had seen in the books on cosmology. Naturally then, I had expected to find a band of brilliant colours or sharp luminous lines as I had seen in other contexts. But here was a diffuse, faint grey strip with a few fuzzy dark lines distributed along its length.

'Surprised, are we?' laughed Humason. 'The light captured on the plate comes from a nebula millions of light-years away from us, you know. That is why the photograph is so faint. And that is why I have to toil for nights and expose the plates for awfully long durations. Why are the lines dark? Well, they are *absorption lines*. You see, suppose white light passes through the vapour of an element which is not hot enough to emit its characteristic radiation. Then that vapour absorbs precisely those wavelengths that the element would emit if it were hot. That is how absorption lines are produced instead of emission lines. Now, the lines you see are those of the element calcium—not the emission lines, but absorption lines. Calcium is very important stuff if you ask me. Why? Because, calcium is essential to the bones. And I need strong bones to tame the mighty telescope over there, got it?'

It was obvious that Humason was an expert in the matter of spectra and in taking their photographs. He was very lucid and animated the way he described his work.

'Now, here is a photograph of calcium lines taken directly in the laboratory. Hope you will be satisfied,' Humason pulled out another plate. This was much better although it was not coloured. Against a dark background white lines stood out sharp and clear. Humason superposed the two plates with edges coinciding exactly. 'Take a look. The absorption lines on the first photograph are all shifted by the same amount in comparison to the *emission lines* captured in the laboratory. You see it? Excellent! Aha, that is the Doppler Effect for you. What does that mean? They say it means that the nebula is running away from us, quite fast too. I capture the spectrum, we measure the redshift, and my boss, Mr Hubble here, calculates the velocity corresponding to that shift using his Doppler formula. Also we know the distance to the nebula by observing the Cepheid variables it contains. Now I know you know how it is done. Now you know I know you know.' Humason gave his hearty laugh and continued. 'Mr Hubble will tell you that this nebula, in the constellation Virgo, is at a distance of some 50 million light-years from us and happens to be flying away at a speed of around 750 miles per second. There you have it—redshift showing how a distant nebula is rushing away from us. Whew, that was a long tiring explanation.' Humason took another swig of his Panther Juice.

'Milt gets easily tired talking, at least that is what he claims so he can keep guzzling his Panther Juice,' chuckled Hubble. 'The photograph Milt is taking right now is of the nebula in the constellation called Corona Borealis, which is at a distance of, say, 940 million light-years from us. And, we expect its velocity to be about 13,400 miles per second. Far enough and fast enough, what ho? As I told you, we can detect redshift of light, if the speed of the source is comparable to that of light. And that is precisely what is happening with those nebulae up there in the sky. We hope to keep measuring even greater distances and higher speeds.'

'That means more photographing,' Humason groaned. 'Tell our friend about the distance-velocity relation, boss.'

'Ah, that is another thing—really interesting,' Hubble's eyes lighted up. 'We found that the two quantities, namely the distance to a nebula and its redshift, or equivalently the velocity with which the nebula is flying away, are very simply inter-related. Double the distance, the velocity doubles too. Make the distance, say ten times larger, and the velocity becomes ten times greater too. This is known as a *linear relation.*'

Yes, it was well-known as the *Hubble Law* in his honour.

I could see a faint glimmer of pride appearing on Hubble's countenance. That was quite natural, wasn't it?

'The speeds you encounter in astronomy are truly great. And the nebulae take the cake,' Hubble said. 'Here, I clap my hand twice. That makes it a second. In that time, the Earth has gone around the Sun by about 18 miles. Clap, clap, the entire solar system has moved by some 137 miles about the centre of our Milky Way. Clap, clap, again, and the nebula in Corona Borealis has zoomed away by 13,400 miles!'

'Oh, well, as far as I am concerned, I have always been rather happy that my end of it—my part in the work—is, you might say, fundamental. It can never be changed, no matter what the decision is as to what it means,' shrugged Humason. 'Those lines are always where I measured them. And the redshifts, if you want to call them that or velocities or whatever they are going to be called eventually, will always remain the same. So, here I go to get some more lines up over there.'

Humason gave me a broad smile, waved goodbye, and made his way slowly up the ladder back to his photographic plates.

'Yes, he will always be remembered for his spectral lines, good old Milt,' said Hubble his eyes following fondly the receding figure of his dear friend and distinguished collaborator. 'We make our observations all right. But it is important to know what they mean. For that, one has to build models based on mathematical theories. Long ago, Newton did that with his law of gravitation, which described everything that was known at his time. But now we have to explain why the nebulae are flying away at breakneck speeds. Newton's theory cannot do the job. We need a new theory of gravitation for that. Fortunately, such a theory already exists.'

After a momentary pause, Hubble made his dramatic statement. 'The nebulae are flying away from one another. And the reason is this: *the Universe itself is expanding.*'

I was startled by the announcement, although everyone in our own time knows that amazing fact. But, of course, there was a lot more for me to learn about the cosmological expansion, especially its theoretical explanation.

'Patience, my man,' Hubble held up his hand. 'You will learn all about it soon, very soon. I have got work to do you know, quite pressing indeed. So I must take leave of you now. Farewell, old sport!'

With that Hubble started moving out of the Mount Wilson observatory. The observatory faded gradually. In its place, I could see an immense telescope, even larger than the one on Mount Wilson, looking out of the dome of another observatory. I realized what it was—the newly-built two hundred-inch supergiant at Palomar. Strangely enough, seated right inside the telescope at its prime focus was Hubble, who had grown much older. And then even that vision melted away and was replaced by the dark night sky filled with the stars. Strangely enough, I could see everywhere diffuse patches that had grown unusually bright. It was the realm of the nebulae.

The bathwater still resembled the dark night sky with the stars, peppered with the ubiquitous faint patches of light, the nebulae. Those nebulae started moving, gathering speed, flying away from one another, and dissolved away. The bathwater once again assumed its natural appearance.

Ah, the Universe is awesome in its vastness. Yet, it is expanding greedily to become vaster still. There is so much to know about it all. I wished I could stay on in my bathtub and enjoy the rare cosmic visions that kept floating across my mind. But wishes are wishes and reality is reality. With a sigh I got up. *To sleep*, I whispered to myself, *perchance to dream*? Sleep yes, but dream no, I thought. As soon as I hit the bed, the Universe curled up and enveloped me in the form of deep, uninterrupted sleep.

Chapter 14
Space, Time and Gravitation

As always, I knocked on George's office door and entered without waiting for George to respond. George maintains that it is a waste of energy to call out 'Enter', 'Come in,' or whatever. If the door is kept unlocked, it means that everyone is welcome to his office—students, faculty, and visitors, with the only exception of burglars. If he were really busy, then he would keep his door locked.

George was seated in one of the comfortable chairs in front of the low table, on which several books were lying, as well as a few photo-copied articles.

'Listen to this, Alfie,' said George without any pleasantries or preamble as though we were already deep into our dialogue. He was holding in his hand a somewhat large volume, its dusty grey cover needing replacement. Of course, George would never do any such thing claiming that it would destroy the venerable aura surrounding the book.

'Listen, let me read the opening passage for you from this very interesting article appearing in this book,' repeated George and read out the passage he had mentioned. '*Speculations about the Universe in which men live are as old as human thought and art; as old as the view of shining stars on a clear night. Yet it was the general relativity theory, which only thirty years ago, shifted cosmological problems from poetry or speculative philosophy into physics. We can even fix the year in which modern cosmology was born. It was in 1917 when Einstein's paper appeared in the Prussian Academy of Science under the title: Cosmological Considerations in General Relativity Theory*. Beautifully put, what do you think?'

It was indeed an elegant and meaningful passage. But, what was this article about? And who wrote it? I am sure George had noticed my puzzlement.

'Ah, sorry, Alfie, forgot to tell you the source of that opening paragraph,' George hastened to demystify me. 'The article is titled, *On the Structure of Our Universe* and is authored by Leopold Infeld, who had worked with Albert Einstein. The article is one of the several historically important essays including Einstein's own *Autobiographical Notes*. And the volume I am reading from is *Albert Einstein: Philosopher-Scientist*, edited by Paul Arthur Schilpp and brought out in 1949. We shall come back to the article soon enough.'

© Springer International Publishing Switzerland 2015
C.V. Vishveshwara, *Universe Unveiled*, Astronomers' Universe,
DOI 10.1007/978-3-319-08213-4_14

George smiled and added, 'Needless to add, but I shall add anyway, you are most welcome to borrow the precious volume as well as other books I shall lend thee, provided you return them safe and in good time.'

George opened his manuscript for ready reference, collected his thoughts for a few moments, and began his explanation of the fundamental tenets of modern cosmology based on the general theory of relativity.

'What does Infeld say about the importance of general relativity to cosmology?' began George. 'Let me quote him again: *the general relativity theory... shifted cosmological problems from poetry or speculative philosophy into physics.* So let us talk about the general theory of relativity first before thinking of its application to cosmology.'

'All right disciple, let us hark back to our discussion of stellar evolution,' continued George. 'We touched very briefly upon general relativity when we talked about Oppenheimer's studies on the gravitational collapse of a massive star to form a black hole, right? Hark back even further, setting the clock backwards by a few years before you launched upon your life of a vagabond. We discussed relativity theory, both special and general, exhaustively—I mean it exhausted me—in describing black-hole physics in great detail. You can read it all from the copious notes you must have kept at that time. If I remember right, and I do, you even got a book published craftily, didn't you? You can read that too.'

'There was nothing crafty about the publication, George. You gave your *imprimatur*, didn't you?' I protested and added with a grin, 'After making sure I had expurgated all your objectionable expletives.'

'What nonsense, I am incapable of uttering profanities,' asserted George and went on seriously. 'Let us get on with general relativity. We already noted how the most distinguished scientists like J. J. Thomson and Dirac extolled the virtues of the theory. And rightly so. The special theory of relativity itself had been hailed as a revolutionary theory. But general relativity, which now included gravitation within its spacetime description, was the supreme achievement of pure thought based on relentless logic with no appeal to observation. Simply put, it was a unique work of unprecedented genius.'

'There was already Newtonian gravity. Where was the need for a new theory of gravitation then?' I enquired.

'Let us not forget that Newtonian gravity too was a stroke of genius,' commented George. 'It completely transformed physics by bringing in so many factors like setting ideas within mathematical framework, bringing in the notion of force, enunciating universal laws, and so on. All along, there were astronomical observations related to planetary motion in the background. And, as we have repeatedly observed, Newton could explain everything in nature that was known at his time.'

'That is fine with Newtonian physics. But, you haven't still answered my question regarding the need for a new theory of gravitation,' I pointed out.

'All right, let me answer your question,' George said. 'First of all, Newtonian physics did not conform to the tenets of special theory of relativity. So, Einstein had to find a theory that embraced both relativity and gravitation. The starting point for this was the fusion of space and time into a unified entity, namely the spacetime, which had been a natural outcome of Einstein's special theory of relativity.'

'I remember your telling me that it was Hermann Minkowski who blended together space and time,' I said.

'You are right. Minkowski was Einstein's mathematics teacher at the famous Federal Institute of Technology in Zurich, familiarly known as Polytechnic or simply as Poly. Apparently, being more interested in physics rather than in mathematics Einstein used to cut Minkowski's classes,' remarked George.

'And Minkowski used to refer to him as *that lazy dog Einstein*!' I added.

'Precisely,' nodded George. 'Actually, when someone asked what Einstein thought about the spacetime representation of his special relativity, he had replied: *Now that mathematicians have invaded the theory, I no longer understand it*! However, Einstein included Minkowski's name among his important mentors at the Polytechnic later on in his *Autobiographical Notes*. Anyway, Einstein wanted to include gravitation into the spacetime picture. This was a logical extension of special relativity, a theory that had nothing to do with gravitation. The cornerstone of the theory was the fact that all objects fall with the same acceleration.'

'A fact Galileo is believed to have demonstrated by dropping two objects of different masses from the tower of Pisa,' I mentioned.

'Well, it is felt that Galileo in his old age and infirmity could not have gone up the tower to perform that experiment, Alfie,' said George. 'Probably it is just a legend. He could have employed his inclined plane to show the effect. Some people believe that he came to the conclusion by pure logic, by *reduction ad absurdum*. If a heavier mass fell faster than a lighter one, as had been stated by Aristotle, what would happen if you joined the two together? The two would be acting against each other! In any case, the simple phenomenon became an extraordinarily powerful tool in the hands of Einstein. He wrote ecstatically about his realizing the importance of this phenomenon:

When, in 1907, I was working on a comprehensive paper on the special theory of relativity, there occurred to me the happiest thought of my life that, for an observer falling freely from the roof of a house, there exists—at least in his immediate surroundings—no gravitational field.'

'Oh yes, the happiest thought of the great man, I remember that,' I said.

'Let me be brief,' continued George. 'Let us perform Einstein's *gedanken experiment*, or *thought experiment*, involving the *Einstein Elevator*. Hopefully no one will get the idea of repeating it in real life, it is too dangerous you know. Suppose Einstein is inside his elevator and the shaft is cut off at his own behest. He is in free fall and so are all the objects around him—his pipe, match box, his notebook and pencil, so on. He feels no gravity at all, visually or otherwise. So, by free fall, equivalently in an appropriate accelerated frame—namely the Einstein Elevator—gravity has been eliminated. You agree?'

I nodded my agreement.

'All right then. Now think of a rocket or a spaceship far away from the Earth in zero gravitational field,' George went on. 'Let the rocket zoom up with acceleration exactly equal to that on the surface of the Earth. Drop things, apples and oranges, they will all seem to fall as though they are on the Earth, as the floor of the rocket accelerates upwards. So then, you have created gravity by going into an accelerated frame, namely the rocket. You follow?'

I nodded again.

'Here comes Einstein's stroke of genius,' announced George. 'Einstein concluded that gravitation was in essence equivalent to acceleration. Whatever you observe in an accelerated frame, you must observe in a corresponding gravitational field. This is Einstein's famous *equivalence principle*.'

George had remarked how Einstein's general relativity had been the outcome of pure logical reasoning. I was realizing the workings of such a simple but powerful process.

'So far it has been all abstract thinking. How about some practical application? Sorry, Alfie, I have to keep repeating myself. Here comes the man's genius again,' George spread out his hands as if he was helpless in this matter. It was a bit comical though. 'Send a ray of light across a stationary rocket. It streaks in a straight-line. Now let the rocket shoot up with acceleration. The light ray seems to hit a point lower than a point it encountered a moment earlier as it traverses the rocket. This happens every moment of its flight across the rocket accelerating upwards. The net result is that the light-ray appears to bend within the rocket. In other words, light bends in an accelerated frame. And the same thing must happen in a gravitational field according to Einstein's equivalence principle! There you are, a new phenomenon predicted: gravitational bending of light!'

'In Einstein's time there were no spacecraft, I know,' I said. 'But, can this bending of light in a gravitational field be observed in nature or in laboratory?'

'The effect is too minute, Alfie,' answered George. 'It can be observed only on astronomical scales, as we shall see. To continue, when there is no gravity, particles and light travel in straight-lines, the shortest path between two points, or *geodesics* as they are called. On a flat sheet of paper, straight-lines form the geodesics. So, gravity-free spacetime is *flat spacetime*. That means, within the freely-falling Einstein Elevator you have a small patch of flat spacetime. But think of such elevators at various points all around the Earth at different heights. They all fall with different accelerations each containing a bit of flat spacetime. When you put them all together, you get a *curved spacetime*, just as patching together little flat areas on the Earth gives a curved two dimensional surface. In short, gravitation engenders a curved spacetime or gravitation is equivalent to spacetime curvature!'

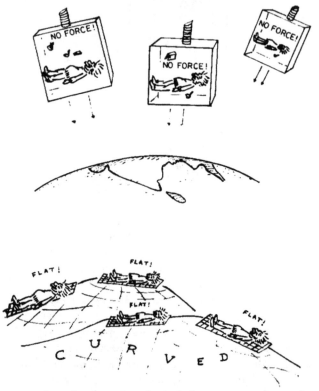

All this seemed so simple and so logical. I was most eager to learn about the further steps taken by Einstein in developing his theory.

'Now, here is an idea that follows that of curved spacetime. First of all, why do particles move along curved paths in a gravitational field? According to Newton, it is because of the gravitational *force*. According to Einstein, there is no gravitational force, but only curved spacetime. In the absence of gravity, spacetime is flat, right? And in a flat spacetime, particles move along straight-lines that are the flat-spacetime *geodesics*. Let us keep in mind that geodesics are those curves along which the distance between two points happens to be the shortest. Now turn on gravity. Spacetime curves. Assume that particles and even light have no choice but to follow geodesics, the most natural curves imbedded in any spacetime. But the spacetime is now curved. *Ipso facto*, particles in a gravitational field, represented by curved spacetime, move along curved trajectories. That is Einstein's famous *geodesic hypothesis*. So, in spacetimes curved by gravity, the trajectories of particles and light are curved too. Conversely, if the trajectories are observed to be curved, then the spacetime in which they move is curved. In essence, Einstein's geodesic hypothesis demonstrates how curved paths lead to a curved spacetime.'

'Geodesic, the shortest path between two points,' I said. 'In a crooked world, the straightest path is crooked too.'

'So is your mind, Alfie,' laughed George. 'That was a lengthy explanation, wasn't it? But then, we cannot interrupt the chain of ideas that had led to the magnificent theory. But let me rest my tortured tongue.'

With that, George sat back and mopped his brow with his handkerchief. But not for long, since he is quite indefatigable when it comes to explaining his favourite subjects. Having had no more than a second or two of rest, George began again.

'Yes, we have touched upon the soul of general relativity,' George resumed. 'Now for some aspects of its corporeal frame. Einstein now held in his palm his beautiful curved spacetime representing gravitation. But, what about the actual theory based on his simple but profound ideas? What about the mathematical frame needed for setting this jewel in? Einstein had no idea how to go about it.'

'Aha, he should have attended Minkowski's math classes more regularly then,' I suggested.

'Well, perhaps Einstein found better help than formal instruction in mathematics,' responded George. 'His classmate at Zurich Poly, Marcel Grossmann, a brilliant mathematician no less, came to his rescue. Earlier, the same Grossmann had rescued Einstein from unemployment by getting him the job at the patent office in Bern through the good offices of his father Grossmann the elder. Now, he pointed out to Einstein what he needed for building up his theory: it was Riemannian geometry. Poor Einstein, lucky him really, had to master this branch of mathematics and then apply it to his own ideas. This took him several years of hard work, meandering through the mathematical maze, taking false turns, retracing his steps, and so on. After all this tortuous journey, he could finally write down the equations that governed the geometry of the spacetime, curved by gravitation. This geometry, in turn, was related directly to the mass distribution that produced the gravitation in the first place. Finally in the year 1915, he unveiled his grand theory, the general theory of relativity that was destined to transform our view of the Universe completely for ever.'

So we had arrived at the most important milestone in our cosmic adventure.

'The theory of theories was ready, the pudding had been cooked, but where was the proof of the pudding?' George raised his questioning eyebrows. 'Well, there were these three phenomena that go under the name of the *classical tests of general relativity*. First, the theory predicted that light or any radiation going from a stronger gravitational field to a weaker one was red-shifted. This is known as *gravitational redshift*. It is not Doppler shift, which is due to the relative motion of a source with respect to a receiver. Then again, let us not confuse the gravitational redshift with the *cosmological redshift* discovered by Hubble and

Humason. We shall talk about it later on. Gravitational redshift could be verified from stellar spectra as well as by eventual terrestrial experiments. Then there was the well known *perihelion precession of the planet Mercury*. The closest point of approach to the Sun, the perihelion, of Mercury did not close but continuously underwent precession by an amount of some five thousand seconds of arc per century. Newtonian gravity could account for most of it as due to the tug of other planets except for a teeny-weeny bit of 43 seconds of arc per century. And what do you think happened?'

'Einstein's general relativity accounted for it,' it was an obvious guess.

'Brilliant,' George grinned. 'After a long calculation, within the rigid framework of his equations with no leeway for any kind of fudging, Einstein could get the miniscule missing amount exactly. In 1916, he wrote to his friend, the renowned physicist Paul Ehrenfest in Holland about his spectacular achievement. Here, let me read the relevant line from the letter: *Imagine my joy at the result that the equations yield the correct perihelion motion for Mercury. I was beside myself with ecstasy for days!*'

This must have been a moment of extraordinary success of Einstein's theory of general relativity.

'Finally, we come to the third and perhaps the most important test of general relativity, namely the bending of light in a curved spacetime,' stated George. 'The story behind the prediction of this phenomenon is quite strange, you know. As early as in 1908, Einstein had calculated the amount by which a ray of light from a distant star would bend as it grazed the surface of the Sun. This was seven years before he had formulated his full-fledged theory. He had just combined Newtonian gravity with his equivalence principle and had obtained a value of 0.83 seconds of arc for the bending. Actually, Einstein had made an arithmetical mistake. The correct value should have been 0.87. In any event, how can one observe this effect? After all, the stars are not visible during the day time. Einstein had this wonderful suggestion that the light bending could be measured during a total solar eclipse when the stars do become visible. And Erwin Finlay-Freundlich, a German astronomer, went all the way to Russia to record this effect in 1914.'

'What happened? Did he measure the light bending?'

'No, the World War broke out and the poor man came back empty handed,' answered George. 'Einstein was most disappointed, to say the least. But it was a blessing in disguise.'

'How is that?' This was surprising.

'Well, after he formulated his final theory, Einstein recalculated and found the light bending was exactly twice the value he had calculated—it was 1.75 seconds of arc,' answered George. 'If Finley-Freundlich had measured this value, if Einstein had confirmed it after putting forward general relativity, maybe one could have felt that the theory had been tailor-made to account for the correct amount of bending. I don't know. In any case, Arthur Eddington verified the correct value for the light bending during the total solar eclipse that occurred in the Portuguese Island of Principe off the western coast of Africa. This was in 1919. The ray of light bent by gravity dazzled the whole world. And Einstein became a household name overnight.'

George smiled as he recollected some amusing anecdote.

'You know, Alfie, it seems someone asked Einstein how he might have felt if Eddington's result had not been in accordance with the general theory of relativity,' related George. 'And apparently Einstein had answered: *Then I would have been sorry for the Lord, for the theory is correct*! What supreme confidence, Alfie.'

George shook his head laughing. I laughed too and waited for George to move on to the next phase of the fantastic story.

George paused drumming his fingers on the armrest of his chair.

'Another phenomenon engendered by general relativity has proved to be of paramount importance, Alfie. It is the existence of *gravitational waves*,' George resumed. 'Believe it or not, these waves were predicted by Einstein himself on the basis of his theory.'

'There seems to be no end to that man's achievements,' I remarked. This was truly remarkable.

'You said it, Alfie,' agreed George. 'Until the advent of general relativity, everyone had been familiar with sound waves and electromagnetic waves. Their production, propagation, and detection were totally clear. Now, the motion of any kind of material objects produces gravitational waves consisting of ripples in the spacetime. These waves carry energy and have peculiar polarization properties quite different from electromagnetic waves.'

'If gravitational waves are generated so easily, have they been observed in the laboratory?'

'No, Alfie, gravitation happens to be an extremely weak force. Unlike electromagnetic waves, gravitational waves of enough strength can neither be produced nor detected in the laboratory. Consequently, here on Earth one may be able to detect only gravitational waves produced in astronomical events,' George answered. 'For decades now, there have been concerted efforts to build gravitational wave detectors and capture those elusive waves. Once this is done, it is hoped that a new window on the Universe would be opened up as has happened with light, radio waves, and X-rays.'

'We have covered enough of general relativity, Alfie,' said George after a while. 'We can proceed now to talk about its application to cosmology that opens up an overwhelming chapter in the study of the Universe. Let us take a break. I have to check my mail gathering dust in the department office you know. It is all junk mail. But I have to clear it to make room for the poor secretaries to fill the mailbox with more junk. On the way back, we can stop at the wending machines and energise ourselves with come cookies and coffee.'

I could see that George needed some respite and energising. I too needed some time to go over what I had heard about general relativity and refresh my memory.

Chapter 15
The Curved Cosmos

We were back from our expedition to the department office and the wending machines suitably energised. Without waiting too long, George began his recounting of the cosmological developments that followed in the wake of the birth of general relativity.

'Ah, where were we? Oh yes, we covered the genesis and the development of general relativity and three classical tests, didn't we?' resumed George. 'Soon after that in 1917, Alfie, Einstein took up the grandest application of his theory to the entire Universe. Let us take note of the status of observations on cosmological scales of those times. Even the size and shape of the Milky Way were unknown, let alone the true expanse of the Universe. No large-scale changes in the sky on truly universal scales had been observed. Is it surprising then that Einstein assumed that the Universe was static? Nevertheless, he realized that his own field equations of general relativity would not allow such a solution. The Universe had to be either expanding or contracting. So what to do? Einstein decided to alter his equations by a teeny bit by adding what is known as the *cosmological term*, characterized by the *cosmological constant* denoted by the Greek letter *Lambda*. With this alteration, Einstein was able to obtain a static cosmological solution. But things changed fast.'

'What happened next? This sounds like a fast-paced suspense story,' I observed.

'Indeed. The next investigator to arrive on the scene around the same time—I don't want to say the scene of the crime, we are talking of cosmology and not crime, you know—was Aleksander Aleksandrovich Friedmann of Russia.'

'Was he an observational astronomer or theoretical cosmologist or what?' I asked.

'He belonged to neither of the first two categories. You could say he happened to be a *what* on your list,' George chuckled. 'Friedmann had been trained in atmospheric science and had a degree in meteorology. In 1917, when Einstein was busy adding his cosmological term to accommodate a static Universe, Friedmann was heading a factory that manufactured aviation instruments. How do you like that?'

'Sorry to keep repeating the fact—but another accidental astronomer added to my list,' I commented. 'But no match to the mule driver!'

© Springer International Publishing Switzerland 2015
C.V. Vishveshwara, *Universe Unveiled*, Astronomers' Universe,
DOI 10.1007/978-3-319-08213-4_15

'By 1919, when general relativity had become big news, Friedmann had moved to Saint Petersburg and held a professorial position. He taught himself the general theory of relativity and turned his attention to its application to cosmology. Friedmann went along with the equations that dictated a dynamic Universe and derived the expanding universal model. And, of course, he was completely vindicated by the actual observational discovery of the universal expansion by Hubble and Humason.'

'What was Einstein's reaction to all this?' I asked the obvious question.

'Well, when Friedmann published his result in 1922, Einstein was reluctant to accept it, claiming that Friedmann had made a mathematical mistake,' answered George. 'It so happened that the mathematical error had been committed by Einstein himself. And in the following year, he gracefully conceded the fact and agreed that Friedmann had a valid cosmological model.'

'And what about the cosmological term Einstein had added to his equations?' I wanted to know.

'Ah, it is a very well-known story. Possibly you have heard it too,' said George. 'When the expanding Universe became a firmly established fact, it seems that Einstein dubbed his introducing the cosmological term *the biggest blunder of my life*! Often, it is pointed out that had Einstein allowed his own original, unsullied equations without the cosmological term to guide him, he could have predicted the existence of an expanding Universe. What a missed opportunity, they shake their heads. Why should one expect that amazing man to be right all the time? Shouldn't one take heart from his mistakes? We make mistakes too, don't we? Well, let us see what Infeld has to say in this matter.'

George picked up the grey volume again and read from Infeld's article. 'Remember, writing in 1949, Infeld is referring to the 1917 paper of Einstein in which Einstein had invoked the cosmological term to generate a static model of the Universe. Here we go.

Although it is difficult to exaggerate the importance of this paper, . . . Einstein's original ideas, as viewed from the perspective of our present day, are antiquated if not even wrong. I believe Einstein would be the first to admit it.

Yet the appearance of this paper is of great importance in the history of theoretical physics. Indeed, it is one more instance showing how a wrong solution of a fundamental problem may be incomparably more important than a correct solution of a trivial, uninteresting problem.

There you are. How perceptive and insightful are Infeld's words! But, my dear Alfie, was the cosmological term a big blunder? Especially, *as viewed from the perspective of our present day*, not Infeld's but our own?'

George paused and gave one of his mysterious smiles. 'The answer, my boy, is for me to know and for you to guess.'

'Come on, George, don't play this mystery game to torture me,' I protested vehemently.

'All in good time, Alfie. After all, the old fox may end up having the last laugh, as we shall see. Let us now get back to some important considerations regarding cosmological models.'

What could I do? I have learnt to be patient after all these years of interaction with George. The waiting has invariably proved to be worthwhile.

'All right, I shall wait,' I said pretending to be somewhat annoyed. 'George, how exactly does the curved spacetime picture or formalism work in the case of the Universe?'

'Excellent question, Alfie,' approved George. 'That is exactly what we are going to discuss next. In order to visualize spacetime, it is customary to think of a two-dimensional *rubber-sheet model.*'

'I remember what you once told me, George. Quite insightful and charming you know,' I recalled. 'You said: *Although we live in a four-dimensional world, our awareness is limited to three dimensions, our imagination to two, and our viewpoints to one.*'

'Did I say that? I don't remember. I must jot down my own quotations,' George seemed to be quite pleased with what I had conveyed to him. 'Well, in order to visualize what goes on in the real curved spacetime, one thinks of what is popularly known as the two-dimensional rubber-sheet model. Take a rubber sheet or a trampoline.'

Before George could continue, I remembered the trampoline in the amusement park I had visited with Ptolemy. It had been so nice to watch the children jumping up and down, especially the toddler sitting in the dimple his weight had created in the trampoline and making a ball either roll down or circle round the dimple.

'Alfie, wake up. Don't get into one of your mysterious reveries,' George admonished me. 'Let me repeat. Take a rubber sheet or a trampoline. When there is no mass in it, it is flat. Place a heavy ball on it, then the ball creates a depression around itself, curving the rubber sheet. Little marbles will roll towards the big ball. A marble, shot normal to the line joining it to the ball, will circle around it like a planet around the Sun. Thus all the gravitational effects can be simulated by the rubber sheet. You can even imagine how light must bend as it moves in this curved two-dimensional space. Now, draw lines on the sheet to make a co-ordinate grid. It could be a Cartesian coordinate system with the lines perpendicular to one another if the rubber sheet is flat. On the other hand, if the sheet is curved, a curvilinear coordinate grid will have to be used, as in the case of the surface of the sphere. Every point on the sheet can be identified by giving its coordinates. Are you with me?'

I nodded to indicate that I was following every word of his explanation.

'Good, now take such a sheet that is curved in general,' George continued. 'Each point on it can be identified by its coordinates, right? If the sheet remains unchanged, not only will the coordinates of the points remain the same, but also the distance between any two points will always be constant. This situation corresponds to Einstein's static Universe. Is that clear?'

'As clear as the crystal spheres of Aristotle,' I smiled.

'Impudent fellow,' George smiled too. 'Now here comes the interesting part. Let the rubber sheet start stretching. The coordinates of every point will remain the same, but the distance between any two of them will keep increasing with time. We now have a two-dimensional model for the expanding Universe as discovered by Friedmann. Got it?'

'Yes, I understand it all,' I confirmed. 'You have been talking about only points, their coordinates, and distances between them on your rubber-sheet model of the Universe. Where did all the stars and the galaxies go?'

'Aha, you noticed the subtle point, didn't you?' George nodded his appreciation. 'Well, that is because Einstein's static model and Friedmann's expanding one were both mathematical constructs. The task of injecting some real physics into the universal models was left to the Belgian Roman Catholic priest, Monseigneur Georges Henri Joseph Édouard Lemaître.'

'In the beginning was the Canon, in the end the Priest,' I remarked.

'The Canon and the cannon he fired will always stay with us, Alfie. We shall come to that shortly. But, let us talk about the priest, who made outstanding fundamental work in cosmology,' George said. 'Lemaître had a rather unusual career, you know. Let me tell you about it briefly. To begin with, he attended a Jesuit secondary school and then he joined the Catholic University in Louvain to study civil engineering. In 1914, he interrupted his studies to serve as an artillery officer in the Belgian army during the First World War, winning the Belgian War Cross with Palms, Alfie. Somewhat like Hubble, I must say. Returning to the University, he received his doctorate in mathematics in 1920. Once again, he stayed away from academics, entering a seminary in Brussels. After a while, he was ordained a priest in 1923. He pursued his studies thereafter at the University of Cambridge, where he was initiated into modern cosmology by no less a person than Arthur Eddington. That is not all. Then he worked with Harlow Shapley at the Harvard College Observatory and spent some time at the Massachusetts Institute of Technology. And finally he returned to Louvain University in 1927, where he worked for the rest of his life until 1966.'

'Quite a career,' I admitted. 'What were his contributions to cosmology?'

'As I told you, theoretical cosmology had been confined essentially to mathematical modelling until Lemaître appeared on the scene,' George explained. 'It was Lemaître who combined the mathematical model with physics and proposed an astronomically realistic expanding Universe. Not only that, he was the first one to derive the Hubble Law, which he published in 1927, two years before Hubble's article appeared. Unfortunately, his paper did not receive the attention it deserved. Now, here comes his novel idea that has now become commonplace. We know that the galaxies are rushing away or, equivalently, the Universe is expanding. If we can go back in time, we would find the galaxies closer to one another than they are now, right? If we continue this backward time travel, galaxies too would come closer and closer to one another. Finally, we would reach a point when all the galaxies, all the matter and radiation in the Universe, would be together. The universal expansion must have started from such a state. Lemaître called this state as that of the *primeval atom*. He even proposed that the Universe had a *fireworks beginning* for its expansion.'

'Ah, that is the Big Bang! Everyone knows about it nowadays.'

'Yes indeed. Once again, unfortunately, Lemaître's work was not properly recognized as it ought to have been during his active career,' reflected George. 'But he will always be remembered as a pioneer of physical cosmology.'

'George, there are some questions that are commonly asked. Maybe you can answer them for me at least,' I ventured.

'Of course, Alfie. Fire away,' invited George.

'First of all, we know that galaxies are flying away with phenomenal speeds as observed from the Earth. If they are indeed rushing away from *us*, are *we* at the centre of the Universe? Does the Universe have a centre?'

'Ah, you have raised one of the central questions of cosmology, no pun intended,' George raised his forefinger as if to illustrate the word *raise*! 'That brings back the canon and his canonical principle, pun intended. In fact, it is known as the *Copernican Principle*. Look at the solar system. There is nothing special about the Earth or any other planet in particular. They are all the same. Consider the Milky Way, our galaxy. There is nothing special about our Sun or any other star. They are all the same. So, one assumes that there is nothing special about our galaxy either. All galaxies are the same. This universal sameness is tantamount to the Copernican Principle, a generalisation starting with the Copernican model of the solar system. To reiterate, on large galactic scales, the Universe possesses similar properties, or the Universe is *homogeneous*. There is one other property of the Universe that is important. Turn your telescope in whichever direction you like, you will find the same large-scale structure, be it the density of galaxies, or the number of radio sources, and so on. This is known as *isotropy*. Of course, isotropy is consistent with homogeneity. We shall see that the isotropy of the Universe raises some important questions in cosmology.'

George took a short breather before proceeding with his further explanation.

'Hubble and Humason discovered that the galaxies were rushing away from us by measuring their redshifts, didn't they?' resumed George. 'First, let us see the reason for this *cosmological redshift*. It is quite simple. Suppose some source in a far off galaxy emits light, which is in the form of a wave. As the wave travels towards us, the Universe keeps expanding. As a result, the wave that is embedded in the spacetime gets continuously stretched out. Therefore its wavelength increases and light turns redder. There you are, you have your cosmological redshift. Simple enough, isn't it? As an approximation, the formula for this redshift goes over to Doppler shift. That is what the H-H pair used in their observations. However, later on with advanced measurements the full formula for the cosmological redshift became indispendable.'

'All right, what comes next?' I asked with increasing interest.

'Let us move on then,' proceeded George. 'According to the Copernican Principle, one must find this recession of galaxies irrespective of where the observation is made from, may be from another galaxy. Perhaps by another H-H pair living in Andromeda, who knows? There is no centre from which the Universe is expanding. The common two-dimensional model to illustrate this basic phenomenon is a balloon that is being inflated. Mark points all over the balloon with your pen to represent the galaxies. Be careful, don't puncture it. Now when you blow air into it, the dots will move away from one another, not from any central point on the balloon.'

I was careful enough not get caught in another reverie revolving around Ptolemy's amusement park, where I had seen George's balloon-analogy happening.

'A three-dimensional analogy is that of a raisin cake, the raisins being the galaxies. As the cake rises in the oven, the raisins move away uniformly from one another,' George added.

'Although I love raisin cakes, let me get back to your balloon,' I said. 'I can easily visualize the balloon expanding into the surrounding space. How about the Universe? What is it expanding *into*?'

'Aha, that is an oft-repeated no-no question,' George shook his head. 'There is nothing for the Universe to expand into. The three-dimensional space of the Universe is all there is. Think about some cosmologist-ants living on the expanding balloon. Unlike us, they are totally unaware of the third dimension. They will just say that their two-dimensional Universe is expanding—into what is a no-no question for them. So, don't go mad, or madder than you are, trying to imagine a situation that does not—and cannot—exist.'

'I recall what J. B. S. Haldane said about the Universe,' I began.

'For a change I know what he said: The Universe is not only stranger than we imagine, but stranger than we *can* imagine! Beautifully put,' George completed the quotation.

'Agreed,' I said. Some people have the talent for expressing profound thoughts in the most telling ways. Haldane was one of them, I suppose. 'Is there more to the homogeneity of the Universe than leading to the fact that there is no special cosmological centre?'

'Yes, much more, and you anticipated what I was going to say next, as you often do,' George went on to the next phase of the exciting story of the Universe. 'The two properties, namely isotropy and homogeneity of the Universe turned out to be powerful tools in the hands of two mathematicians, the American Howard Percy Robertson and Arthur Geoffrey Walker of England. Just on the basis of these two fundamental properties, they showed that there were only three kinds of spatial geometries the Universe could possess. Here, let me show you the two-dimensional versions of these geometries. First you have the *elliptical space*, the surface of a sphere being the important special example. Then comes the *Euclidean space*, analogous to ordinary plane. And finally there is the *hyperbolic space*, saddle shaped or like a valley between two mountains. Forget not, Alfie boy, that the actual space is three-dimensional that cannot be, must not be, visualized. Incidentally cosmological models together bear the generic name in honour of the pioneers in the field: *Friedmann-Lemaître-Robertson-Walker Universe* of *FLRW Universe* for short.'

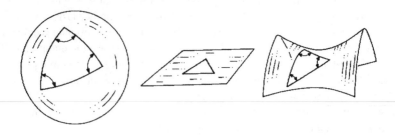

'I realize that reducing the possible cosmological spaces into just three categories is a tremendous feat of simplification,' I was truly impressed, even if the mathematics involved was way beyond me. 'But tell me, which of the three spaces actually corresponds to the Universe we live in.'

'Well that depends on the matter content of our Universe,' answered George. 'Let me explain. Take the average density of the Universe. You can estimate it if you know how many galaxies happen to be there in a given volume. This volume is obviously huge since we are talking about galactic scales. There is a critical cosmic density that generates the Euclidean space. If the actual density of the Universe is more than the critical density, you get the elliptic space. And less than the critical density, then the Universe would be riding the saddle—the three-dimensional cosmic space would assume the shape of the hyperbolic model. Got it?'

'Got it all,' I replied.

'All right then. Now, there is an extremely important correlation between the geometry and the evolution of the Universe, you know.'

'What is that?' The whole cosmic story was getting curioser and curioser as Alice might have put it.

'Both the plane and the saddle are infinite in extent and they keep expanding forever,' George said. 'But they do slow down at different rates. As the Universe expands relentlessly, the stars will die out, the temperature will keep diminishing. And the Universe will die out a cold death infinitely slowly.'

'What a tragedy!' I exclaimed *'This is the way the world ends, not with a bang but a whimper.'*

'Aha, I know that famous quote from T. S. Eliot,' said George triumphantly. 'Let me tell you how the Universe can end with a bang. Let us take the third alternative, the spherical analogue. It expands, slowing down all the time. Again the Universe will cool down with the stars dying out. Then it comes to a halt, and starts collapsing. For a while the Universe will be like what it is now. Then the galaxies come together and merge. Stars will collide and explode. Finally the Universe will end in a *Big Crunch* with all the matter and radiation coalescing together. Then the Universe will explode with a Big Bang and the whole process will repeat again and again. You have now an *oscillating* or a *cyclic Universe.'*

'Well, you have at least a chance of reincarnation in a cyclic Universe. That is some comfort,' I commented. 'By the way, how big is this critical density?'

'You must say how small, Alfie,' George replied. 'It is of the order of 10^{-26} kilograms per cubic metre, an amazingly small number. That translates into, say, 6 hydrogen atoms per cubic metre.'

'Good heavens, such a minute number would decide the spatial curvature of our entire Universe?' This was incredible. 'How do you find out which of the three kinds of the Universe we are living in? Is it by measuring the average density of the Universe and comparing it with the critical density?'

Questions came one after another in a natural sequence.

'That is one way. It is not easy and there may be observational fine-tuning among the three alternatives, you know,' answered George. 'Another way is to find out the rate at which the universal expansion is slowing down. The three models distinctly differ in their slowing down rates.'

'Can the Hubble Law give the answer to this question, George?'

'Yes and no,' smiled George. 'I shall explain those *yes and no* soon, fear not, Alfie. But first, let us take a closer look at that law. You see, Hubble and Humason's measurements were limited to galaxies that are comparatively close to us, even if millions of light years seem to be quite large. There is another complication with the velocities one measures for these galaxies. In general, galaxies have their own peculiar motions. For instance, the one in Andromeda is in fact moving towards us in addition to its cosmological recession. So, if the recessional velocity of a galaxy is not too high, its own peculiar velocity would be comparatively appreciable. As a result, the velocity measured may not follow the Hubble Law exactly. Take a look at this figure in the manuscript. It is the original Hubble plot showing velocities of galaxies and their corresponding distances. Do you see how those points are scattered? But look at the straight line drawn through them by Hubble showing his linear law.'

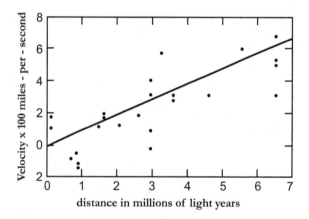

distance in millions of light years

Yes, the points were pretty much scattered. But there was a bold straight line running in the middle of those dots showing the recessional velocities of galaxies as a function of their distances from us.

'We wouldn't allow any of our students to do any such thing with their experimental values and proclaim they have a linear law,' observed George. 'But Hubble was Hubble and he did it! Was it his intuition, conviction, or sheer audacity that drove him to draw that straight line, the simple straight line that governed the universal expansion? Who knows? Of course, as techniques improved, measurements went way beyond Hubble's observations on distances and velocities of galaxies. And the values that were obtained did lie along a straight line and showed that the velocities were directly proportional to the distances.'

'All right, the modern plot is a straight line,' I said. 'Then how can you tell which of the three cosmic models corresponds to the one we happen to inhabit?'

'Aha, we come to the yes-or-no business on hand,' George held up his forefinger. 'Theoretically, one can compute how the Hubble Law graph changes with the three cosmologies. As I told you before, they give distinctly different rates of slowing down. On the other hand, even the modern Hubble plots based on observations do not shed any light on this. Why? Because, they are essentially just straight lines. In order to detect the actual slowing down, one has to make measurements on galaxies at enormously large distances. That also means galaxies far back in time. So, by comparing galactic speeds of way back in time with those that are nearer, or equivalently closer to us in time as well, one could determine the slowing-down rate. From this, one can deduce the curvature of the Universe. That is a tall order, Alfie.'

'So, no one knows what sort of Universe we live in, is that it?' I commented. 'Ignorance is bliss. Ignorance on a cosmological scale is ecstasy.'

'Ignorance comes in different shades, Alfie,' responded George. 'To get an idea of the slowing down of the Universe, or the *deceleration* of its expansion, one has to make observations on galaxies that are beyond 10 billion light-years. At such distances, those galaxies are so faint that uncertainties hamper the distance and redshift measurements. Yet, cosmological observations do indicate that we may be living in a spatially flat Universe.'

After a moment he added, his eyes shining mischievously, 'Well, next time we meet, we shall talk about some dark secrets of the cosmos.'

I wondered what those dark secrets were. As I remarked earlier, this whole business with the Universe read like a suspense story. One should never turn to the last page to find out the ending. Then all the charm of reading the book is lost. Of course, we know that the cosmic saga has no ending. It is an ever continuing quest.

'George, you explained how the three possible types of the Universe differ from one another in their slowing down rates,' I said. 'What about their origin? I mean do they all start expanding in the same way?'

'Of course, all the three Universes begin their lives explosively with a Big Bang.'

'All right, but how do you pin down that moment of creation of the Universe?' I asked. 'In other words, how do you determine the age of the Universe?'

'Good question, Alfie,' nodded George. 'For that, we go back to the Hubble Law once again.'

'Even if we try to shake off that Law, it follows us everywhere like a tail in a spy novel,' I remarked.

'Why don't we work out the age of the Universe with some simple arithmetic, Alfie?' invited George. 'Let us make a simple assumption that the galaxies have been flying away from one another with constant velocities ever since the Big Bang. Take for instance our own Milky Way and any other galaxy. Then what would be the relation between the distance between them and the age of the Universe? Remember the age of the Universe is nothing but the time taken for the galaxy to reach its position now starting from the Big Bang.'

'That is quite simple, George,' I answered. 'The distance would be the product of the recessional velocity and the age of the Universe.'

'Excellent. But then, the Hubble Law tells us that velocity is proportional to the distance. The constant of proportionality is called, what else, the *Hubble Constant*, H_0. Or equivalently, Hubble tells us that the distance between the galaxies is given by the velocity divided by his constant. But, just now you told me that the distance is nothing but velocity times the age of the Universe. Compare the two equations and, my dear Alfie, and tell me what would be the age of the Universe in terms of the Hubble constant.'

'Let me see,' I needed no more than a moment to come up with the answer. *'The age of the Universe T is just the reciprocal of the Hubble Constant H_0!'*

'Splendid. That is simple enough, right?'

'Absolutely simple, George,' I responded. 'Now, two questions present themselves,' I said. 'First, why is there a subscript zero adorning the Hubble Constant?'

'That is because the constant is supposed to have been measured *now*, at the present moment,' answered George. 'The constant need not be a real constant, but a varying quantity and H_0 is its present value. Also remember that the velocities are not constant either since the galaxies are supposed to be slowing down. What we have found out gives the simplest approximation for the age of the Universe as the reciprocal of the Hubble constant. But there are, of course, complications on both observational and theoretical fronts. Don't strain your brain with these things, they might addle it even more than it is. All right, what is your second question?'

'What *is* the age of the Universe? Did Hubble himself find it out too? And what is the modern value?'

'I protest. You have asked three questions, not one,' laughed George. 'Hubble did estimate the age of the Universe from his observations. As we saw, the constant of proportionality he had determined was quite approximate and so was the age of the Universe he obtained. He estimated that the Universe came into existence some two billion years ago. This was quite an embarrassing situation. You know why? Because, by measuring the relative abundances of radioactive elements in the Earth's crust, Lord Rutherford could fix the age of the Earth to be about three billion years. Furthermore, the oldest bacterial fossils have been found to be as old as three and a half billion years. Of course, we know that our Earth is around five billion years old. How could the Earth be older than the Universe? How could an egg be older than the hen that laid it? This was indeed a cosmological crisis!'

'Well, Hubble's estimate of the age of the Universe was much better than that of Archbishop Ussher, wasn't it? Using the chronology found in the Old Testament, he fixed the moment of creation as nightfall preceding 23rd of October 4004 BC,' I observed. 'Of course the Archbishop's finding was far more precise than Hubble's. What do you think?'

'These are high matters I don't want to think about, Alfie,' said George. 'Coming down to Earth, as more accurate measurements were made, the age of the Universe was considered to lie somewhere between ten and twenty billion years. I leave the question of the most modern determination of the age in thrilling suspense till our next meeting.'

'Thrilling suspense indeed, you mean excruciating anticipation,' I grimaced. 'All this is nice. But is there any evidence for the beginning of the Universe in a Big Bang?'

'Yes, there is. The echo of the mighty explosion fills the entire space,' George swept his hand to indicate all of our surroundings. 'Of course that echo is not in the form of sound, but happens to be electromagnetic radiation. It is called the *Cosmic Microwave Background Radiation* or *CMB* for short. Let us first consider the genesis of this background radiation. Right after the Big Bang, the Universe was at an unimaginable temperature of some thousand trillion degrees. This *primordial fireball* consisted of nothing but radiation. Only during a long period of evolution, stretching over billions of years, atoms were created, stars, galaxies and planets formed, and finally life including humans emerged. What was the radiation doing in all this time while the Universe was expanding relentlessly? Think of radiation filling an enclosure that keeps expanding.'

'Aha, we talked about the *blackbody radiation* in the context of the stars, didn't we? The radiation cools down progressively,' I recalled.

'You are absolutely right,' nodded George. 'The phenomenally hot radiation that existed at the beginning of the Universe keeps cooling down. This radiation now permeates the entire space, the whole Universe, lurking in the background as CMB. Its discovery is a strange story in itself.'

George began recounting the story behind the discovery of CMB as he had termed it.

'The year was 1964 and the place Holmdel, New Jersey, USA,' George recalled. 'Two astronomers, Arno Penzias and Robert Wilson, working at the Bell Telephone Laboratories' Crawford Hill Laboratory, had built a large horn-shaped antenna. Their purpose was to observe and study the radiation emanating from a supernova remnant some 10,000 light years away. Their work was hampered by a constant noise that was proving to be an annoying nuisance. This unwanted hum seemed to be coming from every direction in the sky. The two astronomers couldn't figure out the source of this continuous signal. They even cleaned up the white coating two pigeons had deposited within the big antenna and transported the pigeons far away, so that they would not return to the nest they had built within the antenna.'

'Poor pigeons, they had been the true citizens of the Universe. Someone ought to have called the SPCA,' I offered my opinion on the important matter.

'Then CMB might not have been detected, Alfie,' chuckled George. 'To continue the story, the two astronomers realized that they had indeed detected some kind of genuine radiation coming from all directions. But, they had no clue as to its origin. It so happened that a group working at Princeton, almost right in the neighbourhood, not only knew what that radiation was, but were looking for it with their own equipment. Eventually, the nature of the crucial discovery became known to the concerned parties as well as to the rest of the scientific community.'

'A happy ending indeed,' I said. 'Tell me more about this CMB.'

'As we saw, the background radiation had originated from the time of the Big Bang,' continued George. 'As it had been in equilibrium within the primordial fireball, it had the typical black body spectrum. Remember, I showed you the blackbody-radiation plots that displayed the peaks of the curves and the corresponding wavelengths for different temperatures? So, if you assumed that the newly discovered radiation was indeed of the blackbody type, then its temperature could be determined. It turned out to be around three degree Kelvin agreeing with theoretical calculations.'

George paused to give me time to soak in everything he had been describing.

'You know Alfie, often we theoreticians are ahead of our observational colleagues,' George said not without a tinge of pride. 'The possible existence of the radiation, remnant from the Big Bang, had been predicted by George Gamow, although his estimate of its temperature was somewhat higher than the actual value.'

'Ah, George Gamow the author of *Tompkins in Wonderland*,' I said. I had read and thoroughly enjoyed Gamow's writings.

'Gamow was a highly talented, witty, first-rate scientist who made important contributions to nuclear physics and cosmology. Well, more accurate measurements made by other astronomers at different wavelengths showed a perfect fit of observational data to the black-body spectrum corresponding to 2.73 Kelvin. Minor deviations from the black-body spectrum, genuine and not due to errors, are considered to be the signature of the emergence of structures in the early Universe leading to the formation of the galaxies. That is a whole different story, Alfie. I think we have talked enough about CMB.'

George pondered for a long moment.

'Another thing about the distant galaxies,' reflected George. 'Don't forget light takes millions and billions of years to travel from those galaxies to reach us. What does that mean? It means that we are seeing a composite of snapshots of the cosmos taken at different times, millions and billions of years apart. Think about that.'

Absolutely fascinating, I reflected. Everything about the Universe seemed to be strange. I remembered what Francis Bacon had to say about strangeness: *There is no excellent beauty that hath not some strangeness in the proportion.* By that token, the Universe did possess *excellent beauty*.

George pulled out a watch from an inside pocket of his jacket.

'Speaking of cosmic time, we have spent enough of it on a human scale, you know,' observed George looking at his watch.

I could see that the watch had been beautifully crafted and shone with an unusually soft sheen.

'Aha, you noticed my recent acquisition,' George said happily. 'It is a farewell gift from my colleagues at the university I was visiting. They say it is made of seven different metals with a dash of iron from a meteorite added.'

'I have heard that, once upon a time, swords of monarchs and sultans used to contain such meteorite metals to make the owner invincible,' I said. 'Maybe your watch will make you invincible in your fights with the faculty.'

'I hope so,' laughed George. 'Anyway, we have spent enough time today, Alfie. Next time, we shall move on to some really modern stuff including the surprise findings about the universal expansion.'

'May I ask what surprise awaits me then?' I asked knowing fully well what George was going to say.

'Patience, my boy, all in good time,' George's answer was the usual one, not unexpected although disappointing. 'Let us meet again here in my office. Then we can walk over to Bruno's for his surprise dinner. Surprise will be the watchword of the day.'

I stuffed my bag with the books George was lending me, said goodbye to him, and was on my way home, happily whistling.

Chapter 16
The Cosmic Yin and Yang

I found George standing in his office, regarding with concentration two paintings propped up against the wall. Of course, they were reproductions. But nowadays with modern technology these reproductions are exceptionally true to the originals, bringing out their subtle nuances. I could easily identify them: *The Starry Night*, the famous painting by Van Gogh and *Philosophe en Lisant*, *Philosopher Reading*, by Rembrandt, exquisite works by the two masters, poles apart in their styles yet so effective.

'My young colleagues in physics have launched a beautification programme in collaboration with the department of fine arts, you know,' said George. 'They want to put up some nice works of art along the corridor outside. And they have allowed us to select the ones we want across our own offices. And I have chosen these. What do you think?'

'Excellent choice, George,' I endorsed.

'Let us sit down for a while, Alfie, and enjoy those beautiful creations,' offered George.

George continued to gaze upon the paintings. I could see that he was pondering about other matters while admiring the two paintings in front of us.

'Look at Van Gogh's painting, Alfie,' George broke the silence. 'If you go too close to it, all you will see is nothing but the brushstrokes, lots of colourful, interesting details. Only when you stand back, would you see the whole picture with all the glowing stars and the swirling turbulence. It is the same with the Universe, isn't it Alfie? There can be plenty of observational details, but only when you consider them from a distant perspective does the true aspect of the Universe emerge.'

'What about Rembrandt?' I queried.

'Ah, Rembrandt, you don't even see the brush strokes,' commented George. 'The whole canvas is filled with smooth, flowing vibrant colours against the rich dark background. That is what happens when one builds cosmological models. Galaxies are considered to make up the cosmic fluid described by their density, pressure, and velocity. It is just like a gas jar filled with gas—you don't consider the individual atoms and molecules, but just concentrate on the macroscopic properties. And the cosmological models constructed like this yield so much information.'

George paused allowing me to absorb the analogy. I found it not only lovely, but also very appropriate and effective.

'All right, where do we go from whatever we discussed last time?' George began. 'If I remember right, we talked about Hubble's law, which gives a linear relation between the galactic distances and their recessional velocities through the Hubble constant H_0. In order to find out how the galaxies are slowing down, or decelerating, you have to compare their present velocities with their values way back in the past. How do you find out their velocities in the past? You find the velocities of galaxies that are far, far away. Since light from them takes a long time to reach us, we see them as they were in the remote past. In short, one has to make observations on galaxies that are as distant as possible, billions of light years from us. Is that clear, Alfie?'

'Yes, indeed,' I told George.

'As we have seen earlier, both Shapley and Hubble made use of Cepheid variables as their standard candles in respectively charting out the Milky Way and measuring the Universe,' George went on. 'Cepheid variables were quite adequate for the range of distances involved in their measurements. Although they are hundreds and thousands of times brighter than the Sun, they appear too faint if you want to make observations much farther. So, you need much brighter standard sources than the Cepheid variables.'

'Are there such stars with very high luminosities then?'

'Not ordinary or variable stars, but exploding ones,' said George. 'Remember, we considered Type Ia supernovae sometime ago? They are white dwarfs in binaries. When their companion keeps dumping gas on them, their masses keep increasing. When the Chandrasekhar limit is reached, they explode. The luminosities of these white dwarfs bursting out as Ia supernovae are the same since the conditions leading to the explosions are the same. They can therefore serve as standard candles.'

'How bright are they?'

'Well, they are some 100,000 times brighter than the Cepheid variables or around four billion times brighter than the Sun. So, you see, they can act as extraordinarily powerful standard sources for far-off cosmological measurements. But, there are problems.'

'There are always problems, I know. What are they?'

'Well, in the case of the Cepheids, they are always there waiting patiently to be observed. On the other hand, supernova explosions are naturally sporadic and unpredictable. You never know beforehand when they will occur. And where they will occur. You have to keep watching all the galaxies waiting for some fattening white dwarf to explode. But observational astronomers are quite diligent and smart enough to overcome these difficulties, so that these supernovae have indeed turned out to be reliable, bright standard candles.'

I realized that there must have been intense efforts and teamwork behind these findings. But, George was giving me a sweeping panoramic view of the Universe and details were out of place in his narration.

'While there are problems, there are compensations too,' reflected George. 'There have been tremendous advances in technology, as you very well know. The good old photographic plates of Hubble's time have been replaced by electronic detectors combined with computers. And then, one can make observations unhindered by the atmosphere with spacecraft carrying sophisticated telescopes like the *Hubble Space Telescope* or *HST*. What is the end result? Man's reach into the realm of the galaxies has been extended as never imagined before.'

'I remember Robert Browning's quotation, George,' I recalled. '*Ah, but a man's reach should exceed his grasp, or what's a heaven for?* It is so apt in this context, when astronomers keep aiming to extend the range of their observations all the time.'

'I fully agree, Alfie. That quotation happens to be one of my favourites too, you know,' responded George. 'Well, we shall talk about the use of supernovae for cosmic explorations later on. For the moment, let us get back to the determination of Hubble's constant and the age of the Universe. Suppose you repeat Hubble's measurements on the Cepheid variables but now from the Hubble Space Telescope—how nice that the telescope too is Hubble. Now, you can look out to galaxies 25 times farther away than those that were accessible to him originally. There are some adjustments to be made in the data, but let us not worry about these details. The age of the Universe you get this way happens to be 13.6 billion years.'

'So accurate?' This was surprising.

'Yes indeed, observational methods and theoretical calculations keep getting more and more sophisticated, and things get better and better' said George. 'However, you can take 13.6 billion years for the age of the Universe to be quite accurate.

'To quote good old Browning again, *Grow old along with me, the best is yet to be!*' I repeated the well-known line. 'I shall remember the age of the Universe as the lucky number 13 billion years!'

I must have told you before, maybe more than once, that George has the habit of leaning back in his chair, eyes closed, for a few moments before starting upon a new topic. That is what he did now.

'Theoretical models of the Universe, like those with the three different spatial geometries, are like empty houses, Alfie,' George began after his short spell of rest. 'A house becomes a home with its own characteristics only when you fill it with suitable objects of your choice. It is the same with the Universe. Its properties including its evolution depend upon what it contains.'

'You told me that the geometry of the Universe depends on the density of its contents,' I recalled what I had learnt. 'Below the critical density, the cosmological space is saddle shaped, above that density it is like a sphere, and if the actual density is the same as the critical value, then the Universe assumes a flat geometry. And the nature of the cosmic expansion is also decided by the geometry and therefore on the density.'

'Excellent, you have it in a nutshell,' approved George. 'And how does one measure the density of the Universe?'

'Oh, in principle, by adding up the mass of all the matter in a given huge volume—the sparkling stars, the glowing gases, and the shining galaxies. One can do that in different regions of space and get an estimate of the average density.'

'Splendid, again. I like your adjectives—sparkling, glowing, and shining. In other words, we are talking about the *luminous matter* in the Universe. For a long time, it was only the luminous matter that held the attention of the astronomers. But the picture changed, when it was realized that the Universe harboured matter that was invisible, the *dark matter*, as it has come to be known.'

Yes, I had heard that term, the dark matter. One reads about such things in the newspapers nowadays, even if one does not get the real picture. Don't ask me how one gets a picture at all if the matter is dark!

'Dark matter does not emit light or any other electromagnetic radiation like radio waves. So astronomers cannot see it through their telescopes.'

'Then how do you know it exists?'

'Through its gravitational effects, Alfie. Gravity rules the world,' said George almost triumphantly. 'Let me explain the principle involved in hypothesizing the existence of dark matter. Suppose our Sun did not shine. Still, you could deduce its existence from the orbital motion of the planets going around it, can't you? Suppose the matter contained in this dark Sun was spread out diffusely all over the solar system. The motion of any given planet depends upon the gravitational pull of the matter contained within its orbit. From this you can determine the total amount of the dark-Sun matter. Further, let us imagine that this invisible matter is mixed with some luminous matter as well. By studying the planetary motion, you find out the total mass involved. Deduct the visible portion, and, *voila*, you have not only detected the invisible component, but also determined its proportion. Have I made myself clear so far?'

'The answer to a redundant question is an emphatic yes,' I nodded vigorously.

'Fantastic—I am running out of my superlatives,' smiled George. 'Now, we move on from the hypothetical solar system filled with smeared out bright and dark Sun to our real Milky Way. Jan Oort, the famous Dutch astronomer, was the first one to point out the presence of dark matter in the Milky Way. One can compute the velocity of rotation of our galaxy as a function of the distance from the centre. But such a curve drawn from observations reveals that there ought to be more matter present than what is actually observed as luminous matter like stars and glowing gases. So this is the *dark matter*. As a matter of fact, the disk of our Milky Way seems to be cocooned within a large spherical halo of dark matter that extends beyond the visible part.'

As George and I had discussed, the expanding Universe alone was quite strange. But now, this dark matter was making it even stranger.

'That is not all, Alfie,' George went on. 'There was this brilliant Swiss astrophysicist, Fritz Zwicky who worked at the California Institute of Technology or Caltech. He did extraordinarily original work, both theoretical and observational, in many fields including supernovae and neutron stars. Around the same time as Oort,

Zwicky too thought of dark matter to explain the orbital velocities of galaxies within clusters. This was truly a cosmic jump in the dark-matter scenario. Since then, lot of research has been carried out on dark matter. This invisible entity seems to pervade the entire Universe giving rise to many effects. For instance, dark matter can manifest itself through *gravitational lensing.*'

'Gravity acts as a lens?'

'Yes, we saw how gravity bends light as per Einstein's general relativity, didn't we? Now suppose you have a far off bright source like a quasar. Let us have a galaxy between that quasar and ourselves. Then light from the quasar can be bent by the strong gravitational field of the galaxy coming to a focus at our telescope. Then we can see multiple images of the quasar. This has been observed. Now, replace the galaxy by the invisible dark matter. It will act as a lens, just like the galaxy, although you do not see it.'

'Ah, a dark lens. Through a glass darkly, as the Good Book says!'

'That is not all. Dark matter can play an important role in the structure formation in the early Universe leading to the birth of galaxies as well as their evolution. The imprint of dark matter on the cosmic microwave background, CMB, can be observed as small variations in different directions. These are *anisotropies* that may carry information on structures in the early Universe. So you see, Alfie, dark matter is of prime importance in cosmology. As a matter of fact, about 27 percent of the matter in the Universe happens to be dark. Isn't that something?'

'To say the least,' I agreed. 'But what is this dark matter made up of?'

'One of your million-dollar questions as usual,' George shrugged. 'The obvious candidates were regular astronomical objects that are invisible, such as black dwarfs, neutron stars, and of course, black holes. But observations showed that they could form only a small fraction of the dark matter in the Universe. As a result, hypothetical particles have been considered to account for the dark matter, such as Weakly Interacting Massive Particles—WIMPS for short, axions, neutralinos, and so on. Let us not worry about them'

'In short, nobody really knows,' I remarked.

'In short, nobody knows,' echoed George.

Acquisition of cosmic knowledge seemed to have progressed at such different rates over time. Once, as in the case of the passage from myths to natural laws, it took centuries to make real progress. But in modern times, it seems that advances have been made in extraordinarily rapid succession. Or perhaps, bursts of enlightenment are always followed by periods of quiet exploration, who knows?

George was patiently watching me with interest as I was quietly musing to myself.

'Are you ready for more dark at the end of the tunnel, Alfie?' George broke the silence.

'Yes, go ahead and reveal more of the dark secrets, George,' I answered.

'Well, after *dark matter* comes *dark energy*,' proceeded George. 'In 1998, one of the most unexpected observations was made, a real stunner. We saw how the supernovae of Type Ia could be used as powerful standard candles to observe really far off galaxies. We have often remarked that distance in space translates into distance in time, the past that is. So observations were being made on galaxies as they were way back in the past, billions of years ago. Furthermore, their velocities could be determined from their redshifts. If they were slowing down, as everyone expected, the velocities should have been diminishing with time.'

'All this is clear enough. But where is the stunner?'

'They found that the Universe was not slowing down at all, but accelerating,' announced George in his dramatic manner. 'No one had expected this. After all, gravity pulls you back as it does in the case of a stone thrown upwards. How can it accelerate the expansion of the Universe?'

'I wait for the answer to your rhetorical question, George.'

'The answer to my rhetorical question is *dark energy*! While gravity is attractive, dark energy happens to be repulsive in its action. The negative pressure generated by it accelerates the expansion of the Universe. Moreover, observations indicate that this dark energy is homogeneous and constant in time.'

'You have essentially described the action of dark energy,' I observed. 'But what is the root cause of this elusive entity?'

'Elusive entity, that is the right description of dark energy,' said George. 'Let me give you a clue. What are the two important characteristics of dark energy? It is repulsive and, mind you, it is constant. Have we come across anything like?'

'Let me think,' I cudgelled my brains. Enlightenment hit me as it did when the apple is supposed to have fallen on Newton's head. 'Don't tell me it is Einstein's infamous cosmological constant, Lambda!'

'I tell thee that it is indeed the Lambda. But correction, not infamous but famous!' said George. 'We discussed the standard Friedman-Lemaître-Robertson-Walker universal models, didn't we? One can see theoretically that the lambda term can accelerate such a Universe if it is expanding and slow it down if it is contracting. Einstein's biggest blunder is the prime candidate for the dark energy!'

'You told me that dark matter makes up some 27 percent of the mass-energy in the Universe,' I recalled. 'How much of dark energy is there in our Universe?'

'Another surprise, it is about 68 percent,' smiled George.

'Dark matter and dark energy together account for, let me see, 95 percent of the mass-energy content of the Universe, while their luminous counterpart constitutes a paltry 5 percent!' This was incredible. 'This means that the *Yin*, the darker side, predominates way more than the *Yang*, the lighter side, in the Universe. I wonder what kind of coat of arms Niels Bohr would have designed had he known this!'

'Your thoughts go off in crazy directions, don't they, Alfie?' laughed George. He laughs easily as you might have noticed.

'All right, Alfie, we have covered enough of the Universe so far,' said George. 'I would like to wrap up things with one or two details. We talked about the *isotropy* of the Universe, didn't we?'

'Oh yes, the Universe looks the same in every direction of observation,' I recalled. 'Even the cosmic microwave background is by and large isotropic too.'

'That is right,' acknowledged George. 'Even if you look at two regions of the sky on opposite sides, one hundred and eighty degrees apart, they have the same features. This universal isotropy cannot be explained by the standard model of the Universe we have been talking about. Those chunks far apart, as well as the others, haven't had a chance to interact with one another even with light or other kinds of electromagnetic radiation mediating between them.'

'Then how do you account for the isotropy of the cosmos?'

'Well, around 1970s and 1980s, a novel scenario for the evolution was proposed. It is known as *inflation*,' answered George. 'Let me briefly describe this inflation idea. The details are quite complicated. It was proposed that in the very early epoch the Universe expanded at a phenomenally rapid rate. Let me give you a rough idea of how fast this sudden expansion was. The size of the Universe is supposed to have blown up from below that of an elementary particle to the size of, say, a melon in an exceedingly minute fraction of a second. This process helped smooth out spacetime by making regions of space that were once close enough within the highly compact Universe to get mixed up with each other well. This way, the entire Universe would become homogeneous and isotropic as it is observed today. This is like kneading clay or dough thoroughly and then spreading it uniformly in every direction.'

'Sounds pretty remarkable,' I commented.

'Yes indeed,' agreed George. 'Inflation also solves the so-called *flatness problem*. We talked about the three generic shapes of cosmological space, namely those of a sphere, a saddle, and in between a flat plane. Our Universe seems to be essentially flat with the cosmic density being close to the critical value. The smoothening effect of inflation has also been shown to make the Universe flat. Another important feature of inflation is that it brings in quantum effects and particle physics into the fold of cosmology. Quantum fluctuations in the early Universe are supposed to have led to structure formation, which in turn generated galaxies. Obviously that is a very important aspect of the large-scale picture of the Universe today.'

At that point, both of us fell silent. I knew that George was tired. And I had plenty of information to digest.

'This much is enough for today, Alfie,' said George. 'Perhaps, during our dinner we will be able to discuss a bit of the most recent developments in cosmology. Of course, we won't dilute the pleasure Bruno has to offer.'

George got up and stretched.

'The Universe is not static and we shouldn't be either,' remarked George. 'Let us walk over to Bruno's, Alfie. I am sure his surprise dinner will be the grand finale to our cosmic journey.'

We left the physics department and strolled across the campus that had quieted down. Our walk towards Bruno's was a pleasure as our anticipation and appetite steadily built up.

Chapter 17
Food for the Future

I felt that there was something different this time, when George and I entered Bruno's. Usually, one of us would be already seated at our favourite table waiting eagerly for the other to arrive. This time we were both together. It felt a bit strange.

I was surprised to see that there was already someone sitting at our table. Instinctively, he turned around. I was elated to recognize the invader of our privacy. It was our good old friend, the young astronomer Mike Brown, flashing his ever-present boyish grin.

'Surprised, Alfie?' George too grinned. 'Yes, I invited Mike to join us for Bruno's special dinner. I wanted him to come down—how did you put it? Yes—from his mountain of meditation.'

Mike stood up as we strode towards our table. We happily shook hands.

'Hope we didn't keep you waiting too long, Mike,' said George.

'I wish you had, George,' replied Mike. 'I could have enjoyed another one of this delicious exotic drink,' Mike pointed to a tall glass in front of him. It contained a cloudy white liquid, which I could not recognize.

'What is it?' enquired George. 'Is it some kind of wine?'

'Search me, I don't know,' shrugged Mike. 'Definitely non-alcoholic. I asked the waiter. Either he didn't know or didn't want to tell me. Just smiled and left. Maybe Bruno will enlighten us.'

'Non-alcoholic! First, this guy Alfie turns me into a vegetarian. Now, Bruno wants me to be a teetotaller. What goes on here?' George spread out his hands in feigned helplessness.

One of the waiters brought two tall glasses of the cloudy white drink and placed them in front of George and me. He withdrew hastily before we could speak to him, looking at us from the corner of his eye with a hint of a smile on his lips.

I took a sip of the drink. It tasted smooth, gently sweet, soothing, and absolutely delicious. As he sipped, George's face brightened with pleasure.

'This is superb. I don't mind becoming a teetotaller after all,' said George and added quickly. 'Temporarily that is. But where on Earth is Bruno?'

© Springer International Publishing Switzerland 2015
C.V. Vishveshwara, *Universe Unveiled*, Astronomers' Universe,
DOI 10.1007/978-3-319-08213-4_17

As if in answer to George's query, Bruno appeared at the kitchen door, a serious expression on his face. I felt that he was leading someone else. But we could not see the person behind him, as Bruno's big figure covered our view.

Then things happened quickly. I could hear the tinkling laughter of the person following Bruno, taking me back a few years in time. Bruno moved away smiling. I watched George, gazing intently at what was happening in front of him. His expression went through a series of fast transformations—curiosity, incredulity, recognition, finally unbounded exuberant joy.

'Sunny, Sunny! My God, you gave me a heart attack,' George exclaimed holding out his arms.

Sunny, who had made a surprise appearance, ran into his arms. The two hugged each other for a long time. It was like a father greeting his daughter whom he had not seen in a long time.

'Surely you remember Sunny, don't you? Sunny is the short and sweet version of her longer name, Sunitha,' said George proudly. 'It was wonderful to have her working with me for her doctoral degree. And now she is a full-fledged faculty member at one of the top universities.'

Of course, I remembered Sunny and the other two graduate students of George. I had had a great time interacting with them.

Soon after this happy commotion had subsided, our attention turned to the young girl who had followed Sunny. She seemed to be in her early teens. She had lustrous skin, sharp features and her dark eyes shone brightly. She was carrying a guitar apparently with loving care.

'Meet my kid sister, Namratha,' announced Sunny. 'You may call her Nam to save time and energy.'

All three of us greeted Nam warmly.

'Well, Sunny, your sister seems to be fond of her guitar,' observed George.

'Yes sir, she is attached to her guitar or the guitar is attached to her. Both are the same according to relativity,' Sunny laughed. 'She used to be the lead-singer-guitarist in her group called *Nam's Nanosingers*. They even cut a disk, which became quite popular.'

Ah, I remembered the song with the title, *In the Darkness of the Rising Sun* about the atomic bomb that was dropped on Hiroshima. That song was extremely moving and I had included the lyric in the book *Einstein's Enigma*. So, this was Nam of that group! I told George and Mike about this. George, to whom I had quoted the text of the song, was quite impressed. Nam just smiled shyly.

'I am truly impressed, Nam,' acknowledged George. 'Sunny, your sister looks like she is still in high school, but she is already famous, what do you know.'

'Nam dear, why aren't you in school? Won't you be punished for being a truant?' Sunny shook Nam by her shoulders and laughed. She turned to us and said, 'Actually, Nam is well into her doctoral thesis in experimental biology. Nam, tell them about your research.'

'I am working on *prions*—proteins that have become delinquent—using yeast. The results of our research are expected to be important to humans too,' Nam was quite brief. 'My thesis will be titled, *East meets Yeast*.'

'Amazing,' said George admiringly. 'I know Sunny's father who is an astrophysicist like me. How come you chose biology as your subject, Nam?'

'That is because our mother is a biologist,' answered Nam. And then she added with a serious face, 'Sunny and I have divided our parents genetically, our father following Sunny and our mother following me. They happen to be good scientists, as they have inherited our genes.'

'Nam, you are impossible,' Sunny burst out laughing. We couldn't help joining her.

Bruno, who had disappeared for a while, came in carrying a tray with six glasses filled with the delicious drink we had tasted. He joined us at our table.

'At last, the mystery will be solved,' said George. 'What is this unearthly drink of yours, Bruno?'

'Ask, Sunny,' smiled Bruno. 'She got it specially for us.'

'It is in fact a very earthly drink,' explained Sunny. 'It is tender coconut water.'

Ah, my memory went back to my sojourn in India. Yes, I had tasted the soothing drink taken down straight from the top of a coconut tree. A daredevil had climbed up the tree and had brought down a whole bunch of tender coconuts. With his machete, he had trimmed the top and had made a whole in the middle. I had drunk the water that was inside using a straw he had kindly provided for me.

'I must go help Joe with cooking,' Bruno went back to the kitchen after he had finished his drink.

'All right, let us think of the Universe for a bit. It will build up our appetites,' said George. He thought for a moment and added, 'What about Nam? Won't she be bored?'

'If you are capable of explaining your Universe to a barmaid, you can explain it to a biologist,' said Nam dryly.

'Rutherford is supposed to have said the first half of that statement to Niels Bohr. How do you know that? Not many people have even heard about it,' George was quite surprised and so was I.

'It so happens that I have read *Einstein's Enigma,* where it is mentioned,' answered Nam. 'In any case, Sunny will clarify later what I don't understand. Maybe she will do a better job at it.'

'All right Namms, you win,' George was obviously quite happy with the girl. He turned to Mike and said, 'Mike, you have been quite silent, you know. What is new with the Universe?'

'I am a silent observer, George. I am enjoying myself,' replied Mike. 'What is new with the Universe? There is hot stuff coming out from the coldest part of the Earth.'

'See Alfie, what happens to people in your presence? Even Mike has become mysterious,' George unjustly admonished me.

'Not being mysterious, George, but dramatic!' Mike corrected George. 'Recently, there has emerged extraordinary data from Antarctica about the early Universe. So you see, it is indeed hot stuff on the Universe coming from the coldest part of our Earth! This takes some explaining.'

Mike took a couple of sips of his tender coconut water before proceeding with his recounting of the new observations.

'All right guys, I shall try my best to explain briefly the new observations that have excited all cosmologists. But, if you need more details, I am sure you can find them by yourselves,' began Mike. 'The observational equipment at the South Pole has been christened Background Imaging of Cosmic Extragalactic Polarization, or BICEP2, telescope. Cosmologists believe that when the Universe was incredibly young—a mere 10^{-35} second after the Big Bang—it underwent a period of extremely rapid expansion, known as *inflation*. That number indicating the infancy of the Universe translates to ten times trillion-trillion-trillionth of a second. At that time, the volume of the Universe shot up by a factor of up to 10^{80} in a tiny fraction of a second. Again, that number means 1 followed by 80 zeroes! One can't even imagine such numbers, can one? Anyway, about 380,000 years after the Big Bang, the Cosmic Microwave Background or CMB—the thermal remnant of the Big Bang—came into being. Over the years, CMB has been studied extensively.'

'Ah, George has told me about CMB, its discovery and its properties, as well as about inflation,' I said.

'I am sure he has, Alfie. And I am certain Sunny knows it all and will explain it to Nam later on,' responded Mike. 'Well, coming back to inflation, it solved several problems like the isotropy of the Universe and the fact that the cosmic space has a flat geometry. Not only that. The quantum fluctuations inherent to the inflationary phase are believed to have become the seeds for the growth of structure in the Universe—everything from stars to galaxies.'

'All this is nice, Mike,' interjected George. 'But tell us what is relevant for flexing our BICEPs.'

'Patience, man, patience, as you often tell Alfie here,' smiled Mike. 'Scientists also believe that rather extreme gravitational conditions prevailed during the Universe's infancy. I guess you people know about gravitational waves. Well, primordial gravitational waves are thought to have propagated throughout the Universe during the first moments of inflation. This would have produced the so-called cosmic gravitational-wave background, or CGB, similar to CMB that came about later on. This gravitational-wave background would, in turn, have left its own unmistakeably characteristic imprint on the polarization of the CMB. And it is this polarization that BICEP2 has been looking for. The whole experiment was like looking for a needle in a haystack. Such a finding, if firmly established, would be truly profound.'

'This means that gravitational waves act as the brush that paints the portrait of the early Universe on the black canvas of the CMB,' I summed up in my own style.

'Quite a poetic description, Alfie,' Mike nodded his appreciation. 'Of course, as with all important discoveries, this one too ought to be confirmed independently by other experiments. And that has to happen in the future.'

At that point, Sunny and Nam got up without a word and disappeared into the kitchen.

'Why did they go away?' wondered Mike. 'It is all so mysterious.'

'It is a riddle wrapped in a mystery inside an enigma,' I quoted Winston Churchill.

The riddle was soon unwrapped. The procession from the kitchen, now more or less in the reverse order in comparison to the earlier one, emerged from the kitchen. Nam followed by Sunny who led Bruno, Joe, and one of the waiters. Bruno was carrying a bottle of wine along with three glasses. Joe's smile exuded satisfaction. I could not make out what the waiter was carrying on the large tray he was holding up.

Bruno placed the three glasses and the bottle on our table.

'Aha, I am not being forced to be a teetotaller after all,' George's face lit up.

Bruno displayed the bottle for us to see. 'Special wine from India made with Shiraz grapes, courtesy our two dear girls here.'

'Exquisite,' exclaimed George after tasting the wine Bruno had poured into our glasses.

'Yes, there are people who grow grapes and make excellent wine in our own region in the southern part of India,' explained Sunny.

'Move over Bordeaux, here comes Bangalore,' added Nam.

'But why aren't you two girls drinking?' Mike asked.

It was Nam once again who answered Mike's question in a thick outlandish accent. 'We never drink.' Pause. 'Wine!'

'Good heavens, that is Bela Lugosi from the movie classic, *Dracula*,' recalled George laughing.

The waiter who had been watching the goings on with a smile placed on our table three plates holding the first course and withdrew.

I looked at the circular shiny brown dish folded in half, slightly puffed up probably because of some sort of filling. My taste-buds were immediately stimulated.

'This dish is typically south Indian,' Sunny explained. 'It is called *masaalé dosé*. The batter is made with lentils and rice nicely fermented for a day or two. Please, don't call it a pancake.'

'Our *dosé* is thin, crisp, and can be folded, whereas a pancake is thick, soft, and refuses to be folded,' added Nam.

George opened the *dosé* in front of him and looked inside. 'Yes, the outside is glistening with shades of brown and it is delicately crisp. Look at the stuffing inside. Light yellow—looks like potato-based—sprinkled with green peas and pieces of red bell-peppers. A colourful feast for the eyes. And look at the little holes all over the outer shell, surely the result of fermentation.'

I had never heard George waxing so lyrical about food before. He took a piece of this exotic dish and closed his eyes in rapture.

'This is heavenly, out of this world,' exulted George. 'If I were to choose only one dish to eat for the rest of my life, this is it.'

I agreed with George. And I could see from Mike's expression that he did too.

'Why aren't you two girls joining us?' asked George.

'Chefs never eat their own creation,' answered Sunny. 'While helping Joe, who already knew how to make this dish, we nibbled quite a bit.'

Nam took up her guitar and started playing.

'I shall first play some light Italian music in Bruno's honour and a Spanish piece by Granados. I shall then move on to a melody in the Indian *raga* called *Bhairav* that has a touch of the middle-east.'

After we had finished the first course and the table had been cleared away, we relaxed for some time listening to Nam's enchanting music.

After he had finished eating the first course, George asked almost like a greedy child, 'May I have one more?'

Sunny replied in the manner of a stern mother, 'No way, you will lose your appetite for the other courses to come.'

George pulled a long face and told Sunny, 'As Macduff or MacDuck said to Lady Macbeth, cruelty thy name is woman!'

Before I could respond to George's recently acquired habit of deliberately misquoting classics, Nam said without stopping her lovely music, '*Frailty, thy name is woman. Hamlet's soliloquy, Act I, Scene ii*!'

'How does the kid know that?' George was genuinely surprised.

'Among other things, Nam is into the theatre,' said Sunny with pride.

'Let us get back to the BICEP2 search,' George picked up the thread of our earlier discussion after a pause. 'Such observations would bring together so many features of our Universe. We have a whole range of phenomena—CMB the whisper from the Big Bang, gravitational waves that are yet to be observed by terrestrial detectors, inflation with its quantum fluctuations leading to galaxies that form the Universe. In a way, it completes the circle, doesn't it?'

'All those phenomena are macroscopic and classical except the quantum fluctuations,' pointed out Sunny.

'You are absolutely right, Sunny,' acknowledged George. 'Ah, the ever elusive quantum! Physicists have been trying to quantize gravitation for decades now without significant success. Gravity seems to be a tough customer. No one has been able to unify it with the other three forces in nature either—electromagnetic, weak, and strong.'

'But why do you have to quantize gravity?' I asked.

'Why do we need quantum gravity?' George repeated my question. 'Well, Alfie, we have been talking about the Big Bang essentially beginning with the primordial fireball when all the matter and all the radiation in the Universe were confined to a small volume of extremely high temperature and density. But, theoretically one can go back in time even further and hit a point as the origin of the Universe. Mathematically, this point is known as a *singularity* where the curvature of the spacetime becomes infinite. Ironically, all the equations that dictated the evolution of the Universe starting from this point break down precisely around that point. That is disastrous to say the least. Physically, density and temperature too become infinite, since all the matter and radiation are now condensed to a point. That is disaster doubled.'

'So, what is the solution to this cosmic devastation?' I enquired.

'The solution, my dear Alfie, is supposed to be provided by quantum gravity,' answered George not without a hint of a smile. 'Quantum gravity is supposed to smear out the dreaded singularity within a radius of about 10^{-33} centimetre or so. That is the range within which the effects of quantum gravity manifest themselves.'

'Have the theorists achieved this process of smearing out?' was my natural next question.

'I am afraid not,' George shook his head. 'Until we have a definite theory of quantum gravity, this may not be accomplished. But physicists are heroically going about finding such a theory with partial success. Some claim that not only have they smeared out the singularity, but are able to connect our Universe to its phase earlier to the singularity.'

'St Augustine spoke of a scholar who asked the question: *What was God doing before he created Heaven and Earth?* The scholar himself answered the question: *He was preparing a Hell for those who ask such questions,*' I recalled from my memory.

'I now know where some of my friends doing quantum gravity are headed!' George commented with relish.

Ah, now for our second course, I thought as I saw the waiter making his way towards our table. This time, Sunny and Nam did not have to rush to the kitchen since everything seemed to have been pre-arranged.

'Our second offering is *pulao*, variously known as pilaf, pilav, pilau, plov, polu and palaw, depending on where it comes from,' announced Sunny. 'It is grainy *basmati* rice simmered to perfection in mildly spiced vegetable broth with several chunks of colourful vegetables added. It is served with a side dish of *mughlai kofta,* a special kind of fritter in a delicately spiced sauce, a legacy from the golden age of the great Moghuls. It has been specially adapted for vegetarians by using the rarest of rare ingredients, namely lotus root.'

Eyes closed, George savoured the combination of the two culinary delights and sighed deeply with supreme satisfaction.

'If this is the type of food they serve in India, then I shall go settle down in that country,' announced George.

'Please don't, you will be trampled by elephants,' warned Sunny.

'Bitten by king cobras,' added Nam.

'Run over by bullock carts. Although the bullocks have horns, they make no sound.'

'Yogis will force you to lie half naked on a bed of nails.'

'These kids are too much,' guffawed George. Sunny and Nam gave their tinkling laughter in unison. How could Mike and I not join those jolly three?

Once we had finished eating and the table had been cleared, we resumed our discussion again.

'I have a question or two for you guys,' I began addressing both Mike and George. 'Mike explained the BICEP2 experiment. Gravitational waves bring their messages from the early Universe and write them on the blackboard of CMB. That is really nice. But it is all about the past. What about the future? I mean the prospects of future research as well as the future of the Universe itself.'

'Excellent question, Alfie,' Mike nodded. 'Let me go ahead as the harbinger, since nowadays theorists seem to be following in our footsteps. First of all, you have heard about dark matter and dark energy, haven't you?'

'Yes, George has told me those dark secrets of the Universe. But what about Sunny and Nam?' I looked at the two girls who were eagerly following our dialogue.

'Don't worry about us,' assured Sunny. 'I know a bit of those secrets. I shall fill in the cosmic voids for Nam later on.'

'Fine then,' continued Mike. 'At any given time, we cannot think too far ahead in time, can we? Now, some of the most important questions in cosmology centre around the observed acceleration of the Universe and its theoretical explanation. The dark energy responsible for this acceleration needs a lot more exploration. I don't mean more of the same kind of observations that have been made, but vastly improved ones. For instance, the Hubble Space Telescope or HST has been doing a marvellous job in exploring the Universe. There has been a lot of debate as to whether it should be retired or rejuvenated. Hopefully, the capacity of HST will be enhanced and it will be able to aid, even if it is to a limited extent, the observations on distant galaxies. That could throw some light on the dark energy. That last bit of an oxymoron was spontaneous, not pre-planned.'

Mike chuckled before going on.

'Probably, more important than extending the life of HST are new space missions that are being planned to push further our knowledge of dark energy,' continued Mike. 'For instance, HST's worthy descendent is going to be the *James Webb Space Telescope* or *JWST* for short. It is a large orbiting telescope designed to make observations in the infrared frequencies. That is the spectral range in which high-redshift supernovae occurring in really distant galaxies predominantly emit their radiation. Ultimately, JWST is expected to replace HST. That is not all. There are plans afoot to design and look into the possibility of creating what is known as *SNAP*. That acronym stands for *SuperNova Acceleration Probe*. SNAP will be dedicated to the extensive study of dark energy. It is a very ambitious project indeed. The Probe will be equipped with a big telescope comparable to HST, a camera of the highest possible quality, and a spectrograph that can operate in the whole range of wavelengths from infrared to ultraviolet. It is expected to offer important information on some 2000 supernovae.'

'Aha, two thousand is a magic number. That is about the number of galaxies Hubble happened to study,' I observed. 'All right, tell us about the discoveries that are expected to emerge out of all these high-powered space missions.'

'Well, for one thing the acceleration of the universal expansion seems to be constant,' answered Mike. 'With more far-reaching and accurate observations, one can hopefully find out whether it is really a constant or whether it has changed in the past. Another important question is the combined action of the two dark opposing siblings—dark matter and dark energy. Dark matter attracts, while dark energy repels. Consequently, dark matter decelerates the universal expansion, while dark energy accelerates it. Was there a switch from one action to the other, from deceleration to acceleration? This is an important question the answer to which is ardently sought after. In short, the future of cosmology promises to be quite exciting.'

Sunny, who had been listening intently, turned to George. 'So far so good with all the observations to come,' she commented. 'We theorists cannot sit quietly, can we? So tell us what is happening on the theoretical front.'

'I am sure you could have done the job yourself, Sunny,' George wagged his finger. 'All right, age before beauty. Let me do the honours. Now, as Mike pointed out, the acceleration of the universal expansion has been shown to be constant. That is one reason why Einstein's cosmological constant lambda has been considered to be the prime suspect. Remember, Einstein disowned his baby when the Universe was discovered to be expanding and not static. This was in 1917, nearly a century ago. Since that time, the constant has had a varied life. Although Einstein rejected it, Eddington held on to it. As a matter of fact, he attributed the expansion of the Universe discovered by Hubble and Humason to the accelerating effect of the cosmological constant! But, many theorists wanted to abandon the constant, although some of them found it useful in building their cosmic models. Of course, now that constant has made a comeback in a dark disguise. Or, shall we say the ugly duckling has turned out to be a swan?'

'So it is all smooth sailing then?' I asked.

'To repeat the oft-repeated refrain, patience, Alfie, patience,' George laughed. 'Here comes the real big problem. For Einstein, lambda was a mathematical adjustment to accommodate a static Universe within the framework of his general relativistic field equations. But what is its physical origin? Quantum field theorists have tried to generate the constant from the vacuum fluctuations. What we call vacuum is supposed to be seething with virtual particles that are continuously created and annihilated. The Russian physicist Yakov Zel'dovich had shown that these vacuum quantum fluctuations generated a constant energy equivalent to the cosmological constant. But it was 100 magnitude larger than what could be accommodated in a realistic cosmic model. That number happens to be around 0.7 or so. In 1962, Zel'dovich and his younger colleague Igor Novikov wrote: *After a genie has been let out of a bottle. . .legend has it that the genie can be chased back only with great difficulty.* Now with the dark energy stealing away the cork of the bottle, it looks like that the genie cannot be chased back into the bottle at all! How can one reduce the huge theoretical value of the constant to what has been observed? No one knows.'

'In other words, what cannot be cured must be endured, at least for the time being,' I said. 'Mike was talking about the possibility of the acceleration varying with time. What about that?'

'Well, an unvarying acceleration is supposed to be due to the cosmological constant as we have been saying,' George resumed. 'Theorists have been thinking of other alternative explanations for dark energy that accommodate a changing acceleration. For instance, one of them is *quintessence*, an invisible field that repels and therefore accelerates the expansion of the cosmos. Remember, in antiquity this term referred to the ether or the fifth element of nature together with the other four—Earth, fire, water, and air. The new quintessence can lead to varying cosmic acceleration. Another possibility is the presence of an energy called *phantom field* whose density increases with time, causing an exponential cosmic acceleration. This would result in such a speed that it could break the nuclear forces in the atoms and end the Universe in some 20 billion years in what is termed the *Big Rip*!'

'What, in 20 billion years the Universe will be ripped off?' I exclaimed. 'What does that mean?'

'Don't ask me, it is all conjecture. I am just a down-to-earth theorist,' answered George.

'As Sherlock Holmes says: *Ah, my dear Watson, there we come to the realm of conjecture, where the most logical mind may be at fault,*' I quoted.

'Ah, my dear Alfie, write it down for me. I want to add it to my collection of quotations,' said George.

'You mean your bag of misquotations,' I laughed.

Before he could protest, George was distracted by the sight of our smiling waiter marching towards us holding aloft his tray as usual.

There were five bowls, not three, in front of us.

'Oh yes, we are joining you for dessert,' Sunny informed us. 'We would never dream of missing our favourite sweet.'

The small bowl in front of me contained two disks that looked like cookies soaking in thick, creamy milk coloured with saffron. There were slivers of blanched almonds and chopped pistachios floating around in the milk.

'This refreshing delicacy is called *rasamalai*,' Sunny explained. '*Rasa* means juicy and *malai* means cream. Let me not keep you from enjoying it by elaborating on the recipe. Just try it.'

George was already tasting a small piece of rasamalai along with the creamy saffron milk and pieces of nuts.

'Oh, this is divine,' sighed George.

Having tried a bit of my dessert, I fully agreed with George. I instinctively knew that so did Mike.

All of a sudden, Mike turned to me and asked, 'By the way, Alfie, have you found a suitable title for George's book?'

I was taken completely by surprise by Mike's question to say the least. I had not given any thought to the task George had assigned me. Probably, it had been working in my subconscious for all I knew.

'Oh, yes,' I said without hesitation. 'George, how about *Universe Unveiled*?'

'Universe Unveiled, Universe Unveiled,' George repeated the words as though he was turning around my proposed title in his mind. After a moment or two of thinking he said, 'I like it, Alfie. It conveys everything I have tried to put into the book. Thank you, Alfie, thank you.'

Unbidden, words seemed to tumble from my mouth. 'George you could also add the subtitle, *The Cosmos in My Bubble Bath.*'

'What are you talking about, Alfie?' George seemed to be startled by my suggestion. 'Do you mean multiverse consisting of billions or even infinite number of Universes bubbling around? I don't intend to include theories that have no contact with reality.'

'I am with you, George,' agreed Mike. 'I am told that these hypothetical Universes cannot even be observed. As far as we observers are concerned, if they cannot be observed, they do not exist. They are not part of reality.'

'Reality attained through observation,' I pondered. 'With all your hypotheses, theories, and philosophies, do you think you will ever reach ultimate reality?'

Surprisingly, the answer to my question came from Nam who said quietly, 'Ultimate reality is the ultimate illusion!'

'You said it Nam, you said it,' George clapped his hands and laughed like a child.

At that moment, Bruno accompanied by Joe came over.

'Is everything all right?' asked Bruno.

'All right? It has been a superb dinner, Bruno,' complimented George as Mike and I nodded emphatically.

'Inspired by Sunny and Nam,' acknowledged Bruno.

'The problem, Bruno, is how will you match this meal in the future?' wondered George.

'Maybe we will hire Sunny and Nam permanently and turn this into an Indian-Italian restaurant. How do you like that?' smiled Bruno.

'Excellent idea, do that,' George immediately welcomed this novel idea.

'Well, you people look cramped behind the table,' observed Bruno. 'We have created a nice little patio as part of our renovation. You could sit there and talk more comfortably, you know.'

Bruno led us to the lovely little patio outside the dining area. We found it most relaxing to unwind there beneath the cloudless night sky, a gentle breeze blowing in. We enjoyed the ambience, absorbed in one's own thoughts. George sat in silence cradling his chin in his two palms. After a while he spoke up.

'You know, we have been talking about the latest developments in cosmology and even its possible future course,' George began. 'Whatever has been happening with the Universe in the recent past as well as at present has been revealed by fabulous observations supported by brilliant theoretical ideas. But, as we have seen, those phenomena had their origins right at the beginning of the Universe. The Universe has evolved continuously over billions of years. But, the imprints of the initial conditions can never be erased away. And it is the same with cosmology.'

George paused. My mind went back to the day he had handed over the manuscript of his book. Whatever he had told me about cosmology at that time seemed like the prologue to his book. Now, he seemed to be delivering its epilogue.

'All our fascinating discoveries in cosmology had their beginnings centuries ago. Cosmology too has evolved in all that time,' George continued. 'Let us not look down upon the ancient myths. They were the earliest attempts to explain what people saw in the heavens. Even then, the astronomers of antiquity were making their heroic advances into the unknown realms of the night sky. Then came Copernicus. How can anyone describe the complete break from tradition he made, a leap of imagination from appearance to reality? Human thought would never be the same after that. Think of all the pioneering discoveries made by giants like Tycho Brahe, Kepler, and Galileo. Again, how can one estimate the contributions made by Newton? I am not speaking of his individual monumental discoveries, but his ushering in the age of universal laws, his setting science within the framework of mathematics. Of course, he was to find his match in Einstein. Einstein's general relativity, a unique stroke of genius, pervades every aspect of cosmology today. Even if his theory were to be superseded by another one, I don't think the basic tenets of his theory would ever be done away with. Einstein will always be reigning supreme.'

This time George paused for a longer while. He was perhaps pondering over the entire superstructure of cosmology. Every one of us waited in eager anticipation.

'I feel that whatever we have been learning about the Universe is like a story, a ballad, handed down from generation to generation,' resumed George slowly. 'But it is not exactly the same story. It changes continuously even if by minute amounts. The story we tell our children today will have changed by the time they relate that story to their children. That is what makes it so fascinating.'

George nodded a couple of times and went on.

'The study of our Universe is like a tree. We may see immediately only the fresh foliage adorning its top. But beneath it all are the old roots, the sturdy trunk, and the spreading branches. There may be other growths as well supported by that tree. Then again, all our explorations are made up of many strands of coloured threads weaving the grand tapestry of the cosmos. It is like separate streams, each starting from a different point, bringing its own riches, and finally merging with the vast reservoir of knowledge.'

George sat back, hands folded, eyes closed as though he had nothing more to say.

Sunny and Nam exchanged glances. Their minds seemed to work in unison. Nam started to strum her guitar rhythmically producing the same note again and again. I recognized what she was achieving remarkably well. Nam was simulating the repeating sounds of the Indian *ektara*, a single stringed musical instrument used by *Sufi* singers. The two girls sang together a folk song in an Indian language. After each line, they translated its meaning taking turns.

Listen, listen my friend! Listen, my beloved companion!
 You are my mighty tree, I am your creeper
 Entwined around your neck, I smile in happiness
Listen, listen my friend! Listen, my beloved companion!
 My heart says you are the ocean and I your river
 Making waves I wind my way to meet my love
Listen, listen my friend! Listen, my beloved companion!

The haunting soulful song ended with Sunny and Nam singing the last line in a long stretch that must have wafted far into the night carried by the gentle wind. It was a beautiful lyrical reflection of what George had been saying about the nature of the Universe.

The silence was absolute. I realized that Bruno and his crew had lined up behind us listening to the singing of the two lovely young girls.

Sunny and Nam got to their feet, rushed to George and each gave him a big hug. Then they started hugging everybody. And all of us began indulging in this mad and wonderful activity, everyone hugging everyone else, laughing and dancing.

When the tumult had subsided, George spoke up.

'Bruno, my friend, how can I...' George stopped seeing Bruno's upraised hand. 'All right, I won't thank you.'

'Come back soon, all of you,' said Bruno with warmth.

'Where are you two girls headed?' asked George.

'We are going back to our friends with whom we are staying,' answered Sunny. 'Nam and I will drop in at your office tomorrow.'

'Do that girls, do that,' said George and smiled at Mike. 'I suppose I don't have to ask about your plans, Mike.'

'Yes sir, I go back to the Observatory, my mountain of meditation. My wild nightlife begins,' said Mike.

'As for you, Alfie, you must have finished reading the manuscript. Now, get ready to correct the proofs,' smiled George. 'Yes, we shall meet soon, very soon.'

As I walked home after that magical evening, I looked up at the stars above as I always do. They had never seemed brighter or happier.

Chapter 18
Far Horizons

Soon after reaching home, I looked through George's manuscript. Sure enough, the last chapter dealt with whatever George had discussed during the incredible evening. I quickly went over that section just to have all the ideas crystallized in my mind. Then, following George's example, sat back eyes closed, hands folded in my lap, just relaxing. But how long can one relax like that? I got up and picked up a note book in which I usually jot down interesting quotations I have come across.

As I thumbed through my book of quotations, I came upon the lines that had been penned by the poet Archibald MacLeish:

A small planet of a minor star at the edge of an inconsiderable galaxy in the immeasurable distances of space.

How can one even attempt to describe the vastness of our Universe? Awesome, infinite, unimaginable... what words can one use?

I came across several thoughts expressed by Blaise Pascal in his *Pensées* that were so apt in the context of our cosmos. I have often thought about the kind of emotions the expanse of our Universe would inspire. Wonder and awe no doubt, but here is what Pascal felt: *The eternal silence of these infinite spaces terrifies me.* Yes, it can be terrifying, especially when we consider our own place in space and time. Then again, Pascal ponders over this aspect of man's existence as well: *I find myself tied to one corner of this vast expanse, without knowing why I am put in this place rather than in another, nor why the short time which is given me to live is assigned to me at this point rather than at another of the whole eternity which was before me or which shall come after me.* What would we have gleaned about our Universe had we lived in another distant galaxy either way back in the past or far in the future? We do not know. Sitting in our own little corner of the Universe, we have been able to perform the incredible feat of learning so much about our cosmos stretched out in space and time. Underlying all our observations, all our ideas, all our theories is one most important ingredient: human thought. Perhaps, Pascal had the last word on that too: *Man is but a reed, the most feeble in nature; but he is a thinking reed!*

© Springer International Publishing Switzerland 2015
C.V. Vishveshwara, *Universe Unveiled*, Astronomers' Universe,
DOI 10.1007/978-3-319-08213-4_18

That sums up the basis of all that we know about the Universe and all that we will ever come to know.

Well, thinking makes one tired, doesn't it? So, off I go to my bathroom to have all my fatigue, both physical and mental, dissolve away.

The bubbles seemed to be in more profusion than ever. They were everywhere seemingly in competition to massage my body. That was normal, I would say. But strangely enough, there were also a number of them concentrated around my head gently caressing it. Those bubbles were not accompanied by sparkling specks or by swirling spirals. They were perfectly clear and almost transparent. This was a surprise.

'Surprised about the nature and distribution pattern of the bubbles, are we?' softly chuckled my bathtub, Khasbath. 'Well, your last thoughts happened to be about thinking, weren't they? Thinking about thinking, sounds strange, doesn't it, boss? But that is what some philosophers do and get paid for it too, don't they?' Another chuckle before moving on, 'Those bubbles today represent the thought process, boss. Look at them. They are clear like pure consciousness on which one's experience etches the thoughts. Like thought, they can expand. Look, boss, look. They interact, they merge, they produce other bubbles, all like thoughts. Some burst and go out of existence, just like the thoughts you have lost or dismissed.'

There was a long pause before the bathtub spoke again.

'All right, boss, enjoy your bubble bath. Let your thoughts take flight on the wings of fantasy. Yes, it will be an exciting journey.'

In the silence that followed, I closed my eyes and surrendered to the bubbles that would soon transport me to an entirely different plane of perception.

I could hear beautiful music emanating from somewhere nearby. Yes, I could recognize it. It was one of Mozart's violin sonatas. I soon came to the source of that music.

Eyes closed, a rapturous smile playing on his face haloed by wispy white hair sat the violin player. I gasped with astonishment as immediate recognition dawned on me. It was none other than Al, the owner of the little shop selling bathtubs among other things. Strangely, he looked exactly like the great man whom my bathtub keeps referring to as the Master. His first name too began with the two letters *Al*, did it not?

When the music he was playing ended, Al opened his eyes. His smile that greeted me was as always gentle and kind.

'I had told you that we would meet again, had I not? So, we meet again, my friend,' said Al happily. 'Oh yes, that was one of Mozart's violin sonatas I was playing. You know how I feel? I think that the music of Mozart is of such purity and beauty that one feels he merely found it—that it had always existed as part of the inner beauty of the Universe waiting to be revealed. It seems Mozart himself sometimes used to say that he felt like he was not composing so much as taking dictation. Dictation from whom, my young friend? From Nature, from the Universe, from God?'

'Life without playing music is inconceivable for me,' Al declared. 'I live my daydreams in music. I see my life in terms of music. I get most joy in life out of music. I never go anywhere without my violin.'

After a somewhat long pause, Al looked up at me with an amused smile.

'A little while ago, you were reflecting upon the act of thinking, were you not?' Al obviously knew about my ponderings focused on Pascal's *Pensées*. 'You felt that the faculty of thinking was the supreme gift given to man that could unlock the secrets of the Universe, did you not? I perfectly agree with you. Please allow me to tell you my own experience in this important matter. Before I can explain, I must do something important too.'

Al started twirling a lock of hair on his forehead and said in broken English, 'I vill a little t'ink!'

It seems that is how he spoke whenever he was confronted with some difficult problem to solve. On the other hand, I was sure that talking about thinking was no problem to him at all.

Al paused presumably recalling his past experience that had touched upon some aspects of thought.

'Long ago, I was asked to write about my life under the heading, *Autobiographical Notes*. Let me quote the first sentence I wrote, *here I sit in order to write. . .something like my own obituary,*' laughed Al. 'That is how you feel if you were to write about yourself in your old age.'

Al drummed his finger on his chin before continuing.

'It is best if I recount exactly what I wrote about the feeling of wonder one gets when looking at the world with fresh eyes,' said Al and went on to quote further from what he had written in his *Autobiographical Notes*. 'For me it is not dubious that our thinking goes on for the most part without the use of signs, or words, and beyond that to a considerable degree unconsciously. For how, otherwise, would it happen that we *wonder* spontaneously about some experience? This *wondering* seems to occur when an experience comes into conflict with a world of concepts which is already fixed in us. Whenever such a conflict is experienced hard and intensively it reacts back on our thought world in a decisive way. The development of this thought world is in a certain sense a continuous flight from *wonder.*'

A dreamy look came upon Al's face.

'A wonder of such nature I experienced as a child of four or five years, when my father showed me a compass. That this needle behaved in such a determined way did not at all fit into the nature of events, which could find a place in the unconscious world of concepts, I mean the effects connected with direct touch. I can still remember—or at least believe I can remember—that this experience made a deep and lasting impression on me. Something deeply hidden had to be behind things. What man sees before him from infancy causes no reaction of this kind; he is not surprised over the falling bodies, concerning wind and rain, nor concerning the Moon or about the fact that the Moon does not fall down, nor concerning the difference between living and non-living matter.'

There was a long pause before Al spoke again.

'Let me tell you, my friend, what I might have experienced in the case of the Universe,' said Al in a conspiratorial manner. 'I just thought about it now, so I have never mentioned it before. When I considered the Universe as it had been known for ages, ever since the time of Aristotle, I thought nothing happened on a large scale. The Universe was static. So, I put in the cosmological term in my equations. That explained how one could build a model of the Universe in which nothing changed. This model belonged to the thought world. There was no *wonder* in it. Then it was discovered that the Universe was expanding! That was the moment of *wonder*. That magical moment was followed by more observations and more theorizing. Aha, that is what constituted the continuous flight of the thought world from wonder!'

I felt exceptionally privileged to hear the impromptu thoughts of Al that seemed to have been conveyed exclusively to me.

'Enough of all this philosophical talk about thinking,' Al held up his hand. 'To quote the famous words of the Bard, *there are more things in heaven and Earth, my friend, than are dreamt of in your philosophy*! We shall meet again in a short while. Before that, you must go see my dear priestly friend, who is waiting for you.'

Lovingly, caressingly, Al picked up his violin and bow. He closed his eyes, and drew the bow long and soft on the strings of his violin as the silken sound emanating from it seemed to fill the entire Universe once again.

The man I saw now was smiling broadly as he read what looked like a letter. He was stocky, if not fat. He had a pink round face, which looked even rounder with his circular metal-rimmed spectacles on. His cheeks shone like polished apples. On the whole, his countenance made you feel that he was always happy and contented. He wore a tight black suit, waistcoat of heavy wool, and a high stiff collar. It was Georges Lemaître, to whom Al had referred to as his dear priestly friend. I recognized him from the photographs I had seen.

'Ah, you have come here straight from your meeting with the great man. That must have been a wonderful experience.'

Lemaître removed his spectacles, polished them, pinched the bridge of his nose between his thumb and forefinger, and put on the glasses back.

'You know, I had the good fortune of spending some time together with Albert Einstein, whom you met,' recounted Lemaître. 'This was at the California Institute of Technology, or simply Caltech, in the USA. That was a fantastic time for me. Einstein and I used to take frequent walks together deeply absorbed in our discussions. He was so famous that newspaper reporters always followed us at a distance hoping to get some story or the other. They joked that whenever the physicist—Einstein that is—and the priest—that is me—went for a walk, the *little lamb* followed them. The little lamb indeed! That was their short hand for the cosmological constant lambda!'

Lemaître's cheeks shone as he laughed heartily at his recollection and continued.

'Yes, we did discuss the little lamb constantly. Einstein was willing to listen to me although he thought it was not needed at all, since the Universe was not static as he had imagined. Such rejection must have made the poor little thing quite sad.'

A big smile lit up Lemaître's face. I wondered what amused him now. All this was so interesting. This priest was so different from the serious Canon I had visited ages ago.

'In those days I used to smoke cigarettes, you know,' said Lemaître. 'As you perhaps know, Einstein used to smoke a pipe. His wife Elsa did not want him to smoke too much for the sake of his health. He was allowed only a small quantity of tobacco. So what does he do? He takes my cigarettes, opens them up, and fills his pipe with the tobacco he had pulled out from them.'

Lemaître laughed again shaking his head. He removed his glasses, wiped his tears of merriment, put back the glasses, and went on.

'Getting back to the lambda term, Einstein thought it was ugly. Maybe he felt that the little lamb was in reality a black sheep. Here, let me read from the letter he wrote to me.'

Lemaître picked up the sheet of paper he had been looking at with amusement before he started talking to me.

'It is dated September 26, 1947, quite late in the day when so much had already been discussed about cosmology,' Lemaître reflected. 'It begins with dear Professor Lemaître—that is me—thank you very much for your kind letter etc., etc. Here comes the last line of the first paragraph: *It is true that the introduction of the lambda term offers a possibility, it may even be that it is the right one.* Aha, the adamant man bends a bit, doesn't he? But he changes his tune immediately. Let me read what he writes in detail: *Since I have introduced this term I had always a bad conscience. I found it very ugly indeed that the field law of gravitation should be composed of two logically independent terms which are connected by addition. About the justification of such feelings concerning logical simplicity it is difficult to argue. I cannot help but feel it strongly and I am unable to believe that such an ugly thing should be realized in nature.*'

Lemaître put down the letter and shook his head. '*Ugly!* *Ugly* the famous man calls the poor little lamb,' laughed Lemaître. 'Often ugly things are quite useful, aren't they?'

'Well, it was discovered that the Universe was expanding,' continued Lemaître. 'I was able to build a model for it and realized that the Universe must have begun its expansion from a highly condensed state. I thought this was a cold quantum state, I mean a state completely described by quantum theory. However, I believed that the expansion must have started from such a state with some kind of fireworks. Yes, this was the *fireworks-origin of the Universe*. Well, the scientists who came after me have shown that the original state was not only extremely dense but also inconceivably hot, not cold at all as I had imagined. They called it the *primordial fireball*. And the explosion that triggered the expansion of the Universe is known as the *Big Bang*! Quite descriptive, though the violence of such an explosion is unimaginable. What is even more unimaginable is the fact that both space and time came into existence at the origin of the Universe.'

Lemaître's eyes sparkled as he added, 'It was a *day without yesterday*.'

To think of space and time coming into existence at that moment! *A day without yesterday*—that was a lovely description indeed.

Lemaître was about to go on with his exposition. But he was surprised to find that Al or Einstein had quietly materialised beside him. I was astonished by his unexpected appearance too.

'Go on, go on, don't pay any attention to me,' chuckled Einstein as he reached out and picked up Lemaître's pack of cigarettes lying on the table. He pulled out a few of them, opened them up and filled his pipe with the tobacco he had extracted. Then he lit his pipe and puffed on it contentedly, listening and watching with a big smile.

'See, what I told you? He is incorrigible,' Lemaître shook his head.

'Ah, you were talking about the Universe, were you not?' Einstein's eyes were shining. As he continued, he seemed to be speaking to himself forgetting the presence of others. 'Astronomers have found out so much about our cosmos. I think that the most incomprehensible thing about the Universe is that it is comprehensible.'

Einstein pondered for a moment and declared, 'I want to know how God created this world. I want to know His thoughts, the rest are details. I often wonder whether God had any choice in creating the Universe at all.'

'Do you remember what Niels Bohr told you in 1927 during the Fifth Solway Conference?' asked Lemaître with a gentle smile.

'How can I forget what dear Niels said to me at that time? When I heard about Werner Heisenberg's uncertainty principle, I commented that God did not play dice. Einstein, stop telling God what to do, Bohr reprimanded me.' Einstein laughed boisterously.

At that moment, out of nowhere, appeared a strange figure. He was dressed like the legendary sleuth Sherlock Holmes. His clothes were the same as what Holmes is supposed to have worn—a long coat, cape, and a checked cap. Like Holmes, he was holding a pipe in one hand, while in the other he held a magnifying glass through which he was peering at the sky.

Both Einstein and Lemaître were perplexed to say the least.

'*Lieber Gott*, it is none other than our friend, Eddington!' exclaimed Einstein. 'Why is he dressed like that? What is he doing? Is he trying to measure the bending of light in broad day light?'

Then Eddington spoke. 'I am a detective in search of a criminal—the cosmological constant. I know he exists, but I do not know his appearance. For instance, I do not know if he is a little man or a tall man. . . The first move was to search for footprints at the scene of the crime. The search has revealed footprints, or what look like footprints—the recession of the spiral nebulae.'

'I was quoting from the public talk I gave at the International Astronomical Union's meeting in Cambridge, Massachusetts, in 1932,' explained Eddington. 'If the repulsion exerted by the cosmological constant could make the Universe static, why can't it accelerate the Universe at the early stages and lead to the expansion observed by Hubble and Humason?'

'Did you mention the possible accelerating effect of the constant lambda?' Lemaître looked up at Eddington in surprise. 'Well, the modern-day astronomers have made a rather surprising discovery recently that may have some connection to what you just said.'

'What do you mean *recently*?' interjected Einstein. 'All things are relative you know. Especially time. *Ja, ja,* time is often the culprit as I discovered long ago. I shouldn't say *long ago* either, since time is relative. Anyway, tell us about the latest discovery of the astronomers.'

'As I was saying, astronomers have made an unexpected discovery. Actually a stunning one I must admit,' continued Lemaître. 'They were observing extraordinarily bright supernovae occurring in different galaxies that enabled them to estimate the distances to very far-off galaxies. That meant they were looking at those galaxies as they were way back in time. In the process, they found out that the expansion of the Universe is not slowing down at all. *It is actually accelerating!*'

'Good heavens, how can that be?' exclaimed Einstein. He seemed to be pretending to be surprised. But it was not convincing. 'What might be the explanation for this strange behaviour?'

'Well, they think that there must be some agency that is pushing away the Universe making it expand faster than believed earlier,' answered Lemaître. 'Some kind of force or some kind of energy must be responsible for this. They call it *dark energy*.'

'Earlier they had *dark matter*, didn't they?' said Eddington. 'Let us not take it lightly—no, I am not playing with the words *light* and *dark*. Those clever chaps can observe the bending of light by the dark matter. They are following in my footsteps, eh? Anyway, it is called *gravitational lensing*.'

'I see, in the beginning was dark matter and now this dark energy. Don't they know matter and energy are the same?' remarked Einstein and continued.

'I tell you what I think. They call them dark, because they are in the dark. I thought that God said *let there be light*, maybe the Devil countered Him by proclaiming *let there be dark*!' Einstein exploded with laughter and then added, 'Oh, I am sorry, I shouldn't be saying such things in the presence of a priest. In any event, I think it is the *dark age of cosmology*!'

'Here is something that should interest you,' said Lemaître, looking steadily at Einstein. 'One of the candidates for this dark energy is, hold your breath, the cosmological constant lambda!'

'What do you know, the little lamb was not a black sheep after all although black and dark are closely related to each other! *Wunderbar*!' beamed Einstein. 'The biggest blunder of my life strikes again. I am having the last laugh after all.'

With that remark he threw back his head and burst out into his characteristic laughter. His whole body shook as he roared. The entire Universe seemed to vibrate with his booming laughter as it echoed from galaxy to galaxy.

'Well, my friends, Universe is boundless and so is its study,' commented Einstein recovering from his long laughter. 'There are far horizons in our cosmic knowledge. They keep receding as we move towards them. But then, that is what makes all our explorations so exciting, is it not?'

Then Einstein's visage took on a serene expression. He was looking far into space—perhaps beyond speeding swirling galaxies, beyond the Universe in its immense expanse.

Oblivious of the presence of his two colleagues, he whispered to himself, 'Out yonder there is an immense cosmos that stands before us like a great eternal riddle, at least partially accessible to our inspection and thinking. The contemplation of this cosmos beckons like a liberation.'

After a moment or two, he turned to his two companions, 'It is time to go, my friends. I think the others are waiting for us.'

The three of them gave me a warm smile and waved in a gesture of farewell. And they began to walk away chatting, laughing, and animatedly gesticulating. As they moved on some distance, I could see that there were three others who had appeared in front of them. I immediately recognized them: Humason passing his flask of Panther Juice to Hubble, while Shapley looked on with an amused smile. Slowly, steadily, more and more people started materializing and marching ahead. I could identify them all: a dignified lady holding herself erect, Miss Henrietta Swan Leavitt; in front of her, joyously holding hands, was the Herschel family; preceding the three of them were the four giants walking abreast—Tycho Brahe, Kepler, Galileo and, slightly apart from those three and absorbed in thought, Isaac Newton—following their great leader Nicolaus Copernicus; several steps ahead of them, the three astronomers of antiquity, Aristarchus, Aryabhata, and Omar Khayyam engaged in joyful conversation. I was thrilled to see that leading them all was Aristophanes accompanied by his chorus of frogs, hopping and frolicking as usual. I thought I could even hear at that distance the croaking of their quaint refrain.

That procession of some of the greatest contributors to the knowledge of our Universe marched on slowly, steadily, towards the horizon and faded away.

'That was some cosmic parade, boss,' I was woken up from my reverie by those words whispered by my friend, Khasbath.

The clear, transparent bubbles were feverishly moving around in the bathwater. That meant the thought process induced by them was still active.

'You are so fortunate, boss,' continued my bathtub. 'You were in the presence of the Master himself. He told you about his own thinking, didn't he? And you could participate in the great time those three wonderful gentlemen had together—the Master, the astronomer disguised as a detective, and the priestly cosmologist.'

I could not agree more with my bathtub.

'Well, it has been a long, tortuous cosmic journey we have taken, absolutely fabulous,' reflected my bathtub. 'When you started learning about the immensity of the Universe you said that it was—how did you put it?—*not the end, not the beginning of the end, not even the end of the beginning, but the beginning of the beginning*! So it is not the end of the road, boss. The quest goes on.'

Yes, the quest will go on. More and more knowledge of the Universe will be gained. But will I have an opportunity to learn it all as joyfully as I have been doing all these days? I would be missing that pleasure. That thought brought on a wistful feeling.

'Are you feeling sad, boss? Whatever for?' There was a mild reproach in the bathtub's voice. 'It is not over. There is so much to talk about. What about life beyond Earth? Are there living beings, even intelligent ones perhaps, lurking somewhere in the Universe? Maybe in another planetary system, in another galaxy, there is a boss taking bath in his beloved talking bathtub, who knows? Forget the Universe. What about history, philosophy, the workings of the inner spaces of the mind? There is no end to it, is there? No, boss, you will be learning about all this I am sure and we will be talking about whatever you have learnt and more. But for now, you are tired. Enjoy your bath for a while and then go to bed.'

The bathtub fell silent. I felt comforted and relaxed. I luxuriated in the still warm bathwater surrounded by countless bubbles. Then there was a sudden surge of those bubbles, a veritable explosion, and they swarmed all over me, pleasantly messaging every part of my body. Then slowly, gently, they melted away.

Having dried myself, I was about to move out of the bathroom. Then I heard the words distinctly.

'Good night, boss,' said Khasbath clearly. 'Take care and may peace be with you.'

'Good night, Khasbath, and peace be with you too,' I answered.

I went to bed and fell into the most beautiful dreamless sleep ever.

Foundations of Facts and Fantasies

Please allow me to paraphrase what I wrote in *Einstein's Enigma or Black Holes in My Bubble Bath*. Often, written words form the foundation stones and the building blocks of all information and knowledge, imagination and fantasy. Don't you agree? No doubt I learnt a great deal about the Universe from my discussions with George and Mike. But, even they would have gathered their information and knowledge to a considerable extent from their study of books, articles, and original papers. Well, I too have benefited greatly from my own reading of several books and articles. What I learnt from them not only added to my understanding of the Universe, but also stimulated my imagination and, let me admit it, what may pass for my fantasies. I have listed below some of the books and essays I have read more or less in the order in which the contents of the present book unfold.

As you can see, the first book that appears in the list of books is *Einstein's Enigma*. After all, the present book happens to be, in some sense, a sequel to it. As a matter of fact, the Prologue is based on the first chapter of the previous book. Furthermore, there is inevitable overlap between the earlier parts of the two books that deal with the basic aspects of the Universe. I believe that you will find it rewarding and enjoyable to read *Einstein's Enigma*. Also, there are a number of books listed there that may prove to be of interest to you.

As in the case of *Einstein's Enigma*, many of the descriptions given by the past founders of cosmology as well as dialogues among them are based on their original writings. These appear in the fantasies induced by the magical bubble bath. I felt that it would be nice to create some poems and songs for the book: for instance, the lyrics for the chorus of frogs accompanying Aristophanes, translation of Aryabhata's original Sanskrit verse, and Omar Khayyam's impromptu quatrains. Likewise, I drew the cartoons in the book. I hope you found them enjoyable. The song towards the end of Chapter 17, *Food for the Future*, is based on S. D. Burman's haunting Hindi song, *Suno Mere Bandhu Re!* You can listen to it on You Tube.

© Springer International Publishing Switzerland 2015   233
C.V. Vishveshwara, *Universe Unveiled*, Astronomers' Universe,
DOI 10.1007/978-3-319-08213-4

Let me stop here. I am sure you have your own list of books that you have read and enjoyed. It is my ardent hope that you have now found more titles to add to that list.

I am so happy that you joined me on this joyous passage through space and time. As many wished me during the journey, farewell for now and may peace be with you. Till our worldlines meet again then!

List of Some Selected Books

C. V. Vishveshwara, *Einstein's Enigma or Black Holes in My Bubble Bath* (Copernicus Books – An Imprint of Springer Science + Business Media, 2006)

Roy A. Gallant, *The Constellations: How They Came to Be* (Four Winds Press, 1979)

Julius D. W. Staal, *The New Patterns in the Sky – Myths and Legends of the Stars* (McDonald and Woodward, 1988)

Robert Graves, *Greek Gods and Heroes* (Dell Publishing Company, 1972)

Aristophanes, *Four Major Plays* (Airmont Publishing Company, 1969)

George Abell, *Exploration of the Universe* (Holt, Reinhart and Wilson, 1969)

S. Balachandra Rao, *Indian Mathematics and Astronomy* (Bhavan's Gandhi Centre of Science and Human Values, India, 2005)

Omar Khayyam (Translated by Edward Fitzgerald), *Rubaiyat* (Rupa Paperback, India, 2000)

Arthur Koestler, *The Sleepwalkers* (Penguin Books, 1982)

Dava Sobel, *The Planets* (Penguin Books 2006)

Rocky Kolb, *Blind Watchers of the Sky* (Oxford University Press, 1999)

Oliver Lodge, *Pioneers of Science* (Dover, 1960)

Patrick Moore, *The Great Astronomical Revolution* (Albion Publishing, 1994)

Nicolaus Copernicus, *De Revolutionibus Orbium Coelestium*

Charles Glenn Wallis (Translation), *On the Revolutions of Heavenly Spheres* (Prometheus Books, 1995)

Arthur Beer and K. Aa. Strand (Editors), *Copernicus: Yesterday and Today*, Proceedings of the Conference to Commemorate the 500th Anniversary of the birth of Copernicus (Pergamon Press, 1975)

Fred Hoyle, *Nicolaus Copernicus, An Essay on His Life and Work* (Heinemann, London, 1973)

Thomas S. Kuhn, *The Copernican Revolution* (Harvard University Press, 1979)

Dante, *De monorchia*, I. I, translated by P. H. Wicksteed, quoted by John North, *The Medieval Background to Copernicus* in Arthur Beer and K. Aa. Strand (Editors),

© Springer International Publishing Switzerland 2015   235
C.V. Vishveshwara, *Universe Unveiled*, Astronomers' Universe,
DOI 10.1007/978-3-319-08213-4

Copernicus: Yesterday and Today, Proceedings Conference to Commemorate the 500th Anniversary of the birth of Copernicus (Pergamon Press, 1975)

Dante (Translated by Dorothy L. Sayers), *L'Inferno (Hell)* (Penguin Books, 1949)

Arthur Beer and Peter Beer (Editors), 'Kepler – Four Hundred Years', *Vistas in Astronomy* Vol. 18 (Pergamon Press, 1975); In particular: Owen Gingerich, 'Kepler's Place in Astronomy'

Galileo Galilei, *Dialogue Concerning Two Systems of the World: Ptolemaic and Copernican*, Stillman Drake (Translation), (University of California Press, 1967)

Frova A. and Marenzana M., *Thus Spoke Galileo* (Oxford University Press, 1998)

Dava Sobel, *Galileo's Daughter* (Fourth Estate, 1999)

Michael White, *Isaac Newton: The Last Sorcerer* (Fourth Estate, 1998)

William Stukeley, *Memoirs of Sir Isaac Newton's Life* [1752] (Taylor and Francis, 1936)

E. N. Da C. Andrade, *Sir Isaac Newton* (Collins, 1954)

John and Mary Gribbin, *Newton in 90 Minutes* (Constable and Co. 1997 and University Press, India 1997)

John and Mary Gribbin, *Halley in 90 Minutes* (Constable and Co. 1997 and University Press, India 1997)

Isaac Newton, *Philosophiae Naturalis Principia Mathematica. The Principia: Sir Isaac Newton's Mathematical Principles of Natural Philosophy and His System of the World*, Translated into English by Andrew Mott in 1729. Revised and supplied with an historical and explanatory appendix by Florian Cajori (University of California Press, 1946)

Gale E. Christianson, *Edwin Hubble: The Mariner of the Nebulae* (Institute of Physics Publishing, 1995)

Paul Arthur Schilpp (Editor) *Albert Einstein: Philosopher – Scientist* (Tudor Publishing Company, 1951)

Albert Einstein, *Ideas and Opinions* (Bonanza Books, 1988)

Philipp Frank, *Einstein: His Life and Times* (Jonathan Cape, 1948)

Banesh Hoffmann, *Albert Einstein, Creator and Rebel* (Viking Press, 1972)

Ronald W. Clark, *Einstein: The Life and Times* (Avon, 1972)

Alice Calaprice, *The New Quotable Einstein* (Princeton University Press, 2005)

K. Biswas, D.C.V. Mallik, and C.V. Vishveshwara (Editors), *Cosmic Perspectives* (Cambridge University Press, 1989). In particular: C.V. Vihsveshwara, 'Geometry and the Universe'

Herbert Friedman, *Amazing Universe* (National Geographic Society, 1975)

William J. Kaufmann, III, *Universe* (W. H. Freeman and Company, 1988)

Robert P. Kirshner, *The Extravagant Universe* (Princeton University Press, 2004)

Pedro G. Ferreira, *The Perfect Theory* (Little, Brown, 2014)

S. Chandrasekhar, *Truth and Beauty: Aesthetics and Motivations in Science* (University of Chicago Press, 1987)

William Shakespeare, *The Oxford Shakespeare: The Complete Works* (Editors) S. Wells, G. Taylor, J. Jowett, and W. Montgomery (Oxford University Press, 2005)

Peter D. Usher, 'Shakespeare's Cosmic World View', *Mercury*, Vol. 26, No. 1, Jan-Feb (1997)

Robert Frost, *Selected Poems of Robert Frost* (Barnes & Noble Books, 1993)

Sherwood Santos (Translation), *Greek Lyric Poetry* (W. W. Norton, 2005)

Acknowledgements

First of all, I would like to recall my long, happy, and fruitful association with Ramon Khanna, the publishing editor at Springer and my dear friend. A few years ago, when he saw the draft of my earlier book, *Einstein's Enigma or Black Holes in My Bubble Bath,* his response was one of exceptional enthusiasm. An astrophysicist, who has worked on black-hole magneto-hydrodynamics, he was responsible for the publication of that book, taking care of every detail possible. Now, history has repeated itself. When I sent him my proposal for the present book, *Universe Unveiled – The Cosmos in My Bubble Bath*, Ramon responded again with remarkable ardour. As before, he has not only offered me his unstinting support, but has also provided his indispensable advice throughout the entire production of the book. I cannot thank him enough for his personal interest and efforts in this regard. I would also like to thank Ramon Khanna's colleagues at Springer for their generous help throughout, Charlotte Fladt, who has taken care of the administrative work, and Birgit Muench, who has been responsible for overseeing the production process.

Professor Robert Fuller, Former President of Oberlin College and author of several important books, was my mentor at Columbia University, New York, in the early 1960s. His encouragement and help initiated me into general relativity, thereby giving the initial impetus to my lifelong professional career. He was highly complimentary about *Einstein's Enigma*, which was one of the factors that eventually strengthened my plan to write *Universe Unveiled*. Prof Fuller was kind enough to read a preliminary draft of the present book and he made many valuable suggestions. His wonderful reaction to the book is reflected in his comments that appear on the cover. I am deeply indebted to him for our numerous, fruitful interactions beginning with his guidance in the early days of our friendship that has spanned more than five decades.

My friend and colleague, H. R. Madhusudan, Scientific Officer, Jawaharlal Nehru Planetarium, Bangalore, has been of immense help in shaping up the book. He has provided me with important information whenever necessary. Furthermore,

© Springer International Publishing Switzerland 2015

C.V. Vishveshwara, *Universe Unveiled*, Astronomers' Universe,

DOI 10.1007/978-3-319-08213-4

he has spent a great deal of time and energy, with his inimitable zeal, in producing the illustrations. I am highly obliged to him for his indispensable assistance.

B. Srikanth, a gifted graphic artist, has created the beautiful design and artwork for the cover. It is a pleasure to acknowledge his contribution to the attractive appearance of the book. I would like to thank B. S. Shylaja, Director, Jawaharlal Nehru Planetarium, Bangalore, for her astronomical inputs, and G. K. Rajeshwari and R. Pushpa of the Planetarium for their help in many ways.

Now for my family. As always, I have received ready support in all possible ways from my wife Saraswathi. This has been one of the most essential factors in writing the two books. Our older daughter Smitha had appeared in *Einstein's Enigma* in the guise of Sunitha or Sunny. Our second daughter Namitha had also been mentioned in the context of Nam's *Nanosingers*. Both of them reappear in Chapter 17, *Food for the Future*. I have had highly enjoyable discussions with all three of them on various aspects of the two books, which has made writing an exhilarating experience.

As it always happens, I might have left out by oversight the names of others who have helped me during the preparation of the book. My sincere thanks are due to them, even if I do not know who they are at the moment.

And finally, I thank you, dear reader—whoever you are and wherever you are—with all my heart. Without you, what would be the use of this book?

Illustrations and Credits

Illustrations	Credits	Page Numbers
1. Cartoon: Orion and Scorpion	Vishveshwara, C.V. (2006), *Einstein's Enigma or Black Holes in My Bubble Bath*, Springer-Verlag, Berlin, Heidelberg	p. 15
2. Epicycles	Drawn by B. Srikanth	p. 30
3. Heliocentric Model	Nicolaus Copernicus, *De Revolutionibus Orbium Coelestium* (1543)	p. 32
4. Heliocentric retrograde motion	Drawn by B. Srikanth	p. 33
5. Ellipse and Kepler's Second Law	Drawn by B. Srikanth	p. 36
6. Kepler's Planetary Musical Notes	Johannes Kepler, *Harmonices Mundi* (1619)	p. 36
7. Tides	Vishveshwara, C.V. (2006), *Einstein's Enigma or Black Holes in My Bubble Bath*, Springer-Verlag, Berlin, Heidelberg	p. 41
8. *Principia*: Launch of Projectiles	Isaac Newton, *Philosophiae Naturalis Principia Mathematica* (1687)	p. 43
9. Dante's Cosmos	Drawn by B. Srikanth	p. 48
10. Cartoon: Devil in Inferno	Drawn by C.V. Vishveshwara	p. 50
11. Cartoon: Tycho, Kepler, and Jeppe	Vishveshwara, C.V. (2006), *Einstein's Enigma or Black Holes in My Bubble Bath*, Springer-Verlag, Berlin, Heidelberg	p. 60
12. Galileo's drawing of the Moon	Galileo Galilei, *Siderius Nuncius* (1610)	p. 73
13. Galileo's sketch of Jupiter's Moons	Galileo Galilei, *Siderius Nuncius* (1610)	p. 75

(continued)

© Springer International Publishing Switzerland 2015
C.V. Vishveshwara, *Universe Unveiled*, Astronomers' Universe,
DOI 10.1007/978-3-319-08213-4

Illustrations	Credits	Page Numbers
14. Galileo's sketch of Sunspots	Galileo Galilei, *Istoria e Dimostrazioni Intorno Alle Solari e Loro Accidenti Rome* (1613)	p. 78
15. Cartoon: Newton's alchemical laboratory	Drawn by C.V. Vishveshwara	p. 85
16. Sun's interior	Vishveshwara, C.V. (2006), *Einstein's Enigma or Black Holes in My Bubble Bath*, Springer-Verlag, Berlin, Heidelberg	p. 102
17. Blackbody radiation curves	Drawn by B. Srikanth	p. 107
18. Core of a heavy star	Vishveshwara, C.V. (2006), *Einstein's Enigma or Black Holes in My Bubble Bath*, Springer-Verlag, Berlin, Heidelberg	p. 112
19. Pulsar	Vishveshwara, C.V. (2006), *Einstein's Enigma or Black Holes in My Bubble Bath*, Springer-Verlag, Berlin, Heidelberg	p. 114
20. Orion Constellation	Drawn by B. Srikanth	p. 119
21. Herschel's Telescope	Courtesy of Yerkes Observatory	p. 124
22. Herschel's model of the Milky Way	Drawn by B. Srikanth	p. 126
23. Shapley's model of the Milky Way	Drawn by B. Srikanth	p. 131
24. Hubble's photograph of Andromeda	Courtesy of Carnegie Observatories	p. 152
25. Galactic redshifts, distances and velocities	http://hubblesource.stsci.edu/	p. 155
26. Cartoon: Humason the mule driver	Drawn by C.V. Vishveshwara	p. 171
27. Cartoon: Einstein elevator	Vishveshwara, C.V. (2006), *Einstein's Enigma or Black Holes in My Bubble Bath*, Springer-Verlag, Berlin, Heidelberg	p. 180
28. Cartoon: Rocket shooting up	Vishveshwara, C.V. (2006), *Einstein's Enigma or Black Holes in My Bubble Bath*, Springer-Verlag, Berlin, Heidelberg	p. 181
29. Cartoon: Little flats making up curved space	Vishveshwara, C.V. (2006), *Einstein's Enigma or Black Holes in My Bubble Bath*, Springer-Verlag, Berlin, Heidelberg	p. 182
30. Three geometries of the Universe	Vishveshwara, C.V. (2006), *Einstein's Enigma or Black Holes in My Bubble Bath*, Springer-Verlag, Berlin, Heidelberg	p. 192
31. Original Hubble Plot	Hubble, Edwin. (1929), "A Relation Between Distance and Radial Velocity Among Extra-Galactic Nebulae", From the *Proceedings of the National Academy of Sciences Volume 15: March 15, 1929: Number 3*	p. 194

About the Author

Specializing in Einstein's general theory of relativity, C.V. Vishveshwara has worked extensively on the theory of black holes, making major contributions to this exciting field of research since its very beginning.

After receiving his PhD from the University of Maryland, and subsequently serving on the faculties of New York University, Boston University, and University of Pittsburgh, he returned to his hometown of Bangalore, India, where he joined the Raman Research Institute as Professor and retired as Senior Professor from the Indian Institute of Astrophysics.

Besides being the author of a number of technical papers and editor of several books on relativity, astrophysics, and cosmology, Vishveshwara has written popular articles on various topics in science. During his tenure as the founder-director of Jawaharlal Nehru Planetarium in Bangalore, he produced many planetarium programmes. These programmes have proved to be exceedingly popular because of the simple and attractive manner in which difficult concepts have been presented.

Vishveshwara's cartoons have been published in journals and conference proceedings. He has drawn a few new ones for the present book as he did for his highly acclaimed previous book, *Einstein's Enigma or Black Holes in My Bubble Bath*.

© Springer International Publishing Switzerland 2015

C.V. Vishveshwara, *Universe Unveiled*, Astronomers' Universe,

DOI 10.1007/978-3-319-08213-4

CPSIA information can be obtained
at www.ICGtesting.com
Printed in the USA
LVHW080212020622
720308LV00018B/194

Tarzan
the
Terrible

EDGAR RICE BURROUGHS

COSIMOCLASSICS

NEW YORK

TARZAN
the Terrible

Contents

CHAPTER I

The Pithecanthropus

SILENT as the shadows through which he moved, the great beast slunk through the midnight jungle, his yellow-green eyes round and staring, his sinewy tail undulating behind him, his head lowered and flattened, and every muscle vibrant to the thrill of the hunt. The jungle moon dappled an occasional clearing which the great cat was always careful to avoid. Though he moved through thick verdure across a carpet of innumerable twigs, broken branches, and leaves, his passing gave forth no sound that might have been apprehended by dull human ears.

Apparently less cautious was the hunted thing moving even as silently as the lion a hundred paces ahead of the tawny carnivore, for instead of skirting the moon-splashed natural clearings it passed directly across them, and by the tortuous record of its spoor it might indeed be guessed that it sought these avenues of least resistance, as well it might, since, unlike its grim stalker, it walked erect upon two feet—it walked upon two feet and was hairless except for a black thatch upon its head; its arms were well shaped and muscular; its hands powerful and slender with long tapering fingers and thumbs reaching almost to the first joint of the index fingers. Its legs too were shapely but its feet departed from the standards of all races of men, except possibly a few of the lowest races, in that the great toes protruded at right angles from the foot.

1

Pausing momentarily in the full light of the gorgeous African moon the creature turned an attentive ear to the rear and then, his head lifted, his features might readily have been discerned in the moonlight. They were strong, clean cut, and regular—features that would have attracted attention for their masculine beauty in any of the great capitals of the world. But was this thing a man? It would have been hard for a watcher in the trees to have decided as the lion's prey resumed its way across the silver tapestry that Luna had laid upon the floor of the dismal jungle, for from beneath the loin cloth of black fur that girdled its thighs there depended a long hairless, white tail.

In one hand the creature carried a stout club, and suspended at its left side from a shoulder belt was a short, sheathed knife, while a cross belt supported a pouch at its right hip. Confining these straps to the body and also apparently supporting the loin cloth was a broad girdle which glittered in the moonlight as though encrusted with virgin gold, and was clasped in the center of the belly with a huge buckle of ornate design that scintillated as with precious stones.

Closer and closer crept Numa, the lion, to his intended victim, and that the latter was not entirely unaware of his danger was evidenced by the increasing frequency with which he turned his ear and his sharp black eyes in the direction of the cat upon his trail. He did not greatly increase his speed, a long swinging walk where the open places permitted, but he loosened the knife in its scabbard and at all times kept his club in readiness for instant action.

Forging at last through a narrow strip of dense jungle vegetation the man-thing broke through into an almost treeless area of considerable extent. For an instant he hesitated, glancing quickly behind him and then up at the security of the branches of the great trees waving overhead, but some greater urge than fear or caution influenced his decision apparently, for he moved off again across the little plain leaving the safety of the trees behind him. At greater or less intervals leafy sanctuaries dotted the grassy expanse ahead of him and the route he took, leading from one to another, indicated that he had not entirely cast discretion to the winds. But after the second tree had been left behind the distance to the next was considerable, and it was then that Numa walked from the concealing cover of the jungle and, seeing his quarry apparently helpless before him, raised his tail stiffly erect and charged.

Two months—two long, weary months filled with hunger, with thirst, with hardship, with disappointment, and, greater than all, with gnawing pain—had passed since Tarzan of the Apes learned from the diary of the dead German captain that his wife still lived. A brief investigation in which he was enthusiastically aided by the Intelligence Department of the British East African Expedition revealed the fact that an attempt had been made to keep Lady Jane in hiding in the interior, for reasons of which only the German High Command might be cognizant.

In charge of Lieutenant Obergatz and a detachment of native German troops she had been sent across the border into the Congo Free State.

Starting out alone in search of her, Tarzan had succeeded in finding the village in which she had been incarcerated only to learn that she had escaped months before, and that the German officer had disappeared at the same time. From there on the stories of the chiefs and the warriors whom he quizzed, were vague and often contradictory. Even the direction that the fugitives had taken Tarzan could only guess at by piecing together bits of fragmentary evidence gleaned from various sources.

Sinister conjectures were forced upon him by various observations which he made in the village. One was incontrovertible proof that these people were man-eaters; the other, the presence in the village of various articles of native German uniforms and equipment. At great risk and in the face of surly objection on the part of the chief, the ape-man made a careful inspection of every hut in the village from which at least a little ray of hope resulted from the fact that he found no article that might have belonged to his wife.

Leaving the village he had made his way toward the southwest, crossing, after the most appalling hardships, a vast waterless steppe covered for the most part with dense thorn, coming at last into a district that had probably never been previously entered by any white man and which was known only in the legends of the tribes whose country bordered it. Here were precipitous mountains, well-watered plateaus, wide plains, and vast swampy morasses, but neither the plains, nor the plateaus, nor the mountains were accessible to him until after weeks of arduous effort he succeeded in finding a spot where he might cross the morasses—a hideous stretch infested by venomous snakes and other larger dangerous reptiles. On several occasions he glimpsed at distances or by night what might have been titanic reptilian monsters, but as there were hippopotami, rhinoceri, and elephants in great numbers in and about the marsh he was never positive that the forms he saw were not of these.

When at last he stood upon firm ground after crossing the morasses he realized why it was that for perhaps countless ages this territory had defied the courage and hardihood of the heroic races of the outer world that had, after innumerable reverses and unbelievable suffering penetrated to practically every other region, from pole to pole.

From the abundance and diversity of the game it might have appeared that every known species of bird and beast and reptile had sought here a refuge wherein they might take their last stand against the encroaching multitudes of men that had steadily spread themselves over the surface of the earth, wresting the hunting grounds from the lower orders, from the moment that the first ape shed his hair and ceased to walk upon his knuckles. Even the species with which Tarzan was familiar showed here either the results of a divergent line of evolution or an unaltered form that had been transmitted without variation for countless ages.

Too, there were many hybrid strains, not the least interesting of which to Tarzan was a yellow and black striped lion. Smaller than the species with which Tarzan was familiar, but still a most formidable beast, since it possessed in addition to sharp saber-like canines the disposition of a devil. To Tarzan it presented evidence that tigers had once roamed the jungles of Africa, possibly giant saber-tooths of another epoch, and these apparently had crossed with lions with the resultant terrors that he occasionally encountered at the present day.

The true lions of this new, Old World differed but little from those with which he was familiar; in size and conformation they were almost identical, but instead of shedding the leopard spots of cubhood, they retained them through life as definitely marked as those of the leopard.

Two months of effort had revealed no slightest evidence that she he sought had entered this beautiful yet forbidding land. His investigation, however, of the cannibal village and his questioning of other tribes in the neighborhood had convinced him that if Lady Jane still lived it must be in this direction

that he seek her, since by a process of elimination he had reduced the direction of her flight to only this possibility. How
she had crossed the morass he could not guess and yet something within seemed to urge upon him belief that she had
crossed it, and that if she still lived it was here that she must
be sought. But this unknown, untraversed wild was of vast
extent; grim, forbidding mountains blocked his way, torrents
tumbling from rocky fastnesses impeded his progress, and at
every turn he was forced to match wits and muscles with the
great carnivora that he might procure sustenance.

Time and again Tarzan and Numa stalked the same quarry
and now one, now the other bore off the prize. Seldom however did the ape-man go hungry for the country was rich in
game animals and birds and fish, in fruit and the countless
other forms of vegetable life upon which the jungle-bred man
may subsist.

Tarzan often wondered why in so rich a country he found
no evidences of man and had at last come to the conclusion
that the parched, thorn-covered steppe and the hideous morasses had formed a sufficient barrier to protect this country
effectively from the inroads of mankind.

After days of searching he had succeeded finally in discovering a pass through the mountains and, coming down upon
the opposite side, had found himself in a country practically
identical with that which he had left. The hunting was good
and at a water hole in the mouth of a cañon where it de-
bouched upon a tree-covered plain Bara, the deer, fell an
easy victim to the ape-man's cunning.

It was just at dusk. The voices of great four-footed hunters rose now and again from various directions, and as the
cañon afforded among its trees no comfortable retreat the
ape-man shouldered the carcass of the deer and started downward onto the plain. At its opposite side rose lofty trees—a

great forest which suggested to his practised eye a mighty jungle. Toward this the ape-man bent his step, but when midway of the plain he discovered standing alone such a tree as best suited him for a night's abode, swung lightly to its branches and, presently, a comfortable resting place.

Here he ate the flesh of Bara and when satisfied carried the balance of the carcass to the opposite side of the tree where he deposited it far above the ground in a secure place. Returning to his crotch he settled himself for sleep and in another moment the roars of the lions and the howlings of the lesser cats fell upon deaf ears.

The usual noises of the jungle composed rather than disturbed the ape-man but an unusual sound, however imperceptible to the awakened ear of civilized man, seldom failed to impinge upon the consciousness of Tarzan, however deep his slumber, and so it was that when the moon was high a sudden rush of feet across the grassy carpet in the vicinity of his tree brought him to alert and ready activity. Tarzan does not awaken as you and I with the weight of slumber still upon his eyes and brain, for did the creatures of the wild awaken thus, their awakenings would be few. As his eyes snapped open, clear and bright, so, clear and bright upon the nerve centers of his brain, were registered the various perceptions of all his senses.

Almost beneath him, racing toward his tree was what at first glance appeared to be an almost naked white man, yet even at the first instant of discovery the long, white tail projecting rearward did not escape the ape-man. Behind the fleeing figure, and now so close as to preclude the possibility of its quarry escaping, came Numa, the lion, in full charge. Voiceless the prey, voiceless the killer; as two spirits in a dead world the two moved in silent swiftness toward the culminating tragedy of this grim race.

Even as his eyes opened and took in the scene beneath him
—even in that brief instant of perception, followed reason,
judgment, and decision, so rapidly one upon the heels of the
other that almost simultaneously the ape-man was in mid-air,
for he had seen a white-skinned creature cast in a mold similar
to his own, pursued by Tarzan's hereditary enemy. So close
was the lion to the fleeing man-thing that Tarzan had no time
carefully to choose the method of his attack. As a diver leaps
from the springboard headforemost into the waters beneath,
so Tarzan of the Apes dove straight for Numa, the lion; naked
in his right hand the blade of his father that so many times
before had tasted the blood of lions.

A raking talon caught Tarzan on the side, inflicting a long,
deep wound and then the ape-man was on Numa's back and
the blade was sinking again and again into the savage side.
Nor was the man-thing either longer fleeing, or idle. He too,
creature of the wild, had sensed on the instant the truth of
the miracle of his saving, and turning in his tracks, had leaped
forward with raised bludgeon to Tarzan's assistance and
Numa's undoing. A single terrific blow upon the flattened
skull of the beast laid him insensible and then as Tarzan's
knife found the wild heart a few convulsive shudders and a
sudden relaxation marked the passing of the carnivore.

Leaping to his feet the ape-man placed his foot upon the
carcass of his kill and, raising his face to Goro, the moon,
voiced the savage victory cry that had so often awakened the
echoes of his native jungle.

As the hideous scream burst from the ape-man's lips the
man-thing stepped quickly back as in sudden awe, but when
Tarzan returned his hunting knife to its sheath and turned
toward him the other saw in the quiet dignity of his demeanor
no cause for apprehension.

For a moment the two stood appraising each other, and

then the man-thing spoke. Tarzan realized that the creature before him was uttering articulate sounds which expressed in speech, though in a language with which Tarzan was unfamiliar, the thoughts of a man possessing to a greater or less extent the same powers of reason that he possessed. In other words, that though the creature before him had the tail and thumbs and great toes of a monkey, it was, in all other respects, quite evidently a man.

The blood, which was now flowing down Tarzan's side, caught the creature's attention. From the pocket-pouch at his side he took a small bag and approaching Tarzan indicated by signs that he wished the ape-man to lie down that he might treat the wound, whereupon, spreading the edges of the cut apart, he sprinkled the raw flesh with powder from the little bag. The pain of the wound was as nothing to the exquisite torture of the remedy but, accustomed to physical suffering, the ape-man withstood it stoically and in a few moments not only had the bleeding ceased but the pain as well.

In reply to the soft and far from unpleasant modulations of the other's voice, Tarzan spoke in various tribal dialects of the interior as well as in the language of the great apes, but it was evident that the man understood none of these. Seeing that they could not make each other understood, the pithecanthropus advanced toward Tarzan and placing his left hand over his own heart laid the palm of his right hand over the heart of the ape-man. To the latter the action appeared as a form of friendly greeting and, being versed in the ways of uncivilized races, he responded in kind as he realized it was doubtless intended that he should. His action seemed to satisfy and please his new-found acquaintance, who immediately fell to talking again and finally, with his head tipped back, sniffed the air in the direction of the tree above them and then suddenly pointing toward the carcass of Bara, the deer, he touched his stomach in a sign language which even the densest might interpret. With a wave of his hand Tarzan invited his guest to partake of the remains of his savage repast, and the other, leaping nimbly as a little monkey to the lower branches of the tree, made his way quickly to the flesh, assisted always by his long, strong sinuous tail.

The pithecanthropus ate in silence, cutting small strips from the deer's loin with his keen knife. From his crotch in the tree Tarzan watched his companion, noting the preponderance of human attributes which were doubtless accentuated by the paradoxical thumbs, great toes, and tail.

He wondered if this creature was representative of some strange race or if, what seemed more likely, but an atavism. Either supposition would have seemed preposterous enough did he not have before him the evidence of the creature's existence. There he was, however, a tailed man with distinctly arboreal hands and feet. His trappings, gold encrusted and jewel studded, could have been wrought only by skilled artisans; but whether they were the work of this individual or of others like him, or of an entirely different race, Tarzan could not, of course, determine.

His meal finished, the guest wiped his fingers and lips with leaves broken from a nearby branch, looked up at Tarzan with a pleasant smile that revealed a row of strong white teeth, the canines of which were no longer than Tarzan's own, spoke a few words which Tarzan judged were a polite expression of thanks and then sought a comfortable place in the tree for the night.

The earth was shadowed in the darkness which precedes the dawn when Tarzan was awakened by a violent shaking of the tree in which he had found shelter. As he opened his eyes he saw that his companion was also astir, and glancing around quickly to apprehend the cause of the disturbance, the ape-man was astounded at the sight which met his eyes.

The dim shadow of a colossal form reared close beside the tree and he saw that it was the scraping of the giant body against the branches that had awakened him. That such a tremendous creature could have approached so closely without disturbing him filled Tarzan with both wonderment and chagrin. In the gloom the ape-man at first conceived the intruder to be an elephant; yet, if so, one of greater proportions than any he had ever before seen, but as the dim outlines became less indistinct he saw on a line with his eyes and twenty feet above the ground the dim silhouette of a grotesquely serrated back that gave the impression of a creature whose each and every spinal vertebra grew a thick, heavy horn. Only a portion of the back was visible to the ape-man, the rest of the body being lost in the dense shadows beneath the tree, from whence there now arose the sound of giant jaws powerfully crunching flesh and bones. From the odors that rose to the ape-man's sensitive nostrils he presently realized that beneath him was some huge reptile feeding upon the carcass of the lion that had been slain there earlier in the night.

As Tarzan's eyes, straining with curiosity, bored futilely into the dark shadows he felt a light touch upon his shoulder, and, turning, saw that his companion was attempting to attract his attention. The creature, pressing a forefinger to his own lips as to enjoin silence, attempted by pulling on Tarzan's arm to indicate that they should leave at once.

Realizing that he was in a strange country, evidently infested by creatures of titanic size, with the habits and powers of which he was entirely unfamiliar, the ape-man permitted himself to be drawn away. With the utmost caution the pithecanthropus descended the tree upon the opposite side from the great nocturnal prowler, and, closely followed by Tarzan, moved silently away through the night across the plain.

The ape-man was rather loath thus to relinquish an opportunity to inspect a creature which he realized was probably entirely different from anything in his past experience; yet he was wise enough to know when discretion was the better part of valor and now, as in the past, he yielded to that law which dominates the kindred of the wild, preventing them from courting danger uselessly, whose lives are sufficiently filled with danger in their ordinary routine of feeding and mating.

As the rising sun dispelled the shadows of the night, Tarzan found himself again upon the verge of a great forest into which his guide plunged, taking nimbly to the branches of the trees through which he made his way with the celerity of long habitude and hereditary instinct, but though aided by a prehensile tail, fingers, and toes, the man-thing moved through the forest with no greater ease or surety than did the giant ape-man.

It was during this journey that Tarzan recalled the wound in his side inflicted upon him the previous night by the raking talons of Numa, the lion, and examining it was surprised to discover that not only was it painless but along its edges were no indication of inflammation, the results doubtless of the antiseptic powder his strange companion had sprinkled upon it.

They had proceeded for a mile or two when Tarzan's companion came to earth upon a grassy slope beneath a great tree whose branches overhung a clear brook. Here they drank and Tarzan discovered the water to be not only deliciously pure and fresh but of an icy temperature that indicated its rapid descent from the lofty mountains of its origin.

Casting aside his loin cloth and weapons Tarzan entered the little pool beneath the tree and after a moment emerged, greatly refreshed and filled with a keen desire to breakfast. As he came out of the pool he noticed his companion examining him with a puzzled expression upon his face. Taking the ape-man by the shoulder he turned him around so that Tarzan's back was toward him and then, touching the end of Tarzan's spine with his forefinger, he curled his own tail up over his shoulder and, wheeling the ape-man about again, pointed first at Tarzan and then at his own caudal appendage, a look of puzzlement upon his face, the while he jabbered excitedly in his strange tongue.

The ape-man realized that probably for the first time his companion had discovered that he was tailless by nature rather than by accident, and so he called attention to his own great toes and thumbs to further impress upon the creature that they were of different species.

The fellow shook his head dubiously as though entirely unable to comprehend why Tarzan should differ so from him but at last, apparently giving the problem up with a shrug, he laid aside his own harness, skin, and weapons and entered the pool.

His ablutions completed and his meager apparel redonned he seated himself at the foot of the tree and motioning Tarzan to a place beside him, opened the pouch that hung at his right side taking from it strips of dried flesh and a couple of handfuls of thin-shelled nuts with which Tarzan was unfamiliar. Seeing the other break them with his teeth and eat the kernel, Tarzan followed the example thus set him, discovering the meat to be rich and well flavored. The dried flesh also was far from unpalatable, though it had evidently been jerked without salt, a commodity which Tarzan imagined might be rather difficult to obtain in this locality.

As they ate Tarzan's companion pointed to the nuts, the
dried meat, and various other nearby objects, in each instance
repeating what Tarzan readily discovered must be the names
of these things in the creature's native language. The ape-
man could but smile at this evident desire upon the part of
his new-found acquaintance to impart to him instructions
that eventually might lead to an exchange of thoughts between them. Having already mastered several languages
and a multitude of dialects the ape-man felt that he could
readily assimilate another even though this appeared one entirely unrelated to any with which he was familiar.

So occupied were they with their breakfast and the lesson
that neither was aware of the beady eyes glittering down
upon them from above; nor was Tarzan cognizant of any
impending danger until the instant that a huge, hairy body
leaped full upon his companion from the branches above
them.

CHAPTER II

"To the Death!"

I N THE moment of discovery Tarzan saw that the creature was almost a counterpart of his companion in size and conformation, with the exception that his body was entirely clothed with a coat of shaggy black hair which almost concealed his features, while his harness and weapons were similar to those of the creature he had attacked. Ere Tarzan could prevent the creature had struck the ape-man's companion a blow upon the head with his knotted club that felled him, unconscious, to the earth; but before he could inflict further injury upon his defenseless prey the ape-man had closed with him.

Instantly Tarzan realized that he was locked with a creature of almost superhuman strength. The sinewy fingers of a powerful hand sought his throat while the other lifted the bludgeon above his head. But if the strength of the hairy attacker was great, great too was that of his smooth-skinned antagonist. Swinging a single terrific blow with clenched fist to the point of the other's chin, Tarzan momentarily staggered his assailant and then his own fingers closed upon the shaggy throat, as with the other hand he seized the wrist of the arm that swung the club. With equal celerity he shot his right leg behind the shaggy brute and throwing his weight forward hurled the thing over his hip heavily to the ground,

15

at the same time precipitating his own body upon the other's chest.

With the shock of the impact the club fell from the brute's hand and Tarzan's hold was wrenched from its throat. Instantly the two were locked in a deathlike embrace. Though the creature bit at Tarzan the latter was quickly aware that this was not a particularly formidable method of offense or defense, since its canines were scarcely more developed than his own. The thing that he had principally to guard against was the sinuous tail which sought steadily to wrap itself about his throat and against which experience had afforded him no defense.

Struggling and snarling the two rolled growling about the sward at the foot of the tree, first one on top and then the other but each more occupied at present in defending his throat from the other's choking grasp than in aggressive, offensive tactics. But presently the ape-man saw his opportunity and as they rolled about he forced the creature closer and closer to the pool, upon the banks of which the battle was progressing. At last they lay upon the very verge of the water and now it remained for Tarzan to precipitate them both beneath the surface but in such a way that he might remain on top.

At the same instant there came within range of Tarzan's vision, just behind the prostrate form of his companion, the crouching, devil-faced figure of the striped saber-tooth hybrid, eyeing him with snarling, malevolent face.

Almost simultaneously Tarzan's shaggy antagonist discovered the menacing figure of the great cat. Immediately he ceased his belligerent activities against Tarzan and, jabbering and chattering to the ape-man, he tried to disengage himself from Tarzan's hold but in such a way that indicated that as far as he was concerned their battle was over. Appreciating the danger to his unconscious companion and being anxious to protect him from the saber-tooth the ape-man relinquished his hold upon his adversary and together the two rose to their feet.

Drawing his knife Tarzan moved slowly toward the body of his companion, expecting that his recent antagonist would grasp the opportunity for escape. To his surprise, however, the beast, after regaining its club, advanced at his side.

The great cat, flattened upon its belly, remained motionless except for twitching tail and snarling lips where it lay perhaps fifty feet beyond the body of the pithecanthropus. As Tarzan stepped over the body of the latter he saw the eyelids quiver and open, and in his heart he felt a strange sense of relief that the creature was not dead and a realization that without his suspecting it there had arisen within his savage bosom a bond of attachment for this strange new friend.

Tarzan continued to approach the saber-tooth, nor did the shaggy beast at his right lag behind. Closer and closer they came until at a distance of about twenty feet the hybrid charged. Its rush was directed toward the shaggy manlike ape who halted in his tracks with upraised bludgeon to meet the assault. Tarzan, on the contrary, leaped forward and with a celerity second not even to that of the swift-moving cat, he threw himself headlong upon him as might a Rugby tackler on an American gridiron. His right arm circled the beast's neck in front of the right shoulder, his left behind the left foreleg, and so great was the force of the impact that the two rolled over and over several times upon the ground, the cat screaming and clawing to liberate itself that it might turn upon its attacker, the man clinging desperately to his hold.

Seemingly the attack was one of mad, senseless ferocity unguided by either reason or skill. Nothing, however, could have been farther from the truth than such an assumption since every muscle in the ape-man's giant frame obeyed the dictates of the cunning mind that long experience had trained to meet every exigency of such an encounter. The long, powerful legs, though seemingly inextricably entangled with the hind feet of the clawing cat, ever as by a miracle, escaped the raking talons and yet at just the proper instant in the midst of all the rolling and tossing they were where they should be to carry out the ape-man's plan of offense. So that on the instant that the cat believed it had won the mastery of its antagonist it was jerked suddenly upward as the ape-man rose to his feet, holding the striped back close against his body as he rose and forcing it backward until it could but claw the air helplessly.

Instantly the shaggy black rushed in with drawn knife which it buried in the beast's heart. For a few moments Tarzan retained his hold but when the body had relaxed in final dissolution he pushed it from him and the two who had formerly been locked in mortal combat stood facing each other across the body of the common foe.

Tarzan waited, ready either for peace or war. Presently two shaggy black hands were raised; the left was laid upon its own heart and the right extended until the palm touched Tarzan's breast. It was the same form of friendly salutation with which the pithecanthropus had sealed his alliance with the ape-man and Tarzan, glad of every ally he could win in this strange and savage world, quickly accepted the proffered friendship.

At the conclusion of the brief ceremony Tarzan, glancing in the direction of the hairless pithecanthropus, discovered that the latter had recovered consciousness and was sitting erect watching them intently. He now rose slowly and at the same time the shaggy black turned in his direction and addressed him in what evidently was their common language. The hairless one replied and the two approached each other slowly. Tarzan watched interestedly the outcome of their meeting. They halted a few paces apart, first one and then the other speaking rapidly but without apparent excitement, each occasionally glancing or nodding toward Tarzan, indicating that he was to some extent the subject of their conversation.

Presently they advanced again until they met, whereupon was repeated the brief ceremony of alliance which had previously marked the cessation of hostilities between Tarzan and the black. They then advanced toward the ape-man addressing him earnestly as though endeavoring to convey to him some important information. Presently, however, they gave it up as an unprofitable job and, resorting to sign language, conveyed to Tarzan that they were proceeding upon their way together and were urging him to accompany them.

As the direction they indicated was a route which Tarzan had not previously traversed he was extremely willing to accede to their request, as he had determined thoroughly to explore this unknown land before definitely abandoning search for Lady Jane therein.

For several days their way led through the foothills parallel to the lofty range towering above. Often were they menaced by the savage denizens of this remote fastness, and occasionally Tarzan glimpsed weird forms of gigantic proportions amidst the shadows of the nights.

On the third day they came upon a large natural cave in the face of a low cliff at the foot of which tumbled one of the numerous mountain brooks that watered the plain below and fed the morasses in the lowlands at the country's edge. Here the three took up their temporary abode where Tarzan's instruction in the language of his companions progressed more rapidly than while on the march.

The cave gave evidence of having harbored other manlike forms in the past. Remnants of a crude, rock fireplace remained and the walls and ceiling were blackened with the smoke of many fires. Scratched in the soot, and sometimes deeply into the rock beneath, were strange hieroglyphics and the outlines of beasts and birds and reptiles, some of the latter of weird form suggesting the extinct creatures of Jurassic times. Some of the more recently made hieroglyphics Tarzan's companions read with interest and commented upon, and then with the points of their knives they too added to the possibly age-old record of the blackened walls.

Tarzan's curiosity was aroused, but the only explanation at which he could arrive was that he was looking upon possibly the world's most primitive hotel register. At least it gave him a further insight into the development of the strange creatures with which Fate had thrown him. Here were men with the tails of monkeys, one of them as hair covered as any fur-bearing brute of the lower orders, and yet it was evident that they possessed not only a spoken, but a written language. The former he was slowly mastering and at this new evidence of unlooked-for civilization in creatures possessing so many of the physical attributes of beasts, Tarzan's curiosity was still further piqued and his desire quickly to master their tongue strengthened, with the result that he fell to with even greater assiduity to the task he had set himself. Already he knew the names of his companions and the common names of the fauna and flora with which they had most often come in contact.

Ta-den, he of the hairless, white skin, having assumed the rôle of tutor, prosecuted his task with a singleness of purpose that was reflected in his pupil's rapid mastery of Ta-den's mother tongue. Om-at, the hairy black, also seemed to feel that there rested upon his broad shoulders a portion of the burden of responsibility for Tarzan's education, with the result that either one or the other of them was almost constantly coaching the ape-man during his waking hours. The result was only what might have been expected—a rapid assimilation of the teachings to the end that before any of them realized it, communication by word of mouth became an accomplished fact.

Tarzan explained to his companions the purpose of his mission but neither could give him any slightest thread of hope to weave into the fabric of his longing. Never had there been in their country a woman such as he described, nor any tailless man other than himself that they ever had seen.

"I have been gone from A-lur while Bu, the moon, has eaten seven times," said Ta-den. "Many things may happen in seven times twenty-eight days; but I doubt that your woman could have entered our country across the terrible morasses which even you found an almost insurmountable obstacle, and if she had, could she have survived the perils that you already have encountered beside those of which you have yet to learn? Not even our own women venture into the savage lands beyond the cities."

"'A-lur,' Light-city, City of Light," mused Tarzan, translating the word into his own tongue. "And where is A-lur?" he asked. "Is it your city, Ta-den, and Om-at's?"

"It is mine," replied the hairless one; "but not Om-at's. The Waz-don have no cities—they live in the trees of the forests and the caves of the hills—is it not so, *black man?*" he concluded, turning toward the hairy giant beside him.

"Yes," replied Om-at, "We Waz-don are free—only the Ho-don imprison themselves in cities. I would not be a white man!"

Tarzan smiled. Even here was the racial distinction between white man and black man—Ho-don and Waz-don.

Not even the fact that they appeared to be equals in the matter of intelligence made any difference—one was white and one was black, and it was easy to see that the white considered himself superior to the other—one could see it in his quiet smile.

"Where is A-lur?" Tarzan asked again. "You are returning to it?"

"It is beyond the mountains," replied Ta-den. "I do not return to it—not yet. Not until Ko-tan is no more."

"Ko-tan?" queried Tarzan.

"Ko-tan is king," explained the pithecanthropus. "He rules this land. I was one of his warriors. I lived in the palace of Ko-tan and there I met O-lo-a, his daughter. We loved, Like-star-light, and I; but Ko-tan would have none of me. He sent me away to fight with the men of the village of Dak-at, who had refused to pay his tribute to the king, thinking that I would be killed, for Dak-at is famous for his many fine warriors. And I was not killed. Instead I returned victorious with the tribute and with Dak-at himself my prisoner; but Ko-tan was not pleased because he saw that O-lo-a loved me even more than before, her love being strengthened and fortified by pride in my achievement.

"Powerful is my father, Ja-don, the Lion-man, chief of the largest village outside of A-lur. Him Ko-tan hesitated to affront and so he could not but praise me for my success, though he did it with half a smile. But you do not understand! It is what we call a smile that moves only the muscles of the face and effects not the light of the eyes—it means hypocrisy and duplicity. I must be praised and rewarded. What better than that he reward me with the hand of O-lo-a, his daughter? But no, he saves O-lo-a for Bu-lot, son of Mo-sar, the chief whose great-grandfather was king and who thinks that he should be king. Thus would Ko-tan appease the wrath of Mo-sar and win the friendship of those who think with Mo-sar that Mo-sar should be king.

"But what reward shall repay the faithful Ta-den? Greatly do we honor our priests. Within the temples even the chiefs and the king himself bow down to them. No greater honor could Ko-tan confer upon a subject—who wished to be a priest; but I did not so wish. Priests other than the high priest must become eunuchs for they may never marry.

"It was O-lo-a herself who brought word to me that her father had given the commands that would set in motion the machinery of the temple. A messenger was on his way in search of me to summon me to Ko-tan's presence. To have refused the priesthood once it was offered me by the king would have been to have affronted the temple and the gods —that would have meant death; but if I did not appear before Ko-tan I would not have to refuse anything. O-lo-a and I decided that I must not appear. It was better to fly, carrying in my bosom a shred of hope, than to remain and, with my priesthood, abandon hope forever.

"Beneath the shadows of the great trees that grow within the palace grounds I pressed her to me for, perhaps, the last time and then, lest by ill-fate I meet the messenger, I scaled the great wall that guards the palace and passed through the darkened city. My name and rank carried me beyond the city gate. Since then I have wandered far from the haunts of the Ho-don but strong within me is the urge to return if even but to look from without her walls upon the city that holds her most dear to me and again to visit the village of my birth, to see again my father and my mother."

"But the risk is too great?" asked Tarzan.

"It is great, but not too great," replied Ta-den. "I shall go."

"And I shall go with you, if I may," said the ape-man, "for I must see this City of Light, this A-lur of yours, and search there for my lost mate even though you believe that there is little chance that I find her. And you, Om-at, do you come with us?"

"Why not?" asked the hairy one. "The lairs of my tribe lie in the crags above A-lur and though Es-sat, our chief, drove me out I should like to return again, for there is a she there upon whom I should be glad to look once more and who would be glad to look upon me. Yes, I will go with you. Es-sat feared that I might become chief and who knows but that Es-sat was right. But Pan-at-lee! it is she I seek first even before a chieftainship."

"We three, then, shall travel together," said Tarzan.

"And fight together," added Ta-den; "the three as one," and as he spoke he drew his knife and held it above his head.

"The three as one," repeated Om-at, drawing his weapon and duplicating Ta-den's act. "It is spoken!"

"The three as one!" cried Tarzan of the Apes. "To the death!" and his blade flashed in the sunlight.

"Let us go, then," said Om-at; "my knife is dry and cries aloud for the blood of Es-sat."

The trail over which Ta-den and Om-at led and which scarcely could be dignified even by the name of trail was suited more to mountain sheep, monkeys, or birds than to man; but the three that followed it were trained to ways which no ordinary man might essay. Now, upon the lower slopes, it led through dense forests where the ground was so matted with fallen trees and over-rioting vines and brush that the way held always to the swaying branches high above the tangle; again it skirted yawning gorges whose slippery-faced rocks gave but momentary foothold even to the bare feet that lightly touched them as the three leaped chamois-like from one precarious foothold to the next. Dizzy and terrifying was the way that Om-at chose across the summit as he led them

around the shoulder of a towering crag that rose a sheer two
thousand feet of perpendicular rock above a tumbling river.
And when at last they stood upon comparatively level ground
again Om-at turned and looked at them both intently and
especially at Tarzan of the Apes.

"You will both do," he said. "You are fit companions for
Om-at, the Waz-don."

"What do you mean?" asked Tarzan.

"I brought you this way," replied the black, "to learn if
either lacked the courage to follow where Om-at led. It is
here that the young warriors of Es-sat come to prove their
courage. And yet, though we are born and raised upon cliff
sides, it is considered no disgrace to admit that Pastar-ul-ved,
the Father of Mountains, has defeated us, for of those who
try it only a few succeed—the bones of the others lie at the
feet of Pastar-ul-ved."

Ta-den laughed. "I would not care to come this way
often," he said.

"No," replied Om-at; "but it has shortened our journey by
at least a full day. So much the sooner shall Tarzan look
upon the Valley of Jad-ben-Otho. Come!" and he led the
way upward along the shoulder of Pastar-ul-ved until there
lay spread below them a scene of mystery and of beauty—a
green valley girt by towering cliffs of marble whiteness—a
green valley dotted by deep blue lakes and crossed by the
blue trail of a winding river. In the center a city of the whiteness of the marble cliffs—a city which even at so great a distance evidenced a strange, yet artistic architecture. Outside
the city there were visible about the valley isolated groups of
buildings—sometimes one, again two and three and four in
a cluster—but always of the same glaring whiteness, and always in some fantastic form.

About the valley the cliffs were occasionally cleft by deep

gorges, verdure filled, giving the appearance of green rivers rioting downward toward a central sea of green.

"Jad Pele ul Jad-ben-Otho," murmured Tarzan in the tongue of the pithecanthropi; "*The Valley of the Great God* —it is beautiful!"

"Here, in A-lur, lives Ko-tan, the king, ruler over all Pal-ul-don," said Ta-den.

"And here in these gorges live the Waz-don," exclaimed Om-at, "who do not acknowledge that Ko-tan is the ruler over all the *Land-of-man.*"

Ta-den smiled and shrugged. "We will not quarrel, you and I," he said to Om-at, "over that which all the ages have not proved sufficient time in which to reconcile the Ho-don and Waz-don; but let me whisper to you a secret, Om-at. The Ho-don live together in greater or less peace under one ruler so that when danger threatens them they face the enemy with many warriors, for every fighting Ho-don of Pal-ul-don is there. But you Waz-don, how is it with you? You have a dozen kings who fight not only with the Ho-don but with one another. When one of your tribes goes forth upon the fighting trail, even against the Ho-don, it must leave behind sufficient warriors to protect its women and its children from the neighbors upon either hand. When we want eunuchs for the temples or servants for the fields or the homes we march forth in great numbers upon one of your villages. You cannot even flee, for upon either side of you are enemies and though you fight bravely we come back with those who will presently be eunuchs in the temples and servants in our fields and homes. So long as the Waz-don are thus foolish the Ho-don will dominate and their king will be king of Pal-ul-don."

"Perhaps you are right," admitted Om-at. "It is because

our neighbors are fools, each thinking that his tribe is the greatest and should rule among the Waz-don. They will not admit that the warriors of my tribe are the bravest and our shes the most beautiful."

Ta-den grinned. "Each of the others presents precisely the same arguments that you present, Om-at," he said, "which, my friend, is the strongest bulwark of defense possessed by the Ho-don."

"Come!" exclaimed Tarzan; "such discussions often lead to quarrels and we three must have no quarrels. I, of course, am interested in learning what I can of the political and economic conditions of your land; I should like to know something of your religion; but not at the expense of bitterness between my only friends in Pal-ul-don. Possibly, however, you hold to the same god?"

"There indeed we do differ," cried Om-at, somewhat bitterly and with a trace of excitement in his voice.

"Differ!" almost shouted Ta-den; "and why should we not differ? Who could agree with the preposterous——"

"Stop!" cried Tarzan. "Now, indeed, have I stirred up a hornets' nest. Let us speak no more of matters political or religious."

"That is wiser," agreed Om-at; "but I might mention, for your information, that the one and only god has a long tail."

"It is sacrilege," cried Ta-den, laying his hand upon his knife; "Jad-ben-Otho has no tail!"

"Stop!" shrieked Om-at, springing forward; but instantly Tarzan interposed himself between them.

"Enough!" he snapped. "Let us be true to our oaths of friendship that we may be honorable in the sight of God in whatever form we conceive. Him."

"You are right, Tailless One," said Ta-den. "Come, Om-at,

let us look after our friendship and ourselves, secure in the conviction that Jad-ben-Otho is sufficiently powerful to look after himself."

"Done!" agreed Om-at, "but——"

"No 'buts,' Om-at," admonished Tarzan.

The shaggy black shrugged his shoulders and smiled. "Shall we make our way down toward the valley?" he asked. "The gorge below us is uninhabited; that to the left contains the caves of my people. I would see Pan-at-lee once more. Ta-den would visit his father in the valley below and Tarzan seeks entrance to A-lur in search of the mate that would be better dead than in the clutches of the Ho-don priests of Jad-ben-Otho. How shall we proceed?"

"Let us remain together as long as possible," urged Ta-den. "You, Om-at, must seek Pan-at-lee by night and by stealth, for three, even we three, may not hope to overcome Es-sat and all his warriors. At any time may we go to the village where my father is chief, for Ja-don always will welcome the friends of his son. But for Tarzan to enter A-lur is another matter, though there is a way and he has the courage to put it to the test—listen, come close for Jad-ben-Otho has keen ears and this he must not hear," and with his lips close to the ears of his companions Ta-den, the Tall-tree, son of Ja-don, the Lion-man, unfolded his daring plan.

And at the same moment, a hundred miles away, a lithe figure, naked but for a loin cloth and weapons, moved silently across a thorn-covered, waterless steppe, searching always along the ground before him with keen eyes and sensitive nostrils.

CHAPTER III

Pan-at-lee

N IGHT had fallen upon unchartered Pal-ul-don. A slender
moon, low in the west, bathed the white faces of the
chalk cliffs presented to her, in a mellow, unearthly glow.
Black were the shadows in Kor-ul-ja, Gorge-of-lions, where
dwelt the tribe of the same name under Es-sat, their chief.
From an aperture near the summit of the lofty escarpment a
hairy figure emerged—the head and shoulders first—and
fierce eyes scanned the cliff side in every direction.

It was Es-sat, the chief. To right and left and below he
looked as though to assure himself that he was unobserved,
but no other figure moved upon the cliff face, nor did another
hairy body protrude from any of the numerous cave mouths
from the high-flung abode of the chief to the habitations of
the more lowly members of the tribe nearer the cliff's base.
Then he moved outward upon the sheer face of the white
chalk wall. In the half-light of the baby moon it appeared
that the heavy, shaggy black figure moved across the face
of the perpendicular wall in some miraculous manner, but
closer examination would have revealed stout pegs, as large
around as a man's wrist protruding from holes in the cliff
into which they were driven. Es-sat's four handlike members and his long, sinuous tail permitted him to move with
consummate ease whither he chose—a gigantic rat upon a
mighty wall. As he progressed upon his way he avoided the

cave mouths, passing either above or below those that lay in
his path.

The outward appearance of these caves was similar. An
opening from eight to as much as twenty feet long by eight
high and four to six feet deep was cut into the chalklike rock
of the cliff, in the back of this large opening, which formed
what might be described as the front veranda of the home,
was an opening about three feet wide and six feet high, evidently forming the doorway to the interior apartment or
apartments. On either side of this doorway were smaller
openings which it were easy to assume were windows through
which light and air might find their way to the inhabitants.
Similar windows were also dotted over the cliff face between
the entrance porches, suggesting that the entire face of the
cliff was honeycombed with apartments. From many of
these smaller apertures small streams of water trickled down
the escarpment, and the walls above others was blackened
as by smoke. Where the water ran the wall was eroded to a
depth of from a few inches to as much as a foot, suggesting
that some of the tiny streams had been trickling downward
to the green carpet of vegetation below for ages.

In this primeval setting the great pithecanthropus aroused
no jarring discord for he was as much a part of it as the trees
that grew upon the summit of the cliff or those that hid their
feet among the dank ferns in the bottom of the gorge.

Now he paused before an entrance-way and listened and
then, noiselessly as the moonlight upon the trickling waters,
he merged with the shadows of the outer porch. At the
doorway leading into the interior he paused again, listening,
and then quietly pushing aside the heavy skin that covered
the aperture he passed within a large chamber hewn from
the living rock. From the far end, through another doorway,
shone a light, dimly. Toward this he crept with utmost

stealth, his naked feet giving forth no sound. The knotted club that had been hanging at his back from a thong about his neck he now removed and carried in his left hand.

Beyond the second doorway was a corridor running parallel with the cliff face. In this corridor were three more doorways, one at each end and a third almost opposite that in which Es-sat stood. The light was coming from an apartment at the end of the corridor at his left. A sputtering flame rose and fell in a small stone receptacle that stood upon a table or bench of the same material, a monolithic bench fashioned at the time the room was excavated, rising massively from the floor, of which it was a part.

In one corner of the room beyond the table had been left a dais of stone about four feet wide and eight feet long. Upon this were piled a foot or so of softly tanned pelts from which the fur had not been removed. Upon the edge of this dais sat a young female Waz-don. In one hand she held a thin piece of metal, apparently of hammered gold, with serrated edges, and in the other a short, stiff brush. With these she was occupied in going over her smooth, glossy coat which bore a remarkable resemblance to plucked sealskin. Her loin cloth of yellow and black striped *jato*-skin lay on the couch beside her with the circular breastplates of beaten gold, revealing the symmetrical lines of her nude figure in all its beauty and harmony of contour, for even though the creature was jet black and entirely covered with hair yet she was undeniably beautiful.

That she was beautiful in the eyes of Es-sat, the chief, was evidenced by the gloating expression upon his fierce countenance and the increased rapidity of his breathing. Moving quickly forward he entered the room and as he did so the young she looked up. Instantly her eyes filled with terror and as quickly she seized the loin cloth and with a few deft movements adjusted it about her. As she gathered up her breastplates Es-sat rounded the table and moved quickly toward her.

"What do you want?" she whispered, though she knew full well.

"Pan-at-lee," he said, "your chief has come for you."

"It was for this that you sent away my father and my brothers to spy upon the Kor-ul-lul? I will not have you. Leave the cave of my ancestors!"

Es-sat smiled. It was the smile of a strong and wicked man who knows his power—not a pleasant smile at all. "I will leave, Pan-at-lee," he said; "but you shall go with me—to the cave of Es-sat, the chief, to be the envied of the shes of Kor-ul-ja. Come!"

"Never!" cried Pan-at-lee. "I hate you. Sooner would I mate with a Ho-don than with you, beater of women, murderer of babes."

A frightful scowl distorted the features of the chief. "She-*jato!*" he cried. "I will tame you! I will break you! Es-sat, the chief, takes what he will and who dares question his right, or combat his least purpose, will first serve that purpose and then be broken as I break this," and he picked a stone platter from the table and broke it in his powerful hands. "You might have been first and most favored in the cave of the ancestors of Es-sat; but now shall you be last and least and when I am done with you you shall belong to all of the men of Es-sat's cave. Thus for those who spurn the love of their chief!"

He advanced quickly to seize her and as he laid a rough hand upon her she struck him heavily upon the side of his head with her golden breastplates. Without a sound Es-sat, the chief, sank to the floor of the apartment. For a moment Pan-at-lee bent over him, her improvised weapon raised to strike again should he show signs of returning consciousness, her glossy breasts rising and falling with her quickened breathing. Suddenly she stooped and removed Es-sat's knife with its scabbard and shoulder belt. Slipping it over her own shoulder she quickly adjusted her breastplates and keeping a watchful glance upon the figure of the fallen chief, backed from the room.

In a niche in the outer room, just beside the doorway leading to the balcony, were neatly piled a number of rounded pegs from eighteen to twenty inches in length. Selecting five of these she made them into a little bundle about which she twined the lower extremity of her sinuous tail and thus carrying them made her way to the outer edge of the balcony. Assuring herself that there was none about to see, or hinder her, she took quickly to the pegs already set in the face of the cliff and with the celerity of a monkey clambered swiftly aloft to the highest row of pegs which she followed in the direction of the lower end of the gorge for a matter of some hundred yards. Here, above her head, were a series of small round holes placed one above another in three parallel rows. Clinging only with her toes she removed two of the pegs from the bundle carried in her tail and taking one in either hand she inserted them in two opposite holes of the outer rows as far above her as she could reach. Hanging by these new holds she now took one of the three remaining pegs in each of her feet, leaving the fifth grasped securely in her tail. Reaching above her with this member she inserted the fifth peg in one of the holes of the center row and then, alternately hanging by her tail, her feet, or her hands, she moved the pegs upward to new holes, thus carrying her stairway with her as she ascended.

At the summit of the cliff a gnarled tree exposed its time-worn roots above the topmost holes forming the last step from the sheer face of the precipice to level footing. This was the last avenue of escape for members of the tribe hard pressed by enemies from below. There were three such emergency exits from the village and it were death to use them in other than an emergency. This Pan-at-lee well knew; but she knew, too, that it were worse than death to remain where the angered Es-sat might lay hands upon her.

When she had gained the summit, the girl moved quickly through the darkness in the direction of the next gorge which cut the mountain-side a mile beyond Kor-ul-ja. It was the Gorge-of-water, Kor-ul-lul, to which her father and two brothers had been sent by Es-sat ostensibly to spy upon the neighboring tribe. There was a chance, a slender chance, that she might find them; if not there was the deserted Kor-ul-gryf several miles beyond, where she might hide indefinitely from man if she could elude the frightful monster from which the gorge derived its name and whose presence there had rendered its caves uninhabitable for generations.

Pan-at-lee crept stealthily along the rim of the Kor-ul-lul. Just where her father and brothers would watch she did not know. Sometimes their spies remained upon the rim, sometimes they watched from the gorge's bottom. Pan-at-lee was at a loss to know what to do or where to go. She felt very small and helpless alone in the vast darkness of the night. Strange noises fell upon her ears. They came from the lonely reaches of the towering mountains above her, from far away in the invisible valley and from the nearer foothills and once, in the distance, she heard what she thought was the bellow of a bull *gryf*. It came from the direction of the Kor-ul-gryf. She shuddered.

Presently there came to her keen ears another sound. Something approached her along the rim of the gorge. It was coming from above. She halted, listening. Perhaps it was her

father, or a brother. It was coming closer. She strained her
eyes through the darkness. She did not move—she scarcely
breathed. And then, of a sudden, quite close it seemed, there
blazed through the black night two yellow-green spots of fire.

Pan-at-lee was brave, but as always with the primitive, the
darkness held infinite terrors for her. Not alone the terrors
of the known but more frightful ones as well—those of the
unknown. She had passed through much this night and her
nerves were keyed to the highest pitch—raw, taut nerves, they
were, ready to react in an exaggerated form to the slightest
shock.

But this was no slight shock. To hope for a father and a
brother and to see death instead glaring out of the darkness!
Yes, Pan-at-lee was brave, but she was not of iron. With a
shriek that reverberated among the hills she turned and fled
along the rim of Kor-ul-lul and behind her, swiftly, came the
devil-eyed lion of the mountains of Pal-ul-don.

Pan-at-lee was lost. Death was inevitable. Of this there
could be no doubt, but to die beneath the rending fangs of the
carnivore, congenital terror of her kind—it was unthinkable.
But there was an alternative. The lion was almost upon her—
another instant and he would seize her. Pan-at-lee turned
sharply to her left. Just a few steps she took in the new
direction before she disappeared over the rim of Kor-ul-lul.
The baffled lion, planting all four feet, barely stopped upon the
verge, of the abyss. Glaring down into the black shadows
beneath he mounted an angry roar.

Through the darkness at the bottom of Kor-ul-ja, Om-at led
the way toward the caves of his people. Behind him came
Tarzan and Ta-den. Presently they halted beneath a great
tree that grew close to the cliff.

"First," whispered Om-at, "I will go to the cave of Pan-at-lee.

Then will I seek the cave of my ancestors to have speech with my own blood. It will not take long. Wait here—I shall return soon. Afterward shall we go together to Ta-den's people."

He moved silently toward the foot of the cliff up which Tarzan could presently see him ascending like a great fly on a wall. In the dim light the ape-man could not see the pegs set in the face of the cliff. Om-at moved warily. In the lower tier of caves there should be a sentry. His knowledge of his people and their customs told him, however, that in all probability the sentry was asleep. In this he was not mistaken, yet he did not in any way abate his wariness. Smoothly and swiftly he ascended toward the cave of Pan-at-lee while from below Tarzan and Ta-den watched him.

"How does he do it?" asked Tarzan. "I can see no foothold upon that vertical surface and yet he appears to be climbing with the utmost ease."

Ta-den explained the stairway of pegs. "You could ascend easily," he said, "although a tail would be of great assistance."

They watched until Om-at was about to enter the cave of Pan-at-lee without seeing any indication that he had been observed and then, simultaneously, both saw a head appear in the mouth of one of the lower caves. It was quickly evident that its owner had discovered Om-at for immediately he started upward in pursuit. Without a word Tarzan and Ta-den sprang forward toward the foot of the cliff. The pithecanthropus was the first to reach it and the ape-man saw him spring upward for a handhold on the lowest peg above him. Now Tarzan saw other pegs roughly paralleling each other in zigzag rows up the cliff face. He sprang and caught one of these, pulled himself upward by one hand until he could reach a second with his other hand; and when he had ascended far enough to use his feet, discovered that he could make rapid progress. Ta-den was outstripping him, however, for these precarious ladders were no novelty to him and, further, he had an advantage in possessing a tail.

Nevertheless, the ape-man gave a good account of himself, being presently urged to redoubled efforts by the fact that the Waz-don above Ta-den glanced down and discovered his pursuers just before the Ho-don overtook him. Instantly a wild cry shattered the silence of the gorge—a cry that was immediately answered by hundreds of savage throats as warrior after warrior emerged from the entrance to his cave.

The creature who had raised the alarm had now reached the recess before Pan-at-lee's cave and here he halted and turned to give battle to Ta-den. Unslinging his club which had hung down his back from a thong about his neck he stood upon the level floor of the entrance-way effectually blocking Ta-den's ascent. From all directions the warriors of Kor-ul-ja were swarming toward the interlopers. Tarzan, who had reached a point on the same level with Ta-den but a little to the latter's left, saw that nothing short of a miracle could save them. Just at the ape-man's left was the entrance to a cave that either was deserted or whose occupants had not as yet been aroused, for the level recess remained unoccupied. Resourceful was the alert mind of Tarzan of the Apes and quick to respond were the trained muscles. In the time that you or I might give to debating an action he would accomplish it and now, though only seconds separated his nearest antagonist from him, in the brief span of time at his disposal he had stepped into the recess, unslung his long rope and leaning far out shot the sinuous noose, with the precision of long habitude, toward the menacing figure wielding its heavy club above Ta-den. There was a momentary pause of the rope-hand as the noose sped toward its goal, a quick movement of the right wrist that closed it upon its victim as it settled over his head and then a surging tug as, seizing the rope in both hands, Tarzan threw back upon it all the weight of his great frame.

Voicing a terrified shriek, the Waz-don lunged headforemost from the recess above Ta-den. Tarzan braced himself for the coming shock when the creature's body should have fallen the full length of the rope and as it did there was a snap of the vertebrae that rose sickeningly in the momentary silence that had followed the doomed man's departing scream. Unshaken by the stress of the suddenly arrested weight at the end of the rope, Tarzan quickly pulled the body to his side that he might remove the noose from about its neck, for he could not afford to lose so priceless a weapon.

During the several seconds that had elapsed since he cast the rope the Waz-don warriors had remained inert as though paralyzed by wonder or by terror. Now, again, one of them found his voice and his head and straightway, shrieking invectives at the strange intruder, started upward for the ape-man, urging his fellows to attack. This man was the closest to Tarzan. But for him the ape-man could easily have reached Ta-den's side as the latter was urging him to do. Tarzan raised the body of the dead Waz-don above his head, held it poised there for a moment as with face raised to the heavens he screamed forth the horrid challenge of the bull apes of the tribe of Kerchak, and with all the strength of his giant sinews he hurled the corpse heavily upon the ascending warrior. So great was the force of the impact that not only was the Waz-don torn from his hold but two of the pegs to which he clung were broken short in their sockets.

As the two bodies, the living and the dead, hurtled downward toward the foot of the cliff a great cry arose from the Waz-don. "Jad-guru-don! Jad-guru-don!" they screamed, and then: "Kill him! Kill him!"

And now Tarzan stood in the recess beside Ta-den. "Jad-guru-don!" repeated the latter, smiling—"The terrible man! Tarzan the Terrible! They may kill you, but they will never forget you."

"They shall not ki—What have we here?" Tarzan's statement as to what "they" should not do was interrupted by a sudden ejaculation as two figures, locked in deathlike embrace, stumbled through the doorway of the cave to the outer porch. One was Om-at, the other a creature of his own kind but with a rough coat, the hairs of which seemed to grow straight outward from the skin, stiffly, unlike Om-at's sleek covering. The two were quite evidently well matched and equally evident was the fact that each was bent upon murder. They fought almost in silence except for an occasional low growl as one or the other acknowledged thus some new hurt.

Tarzan, following a natural impulse to aid his ally, leaped forward to enter the dispute only to be checked by a grunted admonition from Om-at. "Back!" he said. "This fight is mine, alone."

The ape-man understood and stepped aside.

"It is a *gund-bar*," explained Ta-den, "a *chief-battle*. This fellow must be Es-sat, the chief. If Om-at kills him without assistance Om-at may become chief."

Tarzan smiled. It was the law of his own jungle—the law of the tribe of Kerchak, the bull ape—the ancient law of primitive man that needed but the refining influences of civilization to introduce the hired dagger and the poison cup. Then his attention was drawn to the outer edge of the vestibule. Above it appeared the shaggy face of one of Es-sat's warriors. Tarzan sprang to intercept the man; but Ta-den was there ahead of him. "Back!" cried the Ho-don to the newcomer. "It is *gund-bar*." The fellow looked scrutinizingly at the two fighters, then turned his face downward toward his fellows. "Back!" he cried, "it is *gund-bar* between Es-sat and Om-at." Then he looked back at Ta-den and Tarzan. "Who are you?" he asked.

"We are Om-at's friends," replied Ta-den.

The fellow nodded. "We will attend to you later," he said and disappeared below the edge of the recess.

The battle upon the ledge continued with unabated ferocity, Tarzan and Ta-den having difficulty in keeping out of the way of the contestants who tore and beat at each other with hands and feet and lashing tails. Es-sat was unarmed—Pan-at-lee had seen to that—but at Om-at's side swung a sheathed knife which he made no effort to draw. That would have been contrary to their savage and primitive code for the chief-battle must be fought with nature's weapons.

Sometimes they separated for an instant only to rush upon each other again with all the ferocity and nearly the strength of mad bulls. Presently one of them tripped the other but in that viselike embrace one could not fall alone—Es-sat dragged Om-at with him, toppling upon the brink of the niche. Even Tarzan held his breath. There they surged to and fro perilously for a moment and then the inevitable happened—the two, locked in murderous embrace, rolled over the edge and disappeared from the ape-man's view.

Tarzan voiced a suppressed sigh for he had liked Om-at and then, with Ta-den, approached the edge and looked over. Far below, in the dim light of the coming dawn, two inert forms should be lying stark in death; but, to Tarzan's amazement, such was far from the sight that met his eyes. Instead, there were the two figures still vibrant with life and still battling only a few feet below him. Clinging always to the pegs with two holds—a hand and a foot, or a foot and a tail, they seemed as much at home upon the perpendicular wall as upon the level surface of the vestibule; but now their tactics were slightly altered, for each seemed particularly bent upon dislodging his antagonist from his holds and precipitating him to certain death below. It was soon evident that Om-at, younger and with greater powers of endurance than Es-sat, was gaining an advantage. Now was the chief almost wholly on the defensive. Holding him by the cross belt with one mighty hand Om-at was forcing his foeman straight out from the cliff, and with the other hand and one foot was rapidly breaking first one of Es-sat's holds and then another, alternating his efforts, or rather punctuating them, with vicious blows to the pit of his adversary's stomach. Rapidly was Es-sat weakening and with the knowledge of impending death there came, as there comes to every coward and bully under similar circumstances, a crumbling of the veneer of bravado which had long masqueraded as courage and with it crumbled his code of ethics. Now was Es-sat no longer chief of Kor-ul-ja—instead he was a whimpering craven battling for life. Clutching at Om-at, clutching at the nearest pegs he sought any support that would save him from that awful fall, and as he strove to push aside the hand of death, whose cold fingers he already felt upon his heart, his tail sought Om-at's side and the handle of the knife that hung there.

Tarzan saw and even as Es-sat drew the blade from its sheath he dropped catlike to the pegs beside the battling men. Es-sat's tail had drawn back for the cowardly fatal thrust. Now many others saw the perfidious act and a great cry of rage and disgust arose from savage throats; but as the blade sped toward its goal, the ape-man seized the hairy member that wielded it, and at the same instant Om-at thrust the body of Es-sat from him with such force that its weakened holds were broken and it hurtled downward, a brief meteor of screaming fear, to death.

CHAPTER IV

Tarzan-jad-guru

As TARZAN and Om-at clambered back to the vestibule of Pan-at-lee's cave and took their stand beside Ta-den in readiness for whatever eventuality might follow the death of Es-sat, the sun that topped the eastern hills touched also the figure of a sleeper upon a distant, thorn-covered steppe awakening him to another day of tireless tracking along a faint and rapidly disappearing spoor.

For a time silence reigned in the Kor-ul-ja. The tribesmen waited, looking now down upon the dead thing that had been their chief, now at one another, and now at Om-at and the two who stood upon his either side. Presently Om-at spoke. "I am Om-at," he cried. "Who will say that Om-at is not *gund* of Kor-ul-ja?"

He waited for a taker of his challenge. One or two of the larger young bucks fidgeted restlessly and eyed him; but there was no reply.

"Then Om-at is *gund*," he said with finality. "Now tell me, where are Pan-at-lee, her father, and her brothers?"

An old warrior spoke. "Pan-at-lee should be in her cave. Who should know that better than you who are there now? Her father and her brothers were sent to watch Kor-ul-lul; but neither of these questions arouse any tumult in our breasts. There is one that does: Can Om-at be chief of Kor-ul-ja and yet stand at bay against his own people with a Ho-don and that terrible man at his side—that terrible man who has no tail? Hand the strangers over to your people to be slain as is the way of the Waz-don and then may Om-at be *gund*."

Neither Tarzan nor Ta-den spoke then; they but stood watching Om-at and waiting for his decision, the ghost of a smile upon the lips of the ape-man. Ta-den, at least, knew that the old warrior had spoken the truth—the Waz-don entertain no strangers and take no prisoners of an alien race.

Then spoke Om-at. "Always there is change," he said. "Even the old hills of Pal-ul-don appear never twice alike—the brilliant sun, a passing cloud, the moon, a mist, the changing seasons, the sharp clearness following a storm; these things bring each a new change in our hills. From birth to death, day by day, there is constant change in each of us. Change, then, is one of Jad-ben-Otho's laws.

"And now I, Om-at, your *gund*, bring another change. Strangers who are brave men and good friends shall no longer be slain by the Waz-don of Kor-ul-ja!"

There were growls and murmurings and a restless moving among the warriors as each eyed the others to see who would take the initiative against Om-at, the iconoclast.

"Cease your mutterings," admonished the new *gund*. "I am your chief. My word is your law. You had no part in making me chief. Some of you helped Es-sat to drive me from the cave of my ancestors; the rest of you permitted it. I owe you nothing. Only these two, whom you would have me kill, were loyal to me. I am *gund* and if there be any who doubts it let him speak—he cannot die younger."

Tarzan was pleased. Here was a man after his own heart. He admired the fearlessness of Om-at's challenge and he was a sufficiently good judge of men to know that he had listened to no idle bluff—Om-at would back up his words to the death, if necessary, and the chances were that he would not be the one to die. Evidently the majority of the Kor-ul-jaians entertained the same conviction.

"I will make you a good *gund*," said Om-at, seeing that no one appeared inclined to dispute his rights. "Your wives and daughters will be safe—they were not safe while Es-sat ruled. Go now to your crops and your hunting. I leave to search for Pan-at-lee. Ab-on will be *gund* while I am away—look to him for guidance and to me for an accounting when I return—and may Jad-ben-Otho smile upon you."

He turned toward Tarzan and the Ho-don. "And you, my friends," he said, "are free to go among my people; the cave of my ancestors is yours; do what you will."

"I," said Tarzan, "will go with Om-at to search for Pan-at-lee."

"And I," said Ta-den.

Om-at smiled. "Good!" he exclaimed. "And when we have found her we shall go together upon Tarzan's business and Ta-den's. Where first shall we search?" He turned toward his warriors. "Who knows where she may be?"

None knew other than that Pan-at-lee had gone to her cave with the others the previous evening—there was no clew, no suggestion as to her whereabouts.

"Show me where she sleeps," said Tarzan; "let me see something that belongs to her—an article of her apparel—then, doubtless, I can help you."

Two young warriors climbed closer to the ledge upon which Om-at stood. They were In-sad and O-dan. It was the latter who spoke.

"*Gund* of Kor-ul-ja," he said, "we would go with you to search for Pan-at-lee."

It was the first acknowledgment of Om-at's chieftainship

and immediately following it the tenseness that had prevailed seemed to relax—the warriors spoke aloud instead of in whispers, and the women appeared from the mouths of caves as with the passing of a sudden storm. In-sad and O-dan had taken the lead and now all seemed glad to follow. Some came to talk with Om-at and to look more closely at Tarzan; others, heads of caves, gathered their hunters and discussed the business of the day. The women and children prepared to descend to the fields with the youths and the old men, whose duty it was to guard them.

"O-dan and In-sad shall go with us," announced Om-at, "we shall not need more. Tarzan, come with me and I shall show you where Pan-at-lee sleeps, though why you should wish to know I cannot guess—she is not there. I have looked for myself."

The two entered the cave where Om-at led the way to the apartment in which Es-sat had surprised Pan-at-lee the previous night.

"All here are hers," said Om-at, "except the war club lying on the floor—that was Es-sat's."

The ape-man moved silently about the apartment, the quivering of his sensitive nostrils scarcely apparent to his companion who only wondered what good purpose could be served here and chafed at the delay.

"Come!" said the ape-man, presently, and led the way toward the outer recess.

Here their three companions were awaiting them. Tarzan passed to the left side of the niche and examined the pegs that lay within reach. He looked at them but it was not his eyes that were examining them. Keener than his keen eyes was that marvelously trained sense of scent that had first been developed in him during infancy under the tutorage of his

foster mother, Kala, the she-ape, and further sharpened in
the grim jungles by that master teacher—the instinct of self-
preservation.

From the left side of the niche he turned to the right. Om-at
was becoming impatient.

"Let us be off," he said. "We must search for Pan-at-lee if
we would ever find her."

"Where shall we search?" asked Tarzan.

Om-at scratched his head. "Where?" he repeated. "Why
all Pal-ul-don, if necessary."

"A large job," said Tarzan. "Come," he added, "she went
this way," and he took to the pegs that led aloft toward the
summit of the cliff. Here he followed the scent easily since
none had passed that way since Pan-at-lee had fled. At the
point at which she had left the permanent pegs and resorted
to those carried with her Tarzan came to an abrupt halt. "She
went this way to the summit," he called back to Om-at who
was directly behind him; "but there are no pegs here."

"I do not know how you know that she went this way," said
Om-at; "but we will get pegs. In-sad, return and fetch climbing pegs for five."

The young warrior was soon back and the pegs distributed.
Om-at handed five to Tarzan and explained their use. The
ape-man returned one. "I need but four," he said.

Om-at smiled. "What a wonderful creature you would be
if you were not deformed," he said, glancing with pride at his
own strong tail.

"I admit that I am handicapped," replied Tarzan. "You
others go ahead and leave the pegs in place for me. I am
afraid that otherwise it will be slow work as I cannot hold the
pegs in my toes as you do."

"All right," agreed Om-at; "Ta-den, In-sad, and I will go

first, you follow and O-dan bring up the rear and collect the pegs—we cannot leave them here for our enemies."

"Can't your enemies bring their own pegs?" asked Tarzan.

"Yes; but it delays them and makes easier our defense and —they do not know which of all the holes you see are deep enough for pegs—the others are made to confuse our enemies and are too shallow to hold a peg."

At the top of the cliff beside the gnarled tree Tarzan again took up the trail. Here the scent was fully as strong as upon the pegs and the ape-man moved rapidly across the ridge in the direction of the Kor-ul-lul.

Presently he paused and turned toward Om-at. "Here she moved swiftly, running at top speed, and, Om-at, she was pursued by a lion."

"You can read that in the grass?" asked O-dan as the others gathered about the ape-man.

Tarzan nodded. "I do not think the lion got her," he added; "but that we shall determine quickly. No, he did not get her —look!" and he pointed toward the southwest, down the ridge.

Following the direction indicated by his finger, the others presently detected a movement in some bushes a couple of hundred yards away.

"What is it?" asked Om-at. "Is it she?" and he started toward the spot.

"Wait," advised Tarzan. "It is the lion which pursued her."

"You can see him?" asked Ta-den.

"No, I can smell him."

The others looked their astonishment and incredulity; but of the fact that it was indeed a lion they were not left long in doubt. Presently the bushes parted and the creature stepped out in full view, facing them. It was a magnificent beast, large and beautifully maned, with the brilliant leopard spots of its kind well marked and symmetrical. For a moment it eyed them and then, still chafing at the loss of its prey earlier in the morning, it charged.

The Pal-ul-donians unslung their clubs and stood waiting the onrushing beast. Tarzan of the Apes drew his hunting knife and crouched in the path of the fanged fury. It was almost upon him when it swerved to the right and leaped for Om-at only to be sent to earth with a staggering blow upon the head. Almost instantly it was up and though the men rushed fearlessly in, it managed to sweep aside their weapons with its mighty paws. A single blow wrenched O-dan's club from his hand and sent it hurtling against Ta-den, knocking him from his feet. Taking advantage of its opportunity the lion rose to throw itself upon O-dan and at the same instant Tarzan flung himself upon its back. Strong, white teeth buried themselves in the spotted neck, mighty arms encircled the savage throat and the sinewy legs of the ape-man locked themselves about the gaunt belly.

The others, powerless to aid, stood breathlessly about as the great lion lunged hither and thither, clawing and biting fearfully and futilely at the savage creature that had fastened itself upon him. Over and over they rolled and now the onlookers saw a brown hand raised above the lion's side—a brown hand grasping a keen blade. They saw it fall and rise and fall again—each time with terrific force and in its wake they saw a crimson stream trickling down *ja's* gorgeous coat.

Now from the lion's throat rose hideous screams of hate and rage and pain as he redoubled his efforts to dislodge and punish his tormentor; but always the tousled black head remained half buried in the dark brown mane and the mighty arm rose and fell to plunge the knife again and again into the dying beast.

The Pal-ul-donians stood in mute wonder and admiration. Brave men and mighty hunters they were and as such the first to accord honor to a mightier.

"And you would have had me slay him!" cried Om-at, glancing at In-sad and O-dan.

"Jad-ben-Otho reward you that you did not," breathed In-sad.

And now the lion lunged suddenly to earth and with a few spasmodic quiverings lay still. The ape-man rose and shook himself, even as might *ja*, the leopard-coated lion of Pal-ul-don, had he been the one to survive.

O-dan advanced quickly toward Tarzan. Placing a palm upon his own breast and the other on Tarzan's. "Tarzan the Terrible," he said, "I ask no greater honor than your friendship."

"And I no more than the friendship of Om-at's friends," replied the ape-man simply, returning the other's salute.

"Do you think," asked Om-at, coming close to Tarzan and laying a hand upon the other's shoulder, "that he got her?"

"No, my friend; it was a hungry lion that charged us."

"You seem to know much of lions," said In-sad.

"Had I a brother I could not know him better," replied Tarzan.

"Then where can she be?" continued Om-at.

"We can but follow while the spoor is fresh," answered the ape-man and again taking up his interrupted tracking he led them down the ridge and at a sharp turning of the trail to the left brought them to the verge of the cliff that dropped into the Kor-ul-lul. For a moment Tarzan examined the ground to the right and to the left, then he stood erect and looking at Om-at pointed into the gorge.

For a moment the Waz-don gazed down into the green rift at the bottom of which a tumultuous river tumbled downward along its rocky bed, then he closed his eyes as to a sudden spasm of pain and turned away.

"You—mean—she jumped?" he asked.

"To escape the lion," replied Tarzan. "He was right behind her—look, you can see where his four paws left their impress in the turf as he checked his charge upon the very verge of the abyss."

"Is there any chance—" commenced Om-at, to be suddenly silenced by a warning gesture from Tarzan.

"Down!" whispered the ape-man, "many men are coming. They are running—from down the ridge." He flattened himself upon his belly in the grass, the others following his example.

For some minutes they waited thus and then the others, too, heard the sound of running feet and now a hoarse shout followed by many more.

"It is the war cry of the Kor-ul-lul," whispered Om-at—"the hunting cry of men who hunt men. Presently shall we see them and if Jad-ben-Otho is pleased with us they shall not too greatly outnumber us."

"They are many," said Tarzan, "forty or fifty, I should say; but how many are the pursued and how many the pursuers we cannot even guess, except that the latter must greatly outnumber the former, else these would not run so fast."

"Here they come," said Ta-den.

"It is An-un, father of Pan-at-lee, and his two sons," exclaimed O-dan. "They will pass without seeing us if we do not hurry," he added looking at Om-at, the chief, for a sign.

"Come!" cried the latter, springing to his feet and running rapidly to intercept the three fugitives. The others followed him.

"Five friends!" shouted Om-at as An-un and his sons discovered them.

"*Adenen yo!*" echoed O-dan and In-sad.

The fugitives scare paused as these unexpected reinforcements joined them but they eyed Ta-den and Tarzan with puzzled glances.

"The Kor-ul-lul are many," shouted An-un. "Would that we might pause and fight; but first we must warn Es-sat and our people."

"Yes," said Om-at, "we must warn our people."

"Es-sat is dead," said In-sad.

"Who is chief?" asked one of An-un's sons.

"Om-at," replied O-dan.

"It is well," cried An-un. "Pan-at-lee said that you would come back and slay Es-sat."

Now the enemy broke into sight behind them.

"Come!" cried Tarzan, "let us turn and charge them, raising a great cry. They pursued but three and when they see eight charging upon them they will think that many men have come to do battle. They will believe that there are more even than they see and then one who is swift will have time to reach the gorge and warn your people."

"It is well," said Om-at. "Id-an, you are swift—carry word to the warriors of Kor-ul-ja that we fight the Kor-ul-lul upon the ridge and that Ab-on shall send a hundred men."

Id-an, the son of An-un, sped swiftly toward the cliff-dwellings of the Kor-ul-ja while the others charged the oncoming Kor-ul-lul, the war cries of the two tribes rising and falling in a certain grim harmony. The leaders of the Kor-ul-lul paused at sight of the reinforcements, waiting apparently for those behind to catch up with them and, possibly, also to learn how great a force confronted them. The leaders, swifter runners than their fellows, perhaps, were far in advance while the balance of their number had not yet emerged from the brush; and now as Om-at and his companions fell upon them with a ferocity born of necessity they fell back, so that when their companions at last came in sight of them they appeared to be in full rout. The natural result was that the others turned and fled.

Encouraged by this first success Om-at followed them into the brush, his little company charging valiantly upon his either side, and loud and terrifying were the savage yells with which they pursued the fleeing enemy. The brush, while not growing so closely together as to impede progress, was of such height as to hide the members of the party from one another when they became separated by even a few yards. The result was that Tarzan, always swift and always keen for battle, was soon pursuing the enemy far in the lead of the others—a lack of prudence which was to prove his undoing.

The warriors of Kor-ul-lul, doubtless as valorous as their foemen, retreated only to a more strategic position in the brush, nor were they long in guessing that the number of their pursuers was fewer than their own. They made a stand then where the brush was densest—an ambush it was, and into this ran Tarzan of the Apes. They tricked him neatly. Yes, sad as is the narration of it, they tricked the wily jungle lord. But then they were fighting on their own ground, every foot of which they knew as you know your front parlor, and they were following their own tactics, of which Tarzan knew nothing.

A single black warrior appeared to Tarzan a laggard in the rear of the retreating enemy and thus retreating he lured Tarzan on. At last he turned at bay confronting the ape-man with bludgeon and drawn knife and as Tarzan charged him a score of burly Waz-don leaped from the surrounding brush. Instantly, but too late, the giant Tarmangani realized his peril. There flashed before him a vision of his lost mate and a great and sickening regret surged through him with the realization that if she still lived she might no longer hope, for though she might never know of the passing of her lord the fact of it must inevitably seal her doom.

And consequent to this thought there enveloped him a blind frenzy of hatred for these creatures who dared thwart his purpose and menace the welfare of his wife. With a savage growl he threw himself upon the warrior before him twisting the heavy club from the creature's hand as if he had been a little child, and with his left fist backed by the weight and sinew of his giant frame, he crashed a shattering blow to the center of the Waz-don's face—a blow that crushed the bones and dropped the fellow in his tracks. Then he swung upon the others with their fallen comrade's bludgeon striking to right and left mighty, unmerciful blows that drove down their own weapons until that wielded by the ape-man was splintered and shattered. On either hand they fell before his cudgel; so rapid the delivery of his blows, so catlike his recovery that in the first few moments of the battle he seemed invulnerable to their attack; but it could not last—he was outnumbered twenty to one and his undoing came from a thrown club. It struck him upon the back of the head. For a moment he stood swaying and then like a great pine beneath the woodsman's ax he crashed to earth.

Others of the Kor-ul-lul had rushed to engage the balance of Om-at's party. They could be heard fighting at a short distance and it was evident that the Kor-ul-ja were falling slowly back and as they fell Om-at called to the missing one: "Tarzan the Terrible! Tarzan the Terrible!"

"Jad-guru, indeed," repeated one of the Kor-ul-lul rising from where Tarzan had dropped him. "Tarzan-jad-guru! He was worse than that."

CHAPTER V

In the Kor-ul-gryf

A S TARZAN fell among his enemies a man halted many miles away upon the outer verge of the morass that encircles Pal-ul-don. Naked he was except for a loin cloth and three belts of cartridges, two of which passed over his shoulders, crossing upon his chest and back, while the third encircled his waist. Slung to his back by its leathern sling-strap was an Enfield, and he carried too a long knife, a bow and a quiver of arrows. He had come far, through wild and savage lands, menaced by fierce beasts and fiercer men, yet intact to the last cartridge was the ammunition that had filled his belts the day that he set out.

The bow and the arrows and the long knife had brought him thus far safely, yet often in the face of great risks that could have been minimized by a single shot from the well-kept rifle at his back. What purpose might he have for conserving this precious ammunition? in risking his life to bring the last bright shining missile to his unknown goal? For what, for whom were these death-dealing bits of metal preserved? In all the world only he knew.

When Pan-at-lee stepped over the edge of the cliff above Kor-ul-lul she expected to be dashed to instant death upon the rocks below; but she had chosen this in preference to the rending fangs of *ja*. Instead, chance had ordained that she make the frightful plunge at a point where the tumbling river swung close beneath the overhanging cliff to eddy for a slow moment in a deep pool before plunging madly downward again in a cataract of boiling foam, and water thundering against rocks.

Into this icy pool the girl shot, and down and down beneath the watery surface until, half choked, yet fighting bravely, she battled her way once more to air. Swimming strongly she made the opposite shore and there dragged herself out upon the bank to lie panting and spent until the approaching dawn warned her to seek concealment, for she was in the country of her people's enemies.

Rising, she moved into the concealment of the rank vegetation that grows so riotously in the well-watered *kors* [1] of Pal-ul-don.

Hidden amidst the plant life from the sight of any who might chance to pass along the well-beaten trail that skirted the river Pan-at-lee sought rest and food, the latter growing in abundance all about her in the form of fruits and berries and succulent tubers which she scooped from the earth with the knife of the dead Es-sat.

Ah! if she had but known that he was dead! What trials and risks and terrors she might have been saved; but she thought that he still lived and so she dared not return to Kor-ul-ja. At least not yet while his rage was at white heat. Later, perhaps, her father and brothers returned to their cave, she might risk it; but not now—not now. Nor could she for

---

[1] I have used the Pal-ul-don word for *gorge* with the English plural, which is not the correct native plural form. The latter, it seems to me, is awkward for us and so I have generally ignored it throughout my manuscript, permitting, for example, Kor-ul-ja to answer for both singular and plural. However, for the benefit of those who may be interested in such things I may say that the plurals are formed simply for all words in the Pal-ul-don language by doubling the initial letter of the word, as *k'kor, gorges,* pronounced as though written *kakor,* the *a* having the sound of *a* in *sofa. Lions,* then, would be *j'ja,* or *men d' don.*

long remain here in the neighborhood of the hostile Kor-ul-lul and somewhere she must find safety from beasts before the night set in.

As she sat upon the bole of a fallen tree seeking some solution of the problem of existence that confronted her, there broke upon her ears from up the gorge the voices of shouting men—a sound that she recognized all too well. It was the war cry of the Kor-ul-lul. Closer and closer it approached her hiding place. Then, through the veil of foliage she caught glimpses of three figures fleeing along the trail, and behind them the shouting of the pursuers rose louder and louder as they neared her. Again she caught sight of the fugitives crossing the river below the cataract and again they were lost to sight. And now the pursuers came into view—shouting Kor-ul-lul warriors, fierce and implacable. Forty, perhaps fifty, of them. She waited breathless; but they did not swerve from the trail and passed her, unguessing that an enemy she lay hid within a few yards of them.

Once again she caught sight of the pursued—three Waz-don warriors clambering the cliff face at a point where portions of the summit had fallen away presenting a steep slope that might be ascended by such as these. Suddenly her attention was riveted upon the three. Could it be? O Jad-ben-Otho! had she but known a moment before. When they passed she might have joined them, for they were her father and two brothers. Now it was too late. With bated breath and tense muscles she watched the race. Would they reach the summit? Would the Kor-ul-lul overhaul them? They climbed well, but, oh, so slowly. Now one lost his footing in the loose shale and slipped back! The Kor-ul-lul were ascending—one hurled his club at the nearest fugitive. The Great God was pleased with the brother of Pan-at-lee, for he caused the club to fall short of its target, and to fall, rolling and bounding, back upon its owner carrying him from his feet and precipitating him to the bottom of the gorge.

Standing now, her hands pressed tight above her golden breastplates, Pan-at-lee watched the race for life. Now one, her older brother, reached the summit and clinging there to something that she could not see he lowered his body and his long tail to the father beneath him. The latter, seizing this support, extended his own tail to the son below—the one who had slipped back—and thus, upon a living ladder of their own making, the three reached the summit and disappeared from view before the Kor-ul-lul overtook them. But the latter did not abandon the chase. On they went until they too had disappeared from sight and only a faint shouting came down to Pan-at-lee to tell her that the pursuit continued.

The girl knew that she must move on. At any moment now might come a hunting party, combing the gorge for the smaller animals that fed or bedded there.

Behind her were Es-sat and the returning party of Kor-ul-lul that had pursued her kin; before her, across the next ridge, was the Kor-ul-gryf, the lair of the terrifying monsters that brought the chill of fear to every inhabitant of Pal-ul-don; below her, in the valley, was the country of the Ho-don, where she could look for only slavery, or death; here were the Kor-ul-lul, the ancient enemies of her people and everywhere were the wild beasts that eat the flesh of man.

For but a moment she debated and then turning her face toward the southeast she set out across the gorge of water toward the Kor-ul-gryf—at least there were no men there. As it is now, so it was in the beginning, back to the primitive progenitor of man which is typified by Pan-at-lee and her kind today, of all the hunters that woman fears, man is the most relentless, the most terrible. To the dangers of man she preferred the dangers of the *gryf*.

Moving cautiously she reached the foot of the cliff at the far side of Kor-ul-lul and here, toward noon, she found a comparatively easy ascent. Crossing the ridge she stood at last upon the brink of Kor-ul-gryf—the horror place of the folklore of her race. Dank and mysterious grew the vegetation below; giant trees waved their plumed tops almost level with the summit of the cliff; and over all brooded an ominous silence.

Pan-at-lee lay upon her belly and stretching over the edge scanned the cliff face below her. She could see caves there and the stone pegs which the ancients had fashioned so laboriously by hand. She had heard of these in the firelight tales of her childhood and of how the *gryfs* had come from the morasses across the mountains and of how at last the people had fled after many had been seized and devoured by the hideous creatures, leaving their caves untenanted for no man living knew how long. Some said that Jad-ben-Otho, who has lived forever, was still a little boy. Pan-at-lee shuddered; but there were caves and in them she would be safe even from the *gryfs*.

She found a place where the stone pegs reached to the very summit of the cliff, left there no doubt in the final exodus of the tribe when there was no longer need of safeguarding the deserted caves against invasion. Pan-at-lee clambered slowly down toward the uppermost cave. She found the recess in front of the doorway almost identical with those of her own tribe. The floor of it, though, was littered with twigs and old nests and the droppings of birds, until it was half choked. She moved along to another recess and still another, but all were alike in the accumulated filth. Evidently there was no need in looking further. This one seemed large and commodious. With her knife she fell to work cleaning away the débris by the simple expedient of pushing it over the edge, and always her eyes turned constantly toward the silent gorge where lurked the fearsome creatures of Pal-ul-don. And other eyes there were, eyes she did not see, but that saw her and watched her every move—fierce eyes, greedy eyes, cunning and cruel. They watched her, and a red tongue licked flabby, pendulous lips. They watched her, and a half-human brain laboriously evolved a brutish design.

As in her own Kor-ul-ja, the natural springs in the cliff had been developed by the long-dead builders of the caves so that fresh, pure water trickled now, as it had for ages, within easy access to the cave entrances. Her only difficulty would be in procuring food and for that she must take the risk at least once in two days, for she was sure that she could find fruits and tubers and perhaps small animals, birds, and eggs near the foot of the cliff, the last two, possibly, in the caves themselves. Thus might she live on here indefinitely. She felt now a certain sense of security imparted doubtless by the impregnability of her high-flung sanctuary that she knew to be safe from all the more dangerous beasts, and this one from men, too, since it lay in the abjured Kor-ul-gryf.

Now she determined to inspect the interior of her new home. The sun still in the south, lighted the interior of the first apartment. It was similar to those of her experience—the same beasts and men were depicted in the same crude fashion in the carvings on the walls—evidently there had been little progress in the race of Waz-don during the generations that had come and departed since Kor-ul-gryf had been abandoned by men. Of course Pan-at-lee thought no such thoughts, for evolution and progress existed not for her, or her kind. Things were as they had always been and would always be as they were.

That these strange creatures have existed thus for incalculable ages it can scarce be doubted, so marked are the indications of antiquity about their dwellings—deep furrows worn by naked feet in living rock; the hollow in the jamb of a stone doorway where many arms have touched in passing; the end-less carvings that cover, ofttimes, the entire face of a great cliff and all the walls and ceilings of every cave and each carving wrought by a different hand, for each is the coat of arms, one might say, of the adult male who traced it.

And so Pan-at-lee found this ancient cave homelike and familiar. There was less litter within than she had found without and what there was was mostly an accumulation of dust. Beside the doorway was the niche in which wood and tinder were kept, but there remained nothing now other than mere dust. She had however saved a little pile of twigs from the débris on the porch. In a short time she had made a light by firing a bundle of twigs and lighting others from this fire she explored some of the inner rooms. Nor here did she find aught that was new or strange nor any relic of the departed owners other than a few broken stone dishes. She had been looking for something soft to sleep upon, but was doomed to disappointment as the former owners had evidently made a leisurely departure, carrying all their belongings with them. Below, in the gorge were leaves and grasses and fragrant branches, but Pan-at-lee felt no stomach for descending into that horrid abyss for the gratification of mere creature com-fort—only the necessity for food would drive her there.

And so, as the shadows lengthened and night approached she prepared to make as comfortable a bed as she could by gathering the dust of ages into a little pile and spreading it between her soft body and the hard floor—at best it was only better than nothing. But Pan-at-lee was very tired. She had not slept since two nights before and in the interval she had experienced many dangers and hardships. What wonder then that despite the hard bed, she was asleep almost immediately she had composed herself for rest.

She slept and the moon rose, casting its silver light upon the cliff's white face and lessening the gloom of the dark forest and the dismal gorge. In the distance a lion roared. There was a long silence. From the upper reaches of the gorge came a deep bellow. There was a movement in the trees at the cliff's foot. Again the bellow, low and ominous. It was answered from below the deserted village. Something dropped from the foliage of a tree directly below the cave in which Pan-at-lee slept—it dropped to the ground among the dense shadows. Now it moved, cautiously. It moved toward the foot of the cliff, taking form and shape in the moonlight. It moved like the creature of a bad dream—slowly, sluggishly. It might have been a huge sloth—it might have been a man, with so grotesque a brush does the moon paint—master cubist.

Slowly it moved up the face of the cliff—like a great grubworm it moved; but now the moon-brush touched it again and it had hands and feet and with them it clung to the stone pegs and raised itself laboriously aloft toward the cave where Pan-at-lee slept. From the lower reaches of the gorge came again the sound of bellowing, and it was answered from above the village.

Tarzan of the Apes opened his eyes. He was conscious of a pain in his head, and at first that was about all. A moment later grotesque shadows, rising and falling, focused his arousing perceptions. Presently he saw that he was in a cave. A dozen Waz-don warriors squatted about, talking. A rude stone cresset containing burning oil lighted the interior and as the flame rose and fell the exaggerated shadows of the warriors danced upon the walls behind them.

"We brought him to you alive, *Gund*," he heard one of them saying, "because never before was Ho-don like him seen. He has no tail—he was born without one, for there is no scar to

mark where a tail had been cut off. The thumbs upon his
hands and feet are unlike those of the races of Pal-ul-don. He
is more powerful than many men put together and he attacks
with the fearlessness of *ja*. We brought him alive, that you
might see him before he is slain."

The chief rose and approached the ape-man, who closed his
eyes and feigned unconsciousness. He felt hairy hands upon
him as he was turned over, none too gently. The *gund* examined him from head to foot, making comments, especially
upon the shape and size of his thumbs and great toes.

"With these and with no tail," he said, "it cannot climb."

"No," agreed one of the warriors, "it would surely fall even
from the cliff pegs."

"I have never seen a thing like it," said the chief. "It is
neither Waz-don nor Ho-don. I wonder from whence it
came and what it is called."

"The Kor-ul-ja shouted aloud, 'Tarzan-jad-gurul' and we
thought that they might be calling this one," said a warrior.
"Shall we kill it now?"

"No," replied the chief, "we will wait until it's life returns
into its head that I may question it. Remain here, In-tan, and
watch it. When it can again hear and speak call me."

He turned and departed from the cave, the others, except
In-tan, following him. As they moved past him and out of the
chamber Tarzan caught snatches of their conversation which
indicated that the Kor-ul-ja reinforcements had fallen upon
their little party in great numbers and driven them away.
Evidently the swift feet of Id-an had saved the day for the
warriors of Om-at. The ape-man smiled, then he partially
opened an eye and cast it upon In-tan. The warrior stood at
the entrance to the cave looking out—his back was toward his
prisoner. Tarzan tested the bonds that secured his wrists.
They seemed none too stout and they had tied his hands in

front of him! Evidence indeed that the Waz-don took few
prisoners—if any.

Cautiously he raised his wrists until he could examine the
thongs that confined them. A grim smile lighted his features.
Instantly he was at work upon the bonds with his strong teeth,
but ever a wary eye was upon In-tan, the warrior of Kor-ul-lul.
The last knot had been loosened and Tarzan's hands were free
when In-tan turned to cast an appraising eye upon his ward.
He saw that the prisoner's position was changed—he no longer
lay upon his back as they had left him, but upon his side and
his hands were drawn up against his face. In-tan came closer
and bent down. The bonds seemed very loose upon the prisoner's wrists. He extended his hand to examine them with his
fingers and instantly the two hands leaped from their bonds—
one to seize his own wrist, the other his throat. So unexpected
the catlike attack that In-tan had not even time to cry out
before steel fingers silenced him. The creature pulled him
suddenly forward so that he lost his balance and rolled over
upon the prisoner and to the floor beyond to stop with Tarzan
upon his breast. In-tan struggled to release himself—struggled
to draw his knife; but Tarzan found it before him. The Waz-
don's tail leaped to the other's throat, encircling it—he too
could choke; but his own knife, in the hands of his antagonist,
severed the beloved member close to its root.

The Waz-don's struggles became weaker—a film was ob-
scuring his vision. He knew that he was dying and he was
right. A moment later he was dead. Tarzan rose to his feet
and placed one foot upon the breast of his dead foe. How
the urge seized him to roar forth the victory cry of his kind!
But he dared not. He discovered that they had not removed
his rope from his shoulders and that they had replaced his
knife in its sheath. It had been in his hand when he was
felled. Strange creatures! He did not know that they held

a superstitious fear of the weapons of a dead enemy, believing that if buried without them he would forever haunt his slayers in search of them and that when he found them he would kill the man who killed him. Against the wall leaned his bow and quiver of arrows.

Tarzan stepped toward the doorway of the cave and looked out. Night had just fallen. He could hear voices from the nearer caves and there floated to his nostrils the odor of cooking food. He looked down and experienced a sensation of relief. The cave in which he had been held was in the lowest tier—scarce thirty feet from the base of the cliff. He was about to chance an immediate descent when there occurred to him a thought that brought a grin to his savage lips—a thought that was born of the name the Waz-don had given him —Tarzan-jad-guru—Tarzan the Terrible—and a recollection of the days when he had delighted in baiting the blacks of the distant jungle of his birth. He turned back into the cave where lay the dead body of In-tan. With his knife he severed the warrior's head and carrying it to the outer edge of the recess tossed it to the ground below, then he dropped swiftly and silently down the ladder of pegs in a way that would have surprised the Kor-ul-lul who had been so sure that he could not climb.

At the bottom he picked up the head of In-tan and disappeared among the shadows of the trees carrying the grisly trophy by its shock of shaggy hair. Horrible? But you are judging a wild beast by the standards of civilization. You may teach a lion tricks, but he is still a lion. Tarzan looked well in a Tuxedo, but he was still a Tarmangani and beneath his pleated shirt beat a wild and savage heart.

Nor was his madness lacking in method. He knew that the hearts of the Kor-ul-lul would be filled with rage when they discovered the thing that he had done and he knew, too, that mixed with the rage would be a leaven of fear and it was fear of him that had made Tarzan master of many jungles—one does not win the respect of the killers with bonbons.

Below the village Tarzan returned to the foot of the cliff searching for a point where he could make the ascent to the ridge and thus back to the village of Om-at, the Kor-ul-ja. He came at last to a place where the river ran so close to the rocky wall that he was forced to swim it in search of a trail upon the opposite side and here it was that his keen nostrils detected a familiar spoor. It was the scent of Pan-at-lee at the spot where she had emerged from the pool and taken to the safety of the jungle.

Immediately the ape-man's plans were changed. Pan-at-lee lived, or at least she had lived after the leap from the cliff's summit. He had started in search of her for Om-at, his friend, and for Om-at he would continue upon the trail he had picked up thus fortuitously by accident. It led him into the jungle and across the gorge and then to the point at which Pan-at-lee had commenced the ascent of the opposite cliffs. Here Tarzan abandoned the head of In-tan, tying it to the lower branch of a tree, for he knew that it would handicap him in his ascent of the steep escarpment. Apelike he ascended, following easily the scent spoor of Pan-at-lee. Over the summit and across the ridge the trail lay, plain as a printed page to the delicate senses of the jungle-bred tracker.

Tarzan knew naught of the Kor-ul-gryf. He had seen, dimly in the shadows of the night, strange, monstrous forms and Ta-den and Om-at had spoken of great creatures that all men feared; but always, everywhere, by night and by day, there were dangers. From infancy death had stalked, grim and terrible, at his heels. He knew little of any other existence. To cope with danger was his life and he lived his life as simply and as naturally as you live yours amidst the dangers of the crowded city streets. The black man who goes abroad in the jungle by night is afraid, for he has spent his life since infancy surrounded by numbers of his own kind and safeguarded, especially at night, by such crude means as lie within his powers. But Tarzan had lived as the lion lives and the panther and the elephant and the ape—a true jungle creature dependent solely upon his prowess and his wits, playing a lone hand against creation. Therefore he was surprised at nothing and feared nothing and so he walked through the strange night as undisturbed and unapprehensive as the farmer to the cow lot in the darkness before the dawn.

Once more Pan-at-lee's trail ended at the verge of a cliff; but this time there was no indication that she had leaped over the edge and a moment's search revealed to Tarzan the stone pegs upon which she had made her descent. As he lay upon his belly leaning over the top of the cliff examining the pegs his attention was suddenly attracted by something at the foot of the cliff. He could not distinguish its identity, but he saw that it moved and presently that it was ascending slowly, apparently by means of pegs similar to those directly below him. He watched it intently as it rose higher and higher until he was able to distinguish its form more clearly, with the result that he became convinced that it more nearly resembled some form of great ape than a lower order. It had a tail, though, and in other respects it did not seem a true ape.

Slowly it ascended to the upper tier of caves, into one of which it disappeared. Then Tarzan took up again the trail of Pan-at-lee. He followed it down the stone pegs to the nearest cave and then further along the upper tier. The apeman raised his eyebrows when he saw the direction in which it led, and quickened his pace. He had almost reached the third cave when the echoes of Kor-ul-gryf were awakened by a shrill scream of terror.

CHAPTER VI

The Tor-o-don

PAN-AT-LEE slept—the troubled sleep, of physical and nervous exhaustion, filled with weird dreamings. She dreamed that she slept beneath a great tree in the bottom of the Kor-ul-gryf and that one of the fearsome beasts was creeping upon her but she could not open her eyes nor move. She tried to scream but no sound issued from her lips. She felt the thing touch her throat, her breast, her arm, and there it closed and seemed to be dragging her toward it. With a superhuman effort of will she opened her eyes. In the instant she knew that she was dreaming and that quickly the hallucination of the dream would fade—it had happened to her many times before. But it persisted. In the dim light that filtered into the dark chamber she saw a form beside her, she felt hairy fingers upon her and a hairy breast against which she was being drawn. Jad-ben-Otho! this was no dream. And then she screamed and tried to fight the thing from her; but her scream was answered by a low growl and another hairy hand seized her by the hair of the head. The beast rose now upon its hind legs and dragged her from the cave to the moonlit recess without and at the same instant she saw the figure of what she took to be a Ho-don rise above the outer edge of the niche.

The beast that held her saw it too and growled ominously but it did not relinguish its hold upon her hair. It crouched

67

as though waiting an attack, and it increased the volume and frequency of its growls until the horrid sounds reverberated through the gorge, drowning even the deep bellowings of the beasts below, whose mighty thunderings had broken out anew with the sudden commotion from the high-flung cave. The beast that held her crouched and the creature that faced it crouched also, and growled—as hideously as the other. Pan-at-lee trembled. This was no Ho-don and though she feared the Ho-don she feared this thing more, with its catlike crouch and its beastly growls. She was lost—that Pan-at-lee knew. The two things might fight for her, but whichever won she was lost. Perhaps, during the battle, if it came to that, she might find the opportunity to throw herself over into the Kor-ul-gryf.

The thing that held her she had recognized now as a Tor-o-don, but the other thing she could not place, though in the moonlight she could see it very distinctly. It had no tail. She could see its hands and its feet, and they were not the hands and feet of the races of Pal-ul-don. It was slowly closing upon the Tor-o-don and in one hand it held a gleaming knife. Now it spoke and to Pan-at-lee's terror was added an equal weight of consternation.

"When it leaves go of you," it said, "as it will presently to defend itself, run quickly behind me, Pan-at-lee, and go to the cave nearest the pegs you descended from the cliff top. Watch from there. If I am defeated you will have time to escape this slow thing; if I am not I will come to you there. I am Om-at's friend and yours."

The last words took the keen edge from Pan-at-lee's terror; but she did not understand. How did this strange creature know her name? How did it know that she had descended the pegs by a certain cave? It must, then, have been here when she came. Pan-at-lee was puzzled.

"Who are you?" she asked, "and from whence do you come?"

"I am Tarzan," he replied, "and just now I came from Om-at, *gund* of Kor-ul-ja, in search of you."

Om-at, *gund* of Kor-ul-ja! What wild talk was this? She would have questioned him further, but now he was approaching the Tor-o-don and the latter was screaming and growling so loudly as to drown the sound of her voice. And then it did what the strange creature had said that it would do—it released its hold upon her hair as it prepared to charge. Charge it did and in those close quarters there was no room to fence for openings. Instantly the two beasts locked in deadly embrace, each seeking the other's throat. Pan-at-lee watched, taking no advantage of the opportunity to escape which their preoccupation gave her. She watched and waited, for into her savage little brain had come the resolve to pin her faith to this strange creature who had unlocked her heart with those four words—"I am Om-at's friend!" And so she waited, with drawn knife, the opportunity to do her bit in the vanquishing of the Tor-o-don. That the newcomer could do it unaided she well knew to be beyond the realms of possibility, for she knew well the prowess of the *beastlike man* with whom it fought. There were not many of them in Pal-ul-don, but what few there were were a terror to the women of the Waz-don and the Ho-don, for the old Tor-o-don bulls roamed the mountains and the valleys of Pal-ul-don between rutting seasons and woe betide the women who fell in their paths.

With his tail the Tor-o-don sought one of Tarzan's ankles, and finding it, tripped him. The two fell heavily, but so agile was the ape-man and so quick his powerful muscles that even in falling he twisted the beast beneath him, so that Tarzan fell on top and now the tail that had tripped him sought his throat as had the tail of In-tan, the Kor-ul-lul. In the effort of turning his antagonist's body during the fall Tarzan had had to relinquish his knife that he might seize the shaggy body with both hands and now the weapon lay out of reach at the very edge of the recess. Both hands were occupied for the moment in fending off the clutching fingers that sought to seize him and drag his throat within reach of his foe's formidable fangs and now the tail was seeking its deadly hold with a formidable persistence that would not be denied.

Pan-at-lee hovered about, breathless, her dagger ready, but there was no opening that did not also endanger Tarzan, so constantly were the two duelists changing their positions. Tarzan felt the tail slowly but surely insinuating itself about his neck though he had drawn his head down between the muscles of his shoulders in an effort to protect this vulnerable part. The battle seemed to be going against him for the giant beast against which he strove would have been a fair match in weight and strength for Bolgani, the gorilla. And knowing this he suddenly exerted a single super-human effort, thrust far apart the giant hands and with the swiftness of a striking snake buried his fangs in the jugular of the Tor-o-don. At the same instant the creature's tail coiled about his own throat and then commenced a battle royal of turning and twisting bodies as each sought to dislodge the fatal hold of the other, but the acts of the ape-man were guided by a human brain and thus it was that the rolling bodies rolled in the direction that Tarzan wished—toward the edge of the recess.

The choking tail had shut the air from his lungs, he knew that his gasping lips were parted and his tongue protruding; and now his brain reeled and his sight grew dim; but not before he reached his goal and a quick hand shot out to seize the knife that now lay within reach as the two bodies tottered perilously upon the brink of the chasm.

With all his remaining strength the ape-man drove home the blade—once, twice, thrice, and then all went black before him

as he felt himself, still in the clutches of the Tor-o-don, topple
from the recess.

Fortunate it was for Tarzan that Pan-at-lee had not obeyed
his injunction to make good her escape while he engaged the
Tor-o-don, for it was to this fact that he owed his life. Close
beside the struggling forms during the brief moments of the
terrific climax she had realized every detail of the danger to
Tarzan with which the emergency was fraught and as she saw
the two rolling over the outer edge of the niche she seized
the ape-man by an ankle at the same time throwing herself
prone upon the rocky floor. The muscles of the Tor-o-don relaxed in death with the last thrust of Tarzan's knife and with
its hold upon the ape-man released it shot from sight into the
gorge below.

It was with infinite difficulty that Pan-at-lee retained her
hold upon the ankle of her protector, but she did so and then,
slowly, she sought to drag the dead weight back to the safety
of the niche. This, however, was beyond her strength and
she could but hold on tightly, hoping that some plan would
suggest itself before her powers of endurance failed. She
wondered if, after all, the creature was already dead, but that
she could not bring herself to believe—and if not dead how
long it would be before he regained consciousness. If he did
not regain it soon he never would regain it, that she knew, for
she felt her fingers numbing to the strain upon them and slipping, slowly, slowly, from their hold. It was then that Tarzan
regained consciousness. He could not know what power up-
held him, but he felt that whatever it was it was slowly re-
leasing its hold upon his ankle. Within easy reach of his
hands were two pegs and these he seized upon just as Pan-
at-lee's fingers slipped from their hold.

As it was he came near to being precipitated into the gorge

—only his great strength saved him. He was upright now and his feet found other pegs. His first thought was of his foe. Where was he? Waiting above there to finish him? Tarzan looked up just as the frightened face of Pan-at-lee appeared over the threshold of the recess.

"You live?" she cried.

"Yes," replied Tarzan. "Where is the shaggy one?"

Pan-at-lee pointed downward. "There," she said, "dead."

"Good!" exclaimed the ape-man, clambering to her side. "You are unharmed?" he asked.

"You came just in time," replied Pan-at-lee; "but who are you and how did you know that I was here and what do you know of Om-at and where did you come from and what did you mean by calling Om-at, *gund?*"

"Wait, wait," cried Tarzan; "one at a time. My, but you are all alike—the shes of the tribe of Kerchak, the ladies of England, and their sisters of Pal-ul-don. Have patience and I will try to tell you all that you wish to know. Four of us set out with Om-at from Kor-ul-ja to search for you. We were attacked by the Kor-ul-lul and separated. I was taken prisoner, but escaped. Again I stumbled upon your trail and followed it, reaching the summit of this cliff just as the hairy one was climbing up after you. I was coming to investigate when I heard your scream—the rest you know."

"But you called Om-at, *gund* of Kor-ul-ja," she insisted. "Es-sat is *gund.*"

"Es-sat is dead," explained the ape-man. "Om-at slew him and now Om-at is *gund.* Om-at came back seeking you. He found Es-sat in your cave and killed him."

"Yes," said the girl, "Es-sat came to my cave and I struck him down with my golden breastplates and escaped."

"And a lion pursued you," continued Tarzan, "and you

leaped from the cliff into Kor-ul-lul, but why you were not killed is beyond me."

"Is there anything beyond you?" exclaimed Pan-at-lee. "How could you know that a lion pursued me and that I leaped from the cliff and not know that it was the pool of deep water below that saved me?"

"I would have known that, too, had not the Kor-ul-lul come then and prevented me continuing upon your trail. But now I would ask you a question—by what name do you call the thing with which I just fought?"

"It was a Tor-o-don," she replied. "I have seen but one before. They are terrible creatures with the cunning of man and the ferocity of a beast. Great indeed must be the warrior who slays one single-handed." She gazed at him in open admiration.

"And now," said Tarzan, "you must sleep, for tomorrow we shall return to Kor-ul-ja and Om-at, and I doubt that you have had much rest these two nights."

Pan-at-lee, lulled by a feeling of security, slept peacefully into the morning while Tarzan stretched himself upon the hard floor of the recess just outside her cave.

The sun was high in the heavens when he awoke; for two hours it had looked down upon another heroic figure miles away—the figure of a godlike man fighting his way through the hideous morass that lies like a filthy moat defending Pal-ul-don from the creatures of the outer world. Now waist deep in the sucking ooze, now menaced by loathsome reptiles, the man advanced only by virtue of Herculean efforts gaining laboriously by inches along the devious way that he was forced to choose in selecting the least precarious footing. Near the center of the morass was open water—slimy, green-hued water. He reached it at last after more than two hours of such effort as would have left an ordinary man spent and dying in the

sticky mud, yet he was less than halfway across the marsh.
Greasy with slime and mud was his smooth, brown hide, and
greasy with slime and mud was his beloved Enfield that had
shone so brightly in the first rays of the rising sun.

He paused a moment upon the edge of the open water and
then throwing himself forward struck out to swim across. He
swam with long, easy, powerful strokes calculated less for
speed than for endurance, for his was, primarily, a test of the
latter, since beyond the open water was another two hours or
more of gruelling effort between it and solid ground. He was,
perhaps, halfway across and congratulating himself upon the
ease of the achievement of this portion of his task when there
arose from the depths directly in his path a hideous reptile,
which, with wide-distended jaws, bore down upon him, hissing shrilly.

Tarzan arose and stretched, expanded his great chest and
drank in deep draughts of the fresh morning air. His clear
eyes scanned the wondrous beauties of the landscape spread
out before them. Directly below lay Kor-ul-gryf, a dense,
somber green of gently moving tree tops. To Tarzan it was
neither grim, nor forbidding—it was jungle, beloved jungle.
To his right there spread a panorama of the lower reaches of
the Valley of Jad-ben-Otho, with its winding streams and its
blue lakes. Gleaming whitely in the sunlight were scattered
groups of dwellings—the feudal strongholds of the lesser chiefs
of the Ho-don. A-lur, the City of Light, he could not see as it
was hidden by the shoulder of the cliff in which the deserted
village lay.

For a moment Tarzan gave himself over to that spiritual
enjoyment of beauty that only the man-mind may attain and
then Nature asserted herself and the belly of the beast called
aloud that it was hungry. Again Tarzan looked down at

Kor-ul-gryf. There was the jungle! Grew there a jungle that
would not feed Tarzan? The ape-man smiled and commenced
the descent to the gorge. Was there danger there? Of course.
Who knew it better than Tarzan? In all jungles lies death,
for life and death go hand in hand and where life teems death
reaps his fullest harvest. Never had Tarzan met a creature of
the jungle with which he could not cope—sometimes by virtue
of brute strength alone, again by a combination of brute
strength and the cunning of the man-mind; but Tarzan had
never met a *gryf*.

He had heard the bellowings in the gorge the night before
after he had lain down to sleep and he had meant to ask Pan-
at-lee this morning what manner of beast so disturbed the
slumbers of its betters. He reached the foot of the cliff and
strode into the jungle and here he halted, his keen eyes and
ears watchful and alert, his sensitive nostrils searching each
shifting air current for the scent spoor of game. Again he advanced deeper into the wood, his light step giving forth no
sound, his bow and arrows in readiness. A light morning
breeze was blowing from up the gorge and in this direction
he bent his steps. Many odors impinged upon his organs of
scent. Some of these he classified without effort, but others
were strange—the odors of beasts and of birds, of trees and
shrubs and flowers with which he was unfamiliar. He sensed
faintly the reptilian odor that he had learned to connect with
the strange, nocturnal forms that had loomed dim and bulky
on several occasions since his introduction to Pal-ul-don.

And then, suddenly he caught plainly the strong, sweet odor
of Bara, the deer. Were the belly vocal, Tarzan's would have
given a little cry of joy, for it loved the flesh of Bara. The
ape-man moved rapidly, but cautiously forward. The prey
was not far distant and as the hunter approached it, he took
silently to the trees and still in his nostrils was the faint rep-

tilian odor that spoke of a great creature which he had never yet seen except as a denser shadow among the dense shadows of the night; but the odor was of such a faintness as suggests to the jungle bred the distance of absolute safety.

And now, moving noiselessly, Tarzan came within sight of Bara drinking at a pool where the stream that waters Kor-ul-gryf crosses an open place in the jungle. The deer was too far from the nearest tree to risk a charge, so the ape-man must depend upon the accuracy and force of his first arrow, which must drop the deer in its tracks or forfeit both deer and shaft. Far back came the right hand and the bow, that you or I might not move, bent easily beneath the muscles of the forest god. There was a singing twang and Bara, leaping high in air, collapsed upon the ground, an arrow through his heart. Tarzan dropped to earth and ran to his kill, lest the animal might even yet rise and escape; but Bara was safely dead. As Tarzan stooped to lift it to his shoulder there fell upon his ears a thunderous bellow that seemed almost at his right elbow, and as his eyes shot in the direction of the sound, there broke upon his vision such a creature as paleontologists have dreamed as having possibly existed in the dimmest vistas of Earth's infancy—a gigantic creature, vibrant with mad rage, that charged, bellowing, upon him.

When Pan-at-lee awoke she looked out upon the niche in search of Tarzan. He was not there. She sprang to her feet and rushed out, looking down into Kor-ul-gryf guessing that he had gone down in search of food and there she caught a glimpse of him disappearing into the forest. For an instant she was panic-stricken. She knew that he was a stranger in Pal-ul-don and that, so, he might not realize the dangers that lay in that gorge of terror. Why did she not call to him to return? You or I might have done so, but no Pal-ul-don, for they know the ways of the *gryf*—they know the weak eyes and the keen ears, and that at the sound of a human voice they come. To have called to Tarzan, then, would but have been to invite disaster and so she did not call. Instead, afraid though she was, she descended into the gorge for the purpose of overhauling Tarzan and warning him in whispers of his danger. It was a brave act, since it was performed in the face of countless ages of inherited fear of the creatures that she might be called upon to face. Men have been decorated for less.

Pan-at-lee, descended from a long line of hunters, assumed that Tarzan would move up wind and in this direction she sought his tracks, which she soon found well marked, since he had made no effort to conceal them. She moved rapidly until she reached the point at which Tarzan had taken to the trees. Of course she knew what had happened; since her own people were semi-arboreal; but she could not track him through the trees, having no such well-developed sense of scent as he.

She could but hope that he had continued on up wind and in this direction she moved, her heart pounding in terror against her ribs, her eyes glancing first in one direction and then another. She had reached the edge of a clearing when two things happened—she caught sight of Tarzan bending over a dead deer and at the same instant a deafening roar sounded almost beside her. It terrified her beyond description, but it brought no paralysis of fear. Instead it galvanized her into instant action with the result that Pan-at-lee swarmed up the nearest tree to the very loftiest branch that would sustain her weight. Then she looked down.

The thing that Tarzan saw charging him when the warning bellow attracted his surprised eyes loomed terrifically monstrous before him—monstrous and awe-inspiring; but it did not terrify Tarzan, it only angered him, for he saw that it was beyond even his powers to combat and that meant that it might cause him to lose his kill, and Tarzan was hungry. There was but a single alternative to remaining for annihilation and that was flight—swift and immediate. And Tarzan fled, but he carried the carcass of Bara, the deer, with him. He had not more than a dozen paces start, but on the other hand the nearest tree was almost as close. His greatest danger lay, he imagined, in the great, towering height of the creature pursuing him, for even though he reached the tree he would have to climb high in an incredibly short time as, unless appearances were deceiving, the thing could reach up and pluck him down from any branch under thirty feet above the ground, and possibly from those up to fifty feet, if it reared up on its hind legs.

But Tarzan was no sluggard and though the *gryf* was incredibly fast despite its great bulk, it was no match for Tarzan, and when it comes to climbing, the little monkeys gaze with envy upon the feats of the ape-man. And so it was that the bellowing *gryf* came to a baffled stop at the foot of the tree and even though he reared up and sought to seize his prey among the branches, as Tarzan had guessed he might, he failed in this also. And then, well out of reach, Tarzan came to a stop and there, just above him, he saw Pan-at-lee sitting, wide-eyed and trembling.

"How came you here?" he asked.

She told him. "You came to warn me!" he said. "It was very brave and unselfish of you. I am chagrined that I should have been thus surprised. The creature was up wind from me and yet I did not sense its near presence until it charged. I cannot understand it."

"It is not strange," said Pan-at-lee. "That is one of the peculiarities of the *gryf*—it is said.that man never knows of its

presence until it is upon him—so silently does it move despite
its great size."

"But I should have smelled it," cried Tarzan, disgustedly.

"Smelled it!" ejaculated Pan-at-lee. "Smelled it?"

"Certainly. How do you suppose I found this deer so
quickly? And I sensed the *gryf*, too, but faintly as at a great
distance." Tarzan suddenly ceased speaking and looked down
at the bellowing creature below them—his nostrils quivered as
though searching for a scent. "Ah!" he exclaimed. "I have it!"

"What?" asked Pan-at-lee.

"I was deceived because the creature gives off practically
no odor," explained the ape-man. "What I smelled was the
faint aroma that doubtless permeates the entire jungle because
of the long presence of many of the creatures—it is the sort of
odor that would remain for a long time, faint as it is.

"Pan-at-lee, did you ever hear of a triceratops? No? Well
this thing that you call a *gryf* is a triceratops and it has been
extinct for hundreds of thousands of years. I have seen its
skeleton in the museum in London and a figure of one restored. I always thought that the scientists who did such
work depended principally upon an overwrought imagination,
but I see that I was wrong. This living thing is not an exact
counterpart of the restoration that I saw; but it is so similar
as to be easily recognizable, and then, too, we must remember
that during the ages that have elapsed since the paleontologist's
specimen lived many changes might have been wrought by
evolution in the living line that has quite evidently persisted
in Pal-ul-don."

"Triceratops, London, paleo—I don't know what you are
talking about," cried Pan-at-lee.

Tarzan smiled and threw a piece of dead wood at the face
of the angry creature below them. Instantly the great bony

hood over the neck was erected and a mad bellow rolled upward from the gigantic body. Full twenty feet at the shoulder
the thing stood, a dirty slate-blue in color except for its yellow
face with the blue bands encircling the eyes, the red hood with
the yellow lining and the yellow belly. The three parallel
lines of bony protuberances down the back gave a further
touch of color to the body, those following the line of the
spine being red, while those on either side are yellow. The
five- and three-toed hoofs of the ancient horned dinosaurs had
become talons in the *gryf*, but the three horns, two large ones
above the eyes and a median horn on the nose, had persisted
through all the ages. Weird and terrible as was its appearance
Tarzan could not but admire the mighty creature looming big
below him, its seventy-five feet of length majestically typifying
those things which all his life the ape-man had admired—
courage and strength. In that massive tail alone was the
strength of an elephant.

The wicked little eyes looked up at him and the horny
beak opened to disclose a full set of powerful teeth.

"Herbivorous!" murmured the ape-man. "Your ancestors
may have been, but not you," and then to Pan-at-lee: "Let
us go now. At the cave we will have deer meat and then—
back to Kor-ul-ja and Om-at."

The girl shuddered. "Go?" she repeated. "We will never
go from here."

"Why not?" asked Tarzan.

For answer she but pointed to the *gryf*.

"Nonsense!" exclaimed the man. "It cannot climb. We can
reach the cliff through the trees and be back in the cave before it knows what has become of us."

"You do not know the *gryf*," replied Pan-at-lee gloomily.

"Wherever we go it will follow and always it will be ready

at the foot of each tree when we would descend. It will never give us up."

"We can live in the trees for a long time if necessary," replied Tarzan, "and sometime the thing will leave."

The girl shook her head. "Never," she said, "and then there are the Tor-o-don. They will come and kill us and after eating a little will throw the balance to the *gryf*—the *gryf* and Tor-o-don are friends, because the Tor-o-don shares his food with the *gryf*."

"You may be right," said Tarzan; "but even so I don't intend waiting here for someone to come along and eat part of me and then feed the balance to that beast below. If I don't get out of this place whole it won't be my fault. Come along now and we'll make a try at it," and so saying he moved off through the tree tops with Pan-at-lee close behind. Below them, on the ground, moved the horned dinosaur and when they reached the edge of the forest where there lay fifty yards of open ground to cross to the foot of the cliff he was there with them, at the bottom of the tree, waiting.

Tarzan looked ruefully down and scratched his head.

CHAPTER VII

Jungle Craft

Presently he looked up and at Pan-at-lee. "Can you cross the gorge through the trees very rapidly?" he questioned.

"Alone?" she asked.

"No," replied Tarzan.

"I can follow wherever you can lead," she said then.

"Across and back again?"

"Yes."

"Then come, and do exactly as I bid." He started back again through the trees, swiftly, swinging monkey-like from limb to limb, following a zigzag course that he tried to select with an eye for the difficulties of the trail beneath. Where the underbrush was heaviest, where fallen trees blocked the way, he led the footsteps of the creature below them; but all to no avail. When they reached the opposite side of the gorge the *gryf* was with them.

"Back again," said Tarzan, and, turning, the two retraced their high-flung way through the upper terraces of the ancient forest of Kor-ul-gryf. But the result was the same—no, not quite; it was worse, for another *gryf* had joined the first and now two waited beneath the tree in which they stopped.

The cliff looming high above them with its innumerable cave mouths seemed to beckon and to taunt them. It was so near, yet eternity yawned between. The body of the Tor-o-don lay at the cliff's foot where it had fallen. It was in plain view of the two in the tree. One of the *gryfs* walked over and sniffed about it, but did not offer to devour it. Tarzan had examined it casually as he had passed earlier in the morning. He guessed that it represented either a very high order of ape or a very low order of man—something akin to the Java man, perhaps; a truer example of the pithecanthropi than either the Ho-don or the Waz-don; possibly the precursor of them both. As his eyes wandered idly over the scene below his active brain was working out the details of the plan that he had made to permit Pan-at-lee's escape from the gorge. His thoughts were interrupted by a strange cry from above them in the gorge.

"Whee-oo! Whee-oo!" it sounded, coming closer.

The *gryfs* below raised their heads and looked in the direction of the interruption. One of them made a low, rumbling sound in its throat. It was not a bellow and it did not indicate anger. Immediately the "Whee-oo!" responded. The *gryfs* repeated the rumbling and at intervals the "Whee-oo!" was repeated, coming ever closer.

Tarzan looked at Pan-at-lee. "What is it?" he asked.

"I do not know," she replied. "Perhaps a strange bird, or another horrid beast that dwells in this frightful place."

"Ah," exclaimed Tarzan; "there it is. Look!"

Pan-at-lee voiced a cry of despair. "A Tor-o-don!"

The creature, walking erect and carrying a stick in one hand, advanced at a slow, lumbering gait. It walked directly toward the *gryfs*, who moved aside, as though afraid. Tarzan watched intently. The Tor-o-don was now quite close to one of the triceratops. It swung its head and snapped at him viciously. Instantly the Tor-o-don sprang in and commenced to belabor the huge beast across the face with his stick. To the ape-man's amazement the *gryf*, that might have annihilated the comparatively puny Tor-o-don instantly in any of a dozen ways, cringed like a whipped cur.

"Whee-oo! Whee-oo!" shouted the Tor-o-don and the *gryf* came slowly toward him. A whack on the median horn brought it to a stop. Then the Tor-o-don walked around behind it, clambered up its tail and seated himself astraddle of the huge back. "Whee-oo!" he shouted and prodded the beast with a sharp point of his stick. The *gryf* commenced to move off.

So rapt had Tarzan been in the scene below him that he had given no thought to escape, for he realized that for him and Pan-at-lee time had in these brief moments turned back countless ages to spread before their eyes a page of the dim and distant past. They two had looked upon the first man and his primitive beasts of burden.

And now the ridden *gryf* halted and looked up at them, bellowing. It was sufficient. The creature had warned its master of their presence. Instantly the Tor-o-don urged the beast close beneath the tree which held them, at the same time leaping to his feet upon the horny back. Tarzan saw the bestial face, the great fangs, the mighty muscles. From the loins of such had sprung the human race—and only from such could it have sprung, for only such as this might have survived the horrid dangers of the age that was theirs.

The Tor-o-don beat upon his breast and growled horribly—hideous, uncouth, beastly. Tarzan rose to his full height upon a swaying branch—straight and beautiful as a demigod—unspoiled by the taint of civilization—a perfect specimen of what the human race might have been had the laws of man not interfered with the laws of nature.

The Present fitted an arrow to his bow and drew the shaft far back. The Past basing its claims upon brute strength sought to reach the other and drag him down; but the loosed

arrow sank deep into the savage heart and the Past sank back
into the oblivion that had claimed his kind.

"Tarzan-jad-guru!" murmured Pan-at-lee, unknowingly giving him out of the fullness of her admiration the same title
that the warriors of her tribe had bestowed upon him.

The ape-man turned to her. "Pan-at-lee," he said, "these
beasts may keep us treed here indefinitely. I doubt if we can
escape together, but I have a plan. You remain here, hiding
yourself in the foliage, while I start back across the gorge in
sight of them and yelling to attract their attention. Unless
they have more brains than I suspect they will follow me.
When they are gone you make for the cliff. Wait for me in the
cave not longer than today. If I do not come by tomorrow's
sun you will have to start back for Kor-ul-ja alone. Here is a
joint of deer meat for you." He had severed one of the deer's
hind legs and this he passed up to her.

"I cannot desert you," she said simply; "it is not the way of
my people to desert a friend and ally. Om-at would never
forgive me."

"Tell Om-at that I commanded you to go," replied Tarzan.

"It is a command?" she asked.

"It is! Good-bye, Pan-at-lee. Hasten back to Om-at—you
are a fitting mate for the chief of Kor-ul-ja." He moved off
slowly through the trees.

"Good-bye, Tarzan-jad-guru!" she called after him. "Fortunate are my Om-at and his Pan-at-lee in owning such a friend."

Tarzan, shouting aloud, continued upon his way and the
great *gryfs*, lured, by his voice, followed beneath. His ruse
was evidently proving successful and he was filled with elation as he led the bellowing beasts farther and farther from
Pan-at-lee. He hoped that she would take advantage of the
opportunity afforded her for escape, yet at the same time he
was filled with concern as to her ability to survive the dangers

which lay between Kor-ul-gryf and Kor-ul-ja. There were lions and Tor-o-dons and the unfriendly tribe of Kor-ul-lul to hinder her progress, though the distance in itself to the cliffs of her people was not great.

He realized her bravery and understood the resourcefulness that she must share in common with all primitive people who, day by day, must contend face to face with nature's law of the survival of the fittest, unaided by any of the numerous artificial protections that civilization has thrown around its brood of weaklings.

Several times during this crossing of the gorge Tarzan endeavored to outwit his keen pursuers, but all to no avail. Double as he would he could not throw them off his track and ever as he changed his course they changed theirs to conform. Along the verge of the forest upon the southeastern side of the gorge he sought some point at which the trees touched some negotiable portion of the cliff, but though he traveled far both up and down the gorge he discovered no such easy avenue of escape. The ape-man finally commenced to entertain an idea of the hopelessness of his case and to realize to the full why the Kor-ul-gryf had been religiously abjured by the races of Pal-ul-don for all these many ages.

Night was falling and though since early morning he had sought diligently a way out of this cul-de-sac he was no nearer to liberty than at the moment the first bellowing *gryf* had charged him as he stooped over the carcass of his kill: but with the falling of night came renewed hope for, in common with the great cats, Tarzan was, to a greater or lesser extent, a nocturnal beast. It is true he could not see by night as well as they, but that lack was largely recompensed for by the keenness of his scent and the highly developed sensitiveness of his other organs of perception. As the blind follow and interpret their *Braille* characters with deft fingers, so Tarzan reads the book of the jungle with feet and hands and eyes and ears and nose; each contributing its share to the quick and accurate translation of the text.

But again he was doomed to be thwarted by one vital weakness—he did not know the *gryf*, and before the night was over he wondered if the things never slept, for wheresoever he moved they moved also, and always they barred his road to liberty. Finally, just before dawn, he relinquished his immediate effort and sought rest in a friendly tree crotch in the safety of the middle terrace.

Once again was the sun high when Tarzan awoke, rested and refreshed. Keen to the necessities of the moment he made no effort to locate his jailers lest in the act he might apprise them of his movements. Instead he sought cautiously and silently to melt away among the foliage of the trees. His first move, however, was heralded by a deep bellow from below.

Among the numerous refinements of civilization that Tarzan had failed to acquire was that of profanity, and possibly it is to be regretted since there are circumstances under which it is at least a relief to pent emotion. And it may be that in effect Tarzan resorted to profanity if there can be physical as well as vocal swearing, since immediately the bellow announced that his hopes had been again frustrated, he turned quickly and seeing the hideous face of the *gryf* below him seized a large fruit from a nearby branch and hurled it viciously at the horned snout. The missile struck full between the creature's eyes, resulting in a reaction that surprised the ape-man; it did not arouse the beast to a show of revengeful rage as Tarzan had expected and hoped; instead the creature gave a single vicious side snap at the fruit as it bounded from his skull and then turned sulkily away, walking off a few steps.

There was that in the act that recalled immediately to Tarzan's mind similar action on the preceding day when the Tor-o-don had struck one of the creatures across the face with his staff, and instantly there sprung to the cunning and courageous brain a plan of escape from his predicament that might have blanched the check of the most heroic.

The gambling instinct is not strong among creatures of the wild; the chances of their daily life are sufficient stimuli for the beneficial excitement of their nerve centers. It has remained for civilized man, protected in a measure from the natural dangers of existence, to invent artificial stimulants in the form of cards and dice and roulette wheels. Yet when necessity bids there are no greater gamblers than the savage denizens of the jungle, the forest, and the hills, for as lightly as you roll the ivory cubes upon the green cloth they will gamble with death—their own lives the stake.

And so Tarzan would gamble now, pitting the seemingly wild deductions of his shrewd brain against all the proofs of the bestial ferocity of his antagonists that his experience of them had adduced—against all the age-old folklore and legend that had been handed down for countless generations and passed on to him through the lips of Pan-at-lee.

Yet as he worked in preparation for the greatest play that man can make in the game of life, he smiled; nor was there any indication of haste or excitement or nervousness in his demeanor.

First he selected a long, straight branch about two inches in diameter at its base. This he cut from the tree with his knife, removed the smaller branches and twigs until he had fashioned a pole about ten feet in length. This he sharpened at the smaller end. The staff finished to his satisfaction he looked down upon the triceratops.

"Whee-oo!" he cried.

Instantly the beasts raised their heads and looked at him. From the throat of one of them came faintly a low rumbling sound.

"Whee-oo!" repeated Tarzan and hurled the balance of the carcass of the deer to them.

Instantly the *gryfs* fell upon it with much bellowing, one of them attempting to seize it and keep it from the other: but finally the second obtained a hold and an instant later it had been torn asunder and greedily devoured. Once again they looked up at the ape-man and this time they saw him descending to the ground.

One of them started toward him. Again Tarzan repeated the weird cry of the Tor-o-don. The *gryf* halted in his track, apparently puzzled, while Tarzan slipped lightly to the earth and advanced toward the nearer beast, his staff raised menacingly and the call of the first-man upon his lips.

Would the cry be answered by the low rumbling of the beast of burden or the horrid bellow of the man-eater? Upon the answer to this question hung the fate of the ape-man.

Pan-at-lee was listening intently to the sounds of the departing *gryfs* as Tarzan led them cunningly from her, and when she was sure that they were far enough away to insure her safe retreat she dropped swiftly from the branches to the ground and sped like a frightened deer across the open space to the foot of the cliff, stepped over the body of the Tor-o-don who had attacked her the night before and was soon climbing rapidly up the ancient stone pegs of the deserted cliff village. In the mouth of a cave near that which she had occupied she kindled a fire and cooked the haunch of venison that Tarzan had left her, and from one of the trickling streams that ran down the face of the escarpment she obtained water to satisfy her thirst.

All day she waited, hearing in the distance, and sometimes close at hand, the bellowing of the *gryfs* which pursued the strange creature that had dropped so miraculously into her life. For him she felt the same keen, almost fanatical loyalty that many another had experienced for Tarzan of the Apes. Beast and human, he had held them to him with bonds that were stronger than steel—those of them that were clean and courageous, and the weak and the helpless; but never could Tarzan claim among his admirers the coward, the ingrate or the scoundrel; from such, both man and beast, he had won fear and hatred.

To Pan-at-lee he was all that was brave and noble and heroic and, too, he was Om-at's friend—the friend of the man she loved. For any one of these reasons Pan-at-lee would have died for Tarzan, for such is the loyalty of the simple-minded children of nature. It has remained for civilization to teach us to weigh the relative rewards of loyalty and its antithesis. The loyalty of the primitive is spontaneous, unreasoning, unselfish and such was the loyalty of Pan-at-lee for the Tarmangani.

And so it was that she waited that day and night, hoping that he would return that she might accompany him back to Om-at, for her experience had taught her that in the face of danger two have a better chance than one. But Tarzan-jad-guru had not come, and so upon the following morning Pan-at-lee set out upon her return to Kor-ul-ja.

She knew the dangers and yet she faced them with the stolid indifference of her race. When they directly confronted and menaced her would be time enough to experience fear or excitement or confidence. In the meantime it was unnecessary to waste nerve energy by anticipating them. She moved therefore through her savage land with no greater show of concern than might mark your sauntering to a corner drug-store for a sundae. But this is your life and that is Pan-at-lee's and even now as you read this Pan-at-lee may be sitting upon the edge of the recess of Om-at's cave while the *ja* and *jato* roar from the gorge below and from the ridge above, and the Kor-ul-lul threaten upon the south and the Ho-don from the Valley of Jad-ben-Otho far below, for Pan-at-lee still lives and preens her silky coat of jet beneath the tropical moonlight of Pal-ul-don.

But she was not to reach Kor-ul-ja this day, nor the next, nor for many days after though the danger that threatened her was neither Waz-don enemy nor savage beast.

She came without misadventure to the Kor-ul-lul and after descending its rocky southern wall without catching the slightest glimpse of the hereditary enemies of her people, she experienced a renewal of confidence that was little short of practical assurance that she would successfully terminate her venture and be restored once more to her own people and the lover she had not seen for so many long and weary moons.

She was almost across the gorge now and moving with an extreme caution abated no wit by her confidence, for wariness is an instinctive trait of the primitive, something which cannot be laid aside even momentarily if one would survive. And so she came to the trail that follows the windings of Kor-ul-lul from its uppermost reaches down into the broad and fertile Valley of Jad-ben-Otho.

And as she stepped into the trail there arose on either side of her from out of the bushes that border the path, as though materialized from thin air, a score of tall, white warriors of the Ho-don. Like a frightened deer Pan-at-lee cast a single startled look at these menacers of her freedom and leaped quickly toward the bushes in an effort to escape; but the warriors were too close at hand. They closed upon her from every side and then, drawing her knife she turned at bay, metamorphosed by the fires of fear and hate from a startled deer to a raging tiger-cat. They did not try to kill her, but only to subdue and capture her; and so it was that more than a single Ho-don warrior felt the keen edge of her blade in his flesh before they had succeeded in overpowering her by numbers. And still she fought and scratched and bit after they had taken the knife from her until it was necessary to tie her hands and fasten a piece of wood between her teeth by means of thongs passed behind her head.

At first she refused to walk when they started off in the direction of the valley but after two of them had seized her by the hair and dragged her for a number of yards she thought better of her original decision and came along with them, though still as defiant as her bound wrists and gagged mouth would permit.

Near the entrance to Kor-ul-lul they came upon another body of their warriors with which were several Waz-don prisoners from the tribe of Kor-ul-lul. It was a raiding party come up from a Ho-don city of the valley after slaves. This Pan-at-lee knew for the occurrence was by no means unusual. During her lifetime the tribe to which she belonged had been sufficiently fortunate, or powerful, to withstand successfully the majority of such raids made upon them, but yet Pan-at-lee had known of friends and relatives who had been carried into slavery by the Ho-don and she knew, too, another thing which gave her hope, as doubtless it did to each of the other captives —that occasionally the prisoners escaped from the cities of the hairless whites.

After they had joined the other party the entire band set forth into the valley and presently, from the conversation of her captors, Pan-at-lee knew that she was headed for A-lur, the City of Light; while in the cave of his ancestors, Om-at, chief of the Kor-ul-ja, bemoaned the loss of both his friend and she that was to have been his mate.

CHAPTER VIII

A-lur

A S THE hissing reptile bore down upon the stranger swim-ming in the open water near the center of the morass on the frontier of Pal-ul-don it seemed to the man that this indeed must be the futile termination of an arduous and danger-filled journey. It seemed, too, equally futile to pit his puny knife against this frightful creature. Had he been attacked on land it is possible that he might as a last resort have used his En-field, though he had come thus far through all these weary, danger-ridden miles without recourse to it, though again and again had his life hung in the balance in the face of the savage denizens of forest, jungle, and steppe. For whatever it may have been for which he was preserving his precious ammuni-tion he evidently held it more sacred even than his life, for as yet he had not used a single round and now the decision was not required of him, since it would have been impossible for him to have unslung his Enfield, loaded and fired with the necessary celerity while swimming.

Though his chance for survival seemed slender, and hope at its lowest ebb, he was not minded therefore to give up without a struggle. Instead he drew his blade and awaited the on-coming reptile. The creature was like no living thing he ever before had seen although possibly it resembled a crocodile in some respects more than it did anything with which he was familiar.

93

As this frightful survivor of some extinct progenitor charged upon him with distended jaws there came to the man quickly a full consciousness of the futility of endeavoring to stay the mad rush or pierce the armor-coated hide with his little knife. The thing was almost upon him now and whatever form of defense he chose must be made quickly. There seemed but a single alternative to instant death, and this he took at almost the instant the great reptile towered directly above him.

With the celerity of a seal he dove headforemost beneath the oncoming body and at the same instant, turning upon his back, he plunged his blade into the soft, cold surface of the slimy belly as the momentum of the hurtling reptile carried it swiftly over him; and then with powerful strokes he swam on beneath the surface for a dozen yards before he rose. A glance showed him the stricken monster plunging madly in pain and rage upon the surface of the water behind him. That it was writhing in its death agonies was evidenced by the fact that it made no effort to pursue him, and so, to the accompaniment of the shrill screaming of the dying monster, the man won at last to the farther edge of the open water to take up once more the almost superhuman effort of crossing the last stretch of clinging mud which separated him from the solid ground of Pal-ul-don.

A good two hours it took him to drag his now weary body through the clinging, stinking muck, but at last, mud covered and spent, he dragged himself out upon the soft grasses of the bank. A hundred yards away a stream, winding its way down from the distant mountains, emptied into the morass, and, after a short rest, he made his way to this and seeking a quiet pool, bathed himself and washed the mud and slime from his weapons, accouterments, and loin cloth. Another hour was spent beneath the rays of the hot sun in wiping, polishing, and oiling his Enfield though the means at hand for drying it consisted principally of dry grasses. It was afternoon before he had satisfied himself that his precious weapon was safe from any harm by dirt, or dampness, and then he arose and took up the search for the spoor he had followed to the opposite side of the swamp.

Would he find again the trail that had led into the opposite side of the morass, to be lost there, even to his trained senses? If he found it not again upon this side of the almost impassable barrier he might assume that his long journey had ended in failure. And so he sought up and down the verge of the stagnant water for traces of an old spoor that would have been invisible to your eyes or mine, even had we followed directly in the tracks of its maker.

As Tarzan advanced upon the *gryfs* he imitated as closely as he could recall them the methods and mannerisms of the Tor-o-don, but up to the instant that he stood close beside one of the huge creatures he realized that his fate still hung in the balance, for the thing gave forth no sign, either menacing or otherwise. It only stood there, watching him out of its cold, reptilian eyes and then Tarzan raised his staff and with a menacing "Whee-oo!" struck the *gryf* a vicious blow across the face.

The creature made a sudden side snap in his direction, a snap that did not reach him, and then turned sullenly away, precisely as it had when the Tor-o-don commanded it. Walking around to its rear as he had seen the shaggy first-man do, Tarzan ran up the broad tail and seated himself upon the creature's back, and then again imitating the acts of the Tor-o-don he prodded it with the sharpened point of his staff, and thus goading it forward and guiding it with blows, first upon one side and then upon the other, he started it down the gorge in the direction of the valley.

At first it had been in his mind only to determine if he could successfully assert any authority over the great monsters, realizing that in this possibility lay his only hope of immediate escape from his jailers. But once seated upon the back of his titanic mount the ape-man experienced the sensation of a new thrill that recalled to him the day in his boyhood that he had first clambered to the broad head of Tantor, the elephant, and this, together with the sense of mastery that was always meat and drink to the lord of the jungle, decided him to put his newly acquired power to some utilitarian purpose.

Pan-at-lee he judged must either have already reached safety or met with death. At least, no longer could he be of service to her, while below Kor-ul-gryf, in the soft green valley, lay A-lur, the City of Light, which, since he had gazed upon it from the shoulder of Pastar-ul-ved, had been his ambition and his goal.

Whether or not its gleaming walls held the secret of his lost mate he could not even guess but if she lived at all within the precincts of Pal-ul-don it must be among the Ho-don, since the hairy black men of this forgotten world took no prisoners. And so to A-lur he would go, and how more effectively than upon the back of this grim and terrible creature that the races of Pal-ul-don held in such awe?

A little mountain stream tumbles down from Kor-ul-gryf to be joined in the foothills with that which empties the waters of Kor-ul-lul into the valley, forming a small river which runs southwest, eventually entering the valley's largest lake at the City of A-lur, through the center of which the stream passes. An ancient trail, well marked by countless generations of naked feet of man and beast, leads down toward A-lur beside the river, and along this Tarzan guided the *gryf*. Once clear of the forest which ran below the mouth of the gorge, Tarzan caught occasional glimpses of the city gleaming in the distance far below him.

The country through which he passed was resplendent with the riotous beauties of tropical verdure. Thick, lush grasses grew waist high upon either side of the trail and the way was broken now and again by patches of open park-like forest, or perhaps a little patch of dense jungle where the trees over-arched the way and trailing creepers depended in graceful loops from branch to branch.

At times the ape-man had difficulty in commanding obedience upon the part of his unruly beast, but always in the end its fear of the relatively puny goad urged it on to obedience. Late in the afternoon as they approached the confluence of the stream they were skirting and another which appeared to come from the direction of Kor-ul-ja the ape-man, emerging from one of the jungle patches, discovered a considerable party of Ho-don upon the opposite bank. Simultaneously they saw him and the mighty creature he bestrode. For a moment they stood in wide-eyed amazement and then, in answer to the command of their leader, they turned and bolted for the shelter of the nearby wood.

The ape-man had but a brief glimpse of them but it was sufficient indication that there were Waz-don with them, doubtless prisoners taken in one of the raids upon the Waz-don villages of which Ta-den and Om-at had told him.

At the sound of their voices the *gryf* had bellowed terrifically and started in pursuit even though a river intervened, but by dint of much proding and beating, Tarzan had succeeded in heading the animal back into the path though thereafter for a long time it was sullen and more intractable than ever.

As the sun dropped nearer the summit of the western hills Tarzan became aware that his plan to enter A-lur upon the

back of a *gryf* was likely doomed to failure, since the stub-
bornness of the great beast was increasing momentarily, doubtless due to the fact that its huge belly was crying out for food.
The ape-man wondered if the Tor-o-dons had any means of
picketing their beasts for the night, but as he did not know
and as no plan suggested itself, he determined that he should
have to trust to the chance of finding it again in the morning.

There now arose in his mind a question as to what would be
their relationship when Tarzan had dismounted. Would it
again revert to that of hunter and quarry or would fear of the
goad continue to hold its supremacy over the natural instinct
of the hunting flesh-eater. Tarzan wondered but as he could
not remain upon the *gryf* forever, and as he preferred dis-
mounting and putting the matter to a final test while it was
still light, he decided to act at once.

How to stop the creature he did not know, as up to this time
his sole desire had been to urge it forward. By experimen-
tating with his staff, however, he found that he could bring it
to a halt by reaching forward and striking the thing upon its
beaklike snout. Close by grew a number of leafy trees, in any
one of which the ape-man could have found sanctuary, but it
had occurred to him that should he immediately take to the
trees it might suggest to the mind of the *gryf* that the creature
that had been commanding him all day feared him, with the
result that Tarzan would once again be held a prisoner by the
triceratops.

And so, when the *gryf* halted, Tarzan slid to the ground,
struck the creature a careless blow across the flank as though in
dismissal and walked indifferently away. From the throat of
the beast came a low rumbling sound and without even a
glance at Tarzan it turned and entered the river where it stood
drinking for a long time.

Convinced that the *gryf* no longer constituted a menace to

him the ape-man, spurred on himself by the gnawing of hunger, unslung his bow and selecting a handful of arrows set forth cautiously in search of food, evidence of the near presence of which was being borne up to him by a breeze from down river.

Ten minutes later he had made his kill, again one of the Pal-ul-don specimens of antelope, all species of which Tarzan had known since childhood as Bara, the deer, since in the little primer that had been the basis of his education the picture of a deer had been the nearest approach to the likeness of the antelope, from the giant eland to the smaller bushbuck of the hunting grounds of his youth.

Cutting off a haunch he cached it in a nearby tree, and throwing the balance of the carcass across his shoulder trotted back toward the spot at which he had left the *gryf*. The great beast was just emerging from the river when Tarzan, seeing it, issued the weird cry of the Tor-o-don. The creature looked in the direction of the sound voicing at the same time the low rumble with which it answered the call of its master. Twice Tarzan repeated his cry before the beast moved slowly toward him, and when it had come within a few paces he tossed the carcass of the deer to it, upon which it fell with greedy jaws.

"If anything will keep it within call," mused the ape-man as he returned to the tree in which he had cached his own portion of his kill, "it is the knowledge that I will feed it." But as he finished his repast and settled himself comfortably for the night high among the swaying branches of his eyrie he had little confidence that he would ride into A-lur the following day upon his prehistoric steed.

When Tarzan awoke early the following morning he dropped lightly to the ground and made his way to the stream. Removing his weapons and loin cloth he entered the cold waters of the little pool, and after his refreshing bath returned to the tree to breakfast upon another portion of Bara, the deer, adding to his repast some fruits and berries which grew in abundance nearby.

His meal over he sought the ground again and raising his voice in the weird cry that he had learned, he called aloud on the cha nce of attracting the *gryf*, but though he waited for some time and continued calling there was no response, and he was finally forced to the conclusion that he had seen the last of his great mount of the preceding day.

And so he set his face toward A-lur, pinning his faith upon his knowledge of the Ho-don tongue, his great strength and his native wit.

Refreshed by food and rest, the journey toward A-lur, made in the cool of the morning along the bank of the joyous river, he found delightful in the extreme. Differentiating him from his fellows of the savage jungle were many characteristics other than those physical and mental. Not the least of these were in a measure spiritual, and one that had doubtless been as strong as another in influencing Tarzan's love of the jungle had been his appreciation of the beauties of nature. The apes cared more for a grubworm in a rotten log than for all the majestic grandeur of the forest giants waving above them. The only beauties that Numa acknowledged were those of his own person as he paraded them before the admiring eyes of his mate; but in all the manifestations of the creative power of nature of which Tarzan was cognizant he appreciated the beauties.

As Tarzan neared the city his interest became centered upon the architecture of the outlying buildings which were hewn from the chalklike limestone of what had once been a group of low hills, similar to the many grass-covered hillocks that dotted the valley in every direction. Ta-den's explanation of the Ho-don methods of house construction accounted for the ofttimes remarkable shapes and proportions of the buildings which, during the ages that must have been required for their construction, had been hewn from the limestone hills, the exteriors chiseled to such architectural forms as appealed to the eyes of the builders while at the same time following roughly the original outlines of the hills in an evident desire to economize both labor and space. The excavation of the apartments within had been similarly governed by necessity.

As he came nearer Tarzan saw that the waste material from these building operations had been utilized in the construction of outer walls about each building or group of buildings resulting from a single hillock, and later he was to learn that it had also been used for the filling of inequalities between the hills and the forming of paved streets throughout the city, the result, possibly, more of the adoption of an easy method of disposing of the quantities of broken limestone than by any real necessity for pavements.

There were people moving about within the city and upon the narrow ledges and terraces that broke the lines of the buildings and which seemed to be a peculiarity of Ho-don architecture, a concession, no doubt, to some inherent instinct that might be traced back to their early cliff-dwelling progenitors.

Tarzan was not surprised that at a short distance he aroused no suspicion or curiosity in the minds of those who saw him, since, until closer scrutiny was possible, there was little to distinguish him from a native either in his general conformation or his color. He had, of course, formulated a plan of action and, having decided, he did not hesitate in the carrying out his plan.

With the same assurance that you might venture upon the main street of a neighboring city Tarzan strode into the Ho-don city of A-lur. The first person to detect his spuriousness was a little child playing in the arched gateway of one of the walled buildings. "No tail! no tail!" it shouted, throwing a

stone at him, and then it suddenly grew dumb and its eyes wide
as it sensed that this creature was something other than a mere
Ho-don warrior who had lost his tail. With a gasp the child
turned and fled screaming into the courtyard of its home.

Tarzan continued on his way, fully realizing that the moment was imminent when the fate of his plan would be decided.
Nor had he long to wait since at the next turning of the winding street he came face to face with a Ho-don warrior. He
saw the sudden surprise in the latter's eyes, followed instantly
by one of suspicion, but before the fellow could speak Tarzan
addressed him.

"I am a stranger from another land," he said; "I would speak
with Ko-tan, your king."

The fellow stepped back, laying his hand upon his knife.
"There are no strangers that come to the gates of A-lur," he
said, "other than as enemies or slaves."

"I come neither as a slave nor an enemy," replied Tarzan.
"I come directly from Jad-ben-Otho. Look!" and he held out
his hands that the Ho-don might see how greatly they differed
from his own, and then wheeled about that the other might
see that he was tailless, for it was upon this fact that his plan
had been based, due to his recollection of the quarrel between
Ta-den and Om-at, in which the Waz-don had claimed that
Jad-ben-Otho had a long tail while the Ho-don had been
equally willing to fight for his faith in the taillessness of his
god.

The warrior's eyes widened and an expression of awe crept
into them, though it was still tinged with suspicion. "Jad-
ben-Otho!" he murmured, and then, "It is true that you are
neither Ho-don nor Waz-don, and it is also true that Jad-ben-
Otho has no tail. Come," he said, "I will take you to Ko-tan,
for this is a matter in which no common warrior may interfere.
Follow me," and still clutching the handle of his knife and

keeping a wary side glance upon the ape-man he led the way through A-lur.

The city covered a large area. Sometimes there was a considerable distance between groups of buildings, and again they were quite close together. There were numerous imposing groups, evidently hewn from the larger hills, often rising to a height of a hundred feet or more. As they advanced they met numerous warriors and women, all of whom showed great curiosity in the stranger, but there was no attempt to menace him when it was found that he was being conducted to the palace of the king.

They came at last to a great pile that sprawled over a considerable area, its western front facing upon a large blue lake and evidently hewn from what had once been a natural cliff. This group of buildings was surrounded by a wall of considerably greater height than any that Tarzan had before seen. His guide led him to a gateway before which waited a dozen or more warriors who had risen to their feet and formed a barrier across the entrance-way as Tarzan and his party appeared around the corner of the palace wall, for by this time he had accumulated such a following of the curious as presented to the guards the appearance of a formidable mob.

The guide's story told, Tarzan was conducted into the courtyard where he was held while one of the warriors entered the palace, evidently with the intention of notifying Ko-tan. Fifteen minutes later a large warrior appeared, followed by several others, all of whom examined Tarzan with every sign of curiosity as they approached.

The leader of the party halted before the ape-man. "Who are you?" he asked, "and what do you want of Ko-tan, the king?"

"I am a friend," replied the ape-man, "and I have come from the country of Jad-ben-Otho to visit Ko-tan of Pal-ul-don."

The warrior and his followers seemed impressed. Tarzan could see the latter whispering among themselves.

"How come you here," asked the spokesman, "and what do you want of Ko-tan?"

Tarzan drew himself to his full height. "Enough!" he cried. "Must the messenger of Jad-ben-Otho be subjected to the treatment that might be accorded to a wandering Waz-don? Take me to the king at once lest the wrath of Jad-ben-Otho fall upon you."

There was some question in the mind of the ape-man as to how far he might carry his unwarranted show of assurance, and he waited therefore with amused interest the result of his demand. He did not, however, have long to wait for almost immediately the attitude of his questioner changed. He whitened, cast an apprehensive glance toward the eastern sky and then extended his right palm toward Tarzan, placing his left over his own heart in the sign of amity that was common among the peoples of Pal-ul-don.

Tarzan stepped quickly back as though from a profaning hand, a feigned expression of horror and disgust upon his face.

"Stop!" he cried, "who would dare touch the sacred person of the messenger of Jad-ben-Otho? Only as a special mark of favor from Jad-ben-Otho may even Ko-tan himself receive this honor from me. Hasten! Already now have I waited too long! What manner of reception the Ho-don of A-lur would extend to the son of my father!"

At first Tarzan had been inclined to adopt the rôle of Jad-ben-Otho himself but it occurred to him that it might prove embarrassing and considerable of a bore to be compelled constantly to portray the character of a god, but with the growing success of his scheme it had suddenly occurred to him that the authority of the son of Jad-ben-Otho would be far greater than that of an ordinary messenger of a god, while at the same time giving him some leeway in the matter of his acts and demeanor, the ape-man reasoning that a young god would not be held so strictly accountable in the matter of his dignity and bearing as an older and greater god.

This time the effect of his words was immediately and painfully noticeable upon all those near him. With one accord they shrank back, the spokesman almost collapsing in evident terror. His apologies, when finally the paralysis of his fear would permit him to voice them, were so abject that the ape-man could scarce repress a smile of amused contempt.

"Have mercy, O Dor-ul-Otho," he pleaded, "on poor old Dak-lot. Precede me and I will show you to where Ko-tan, the king, awaits you, trembling. Aside, snakes and vermin," he cried pushing his warriors to right and left for the purpose of forming an avenue for Tarzan.

"Come!" cried the ape-man peremptorily, "lead the way, and let these others follow."

The now thoroughly frightened Dak-lot did as he was bid, and Tarzan of the Apes was ushered into the palace of Ko-tan, King of Pal-ul-don.

CHAPTER IX

Blood-Stained Altars

THE entrance through which he caught his first glimpse of the interior was rather beautifully carved in geometric designs, and within the walls were similarly treated, though as he proceeded from one apartment to another he found also the figures of animals, birds, and men taking their places among the more formal figures of the mural decorator's art. Stone vessels were much in evidence as well as ornaments of gold and the skins of many animals, but nowhere did he see an indication of any woven fabric, indicating that in that respect at least the Ho-don were still low in the scale of evolution, and yet the proportions and symmetry of the corridors and apartments bespoke a degree of civilization.

The way led through several apartments and long corridors, up at least three flights of stone stairs and finally out upon a ledge upon the western side of the building overlooking the blue lake. Along this ledge, or arcade, his guide led him for a hundred yards, to stop at last before a wide entrance-way leading into another apartment of the palace.

Here Tarzan beheld a considerable concourse of warriors in an enormous apartment, the domed ceiling of which was fully fifty feet above the floor. Almost filling the chamber was a great pyramid ascending in broad steps well up under the dome in which were a number of round apertures which let in the light. The steps of the pyramid were occupied by warriors to the very pinnacle, upon which sat a large, imposing figure of a man whose golden trappings shone brightly in the light of the afternoon sun, a shaft of which poured through one of the tiny apertures of the dome.

"Ko-tan!" cried Dak-lot, addressing the resplendent figure at the pinnacle of the pyramid. "Ko-tan and warriors of Pal-ul-don! Behold the honor that Jad-ben-Otho has done you in sending as his messenger his own son," and Dak-lot, stepping aside, indicated Tarzan with a dramatic sweep of his hand.

Ko-tan rose to his feet and every warrior within sight craned his neck to have a better view of the newcomer. Those upon the opposite side of the pyramid crowded to the front as the words of the old warrior reached them. Skeptical were the expressions on most of the faces; but theirs was a skepticism marked with caution. No matter which way fortune jumped they wished to be upon the right side of the fence. For a moment all eyes were centered upon Tarzan and then gradually they drifted to Ko-tan, for from his attitude would they receive the cue that would determine theirs. But Ko-tan was evidently in the same quandary as they—the very attitude of his body indicated it—it was one of indecision and of doubt.

The ape-man stood erect, his arms folded upon his broad breast, an expression of haughty disdain upon his handsome face; but to Dak-lot there seemed to be indications also of growing anger. The situation was becoming strained. Dak-lot fidgeted, casting apprehensive glances at Tarzan and appealing ones at Ko-tan. The silence of the tomb wrapped the great chamber of the throneroom of Pal-ul-don.

At last Ko-tan spoke. "Who says that he is Dor-ul-Otho?" he asked, casting a terrible look at Dak-lot.

"He does!" almost shouted that terrified noble.

"And so it must be true?" queried Ko-tan.

Could it be that there was a trace of irony in the chief's tone?

Otho forbid! Dak-lot cast a side glance at Tarzan—a glance that he intended should carry the assurance of his own faith; but that succeeded only in impressing the ape-man with the other's pitiable terror.

"O Ko-tan!" pleaded Dak-lot, "your own eyes must convince you that indeed he is the son of Otho. Behold his godlike figure, his hands, and his feet, that are not as ours, and that he is entirely tailless as is his mighty father."

Ko-tan appeared to be perceiving these facts for the first time and there was an indication that his skepticism was faltering. At that moment a young warrior who had pushed his way forward from the opposite side of the pyramid to where he could obtain a good look at Tarzan raised his voice.

"Ko-tan," he cried, "it must be even as Dak-lot says, for I am sure now that I have seen Dor-ul-Otho before. Yesterday as we were returning with the Kor-ul-lul prisoners we beheld him seated upon the back of a great *gryf*. We hid in the woods before he came too near, but I saw enough to make sure that he who rode upon the great beast was none other than the messenger who stands here now."

This evidence seemed to be quite enough to convince the majority of the warriors that they indeed stood in the presence of deity—their faces showed it only too plainly, and a sudden modesty that caused them to shrink behind their neighbors. As their neighbors were attempting to do the same thing, the result was a sudden melting away of those who stood nearest the ape-man, until the steps of the pyramid directly before him lay vacant to the very apex and to Ko-tan. The latter, possibly influenced as much by the fearful attitude of his followers as by the evidence adduced, now altered his tone and his manner in such a degree as might comport with the requirements if the stranger was indeed the Dor-ul-Otho while leaving his dignity a loophole of escape should it appear that he had entertained an impostor.

"If indeed you are the Dor-ul-Otho," he said, addressing Tarzan, "you will know that our doubts were but natural since we have received no sign from Jad-ben-Otho that he intended honoring us so greatly, nor how could we know, even, that the Great God had a son? If you are he, all Pal-ul-don rejoices to honor you; if you are not he, swift and terrible shall be the punishment of your temerity. I, Ko-tan, King of Pal-ul-don, have spoken."

"And spoken well, as a king should speak," said Tarzan, breaking his long silence, "who fears and honors the god of his people. It is well that you insist that I indeed be the Dor-ul-Otho before you accord me the homage that is my due. Jad-ben-Otho charged me specially to ascertain if you were fit to rule his people. My first experience of you indicates that Jad-ben-Otho chose well when he breathed the spirit of a king into the babe at your mother's breast."

The effect of this statement, made so casually, was marked in the expressions and excited whispers of the now awe-struck assemblage. At last they knew how kings were made! It was decided by Jad-ben-Otho while the candidate was still a suckling babe! Wonderful! A miracle! and this divine creature in whose presence they stood knew all about it. Doubtless he even discussed such matters with their god daily. If there had been an atheist among them before, or an agnostic, there was none now, for had they not looked with their own eyes upon the son of god?

"It is well then," continued the ape-man, "that you should assure yourself that I am no impostor. Come closer that you may see that I am not as are men. Furthermore it is not meet that you stand upon a higher level than the son of your god."

There was a sudden scramble to reach the floor of the throne-room, nor was Ko-tan far behind his warriors, though he managed to maintain a certain majestic dignity as he descended the broad stairs that countless naked feet had polished to a gleaming smoothness through the ages. "And now," said Tarzan as the king stood before him, "you can have no doubt that I am not of the same race as you. Your priests have told you that Jad-ben-Otho is tailless. Tailless, therefore, must be the race of gods that spring from his loins. But enough of such proofs as these! You know the power of Jad-ben-Otho; how his lightnings gleaming out of the sky carry death as he wills it; how the rains come at his bidding, and the fruits and the berries and the grains, the grasses, the trees and the flowers spring to life at his divine direction; you have witnessed birth and death, and those who honor their god honor him because he controls these things. How would it fare then with an impostor who claimed to be the son of this all-powerful god? This then is all the proof that you require, for as he would strike you down should you deny me, so would he strike down one who wrongfully claimed kinship with him."

This line of argument being unanswerable must needs be convincing. There could be no questioning of this creature's statements without the tacit admission of lack of faith in the omnipotence of Jad-ben-Otho. Ko-tan was satisfied that he was entertaining deity, but as to just what form his entertainment should take he was rather at a loss to know. His conception of god had been rather a vague and hazy affair, though in common with all primitive people his god was a personal one as were his devils and demons. The pleasures of Jad-ben-Otho he had assumed to be the excesses which he himself enjoyed, but devoid of any unpleasant reaction. It therefore occurred to him that the Dor-ul-Otho would be greatly enter-

tained by eating—eating large quantities of everything that
Ko-tan liked best and that he had found most injurious; and
there was also a drink that the women of the Ho-don made
by allowing corn to soak in the juices of succulent fruits, to
which they had added certain other ingredients best known
to themselves. Ko-tan knew by experience that a single
draught of this potent liquor would bring happiness and surcease from worry, while several would cause even a king to do
things and enjoy things that he would never even think of
doing or enjoying while not under the magical influence of
the potion, but unfortunately the next morning brought suffering in direct ratio to the joy of the preceding day. A god,
Ko-tan reasoned, could experience all the pleasure without
the headache, but for the immediate present he must think of
the necessary dignities and honors to be accorded his immortal
guest.

No foot other than a king's had touched the surface of the
apex of the pyramid in the throneroom at A-lur during all the
forgotten ages through which the kings of Pal-ul-don had
ruled from its high eminence. So what higher honor could
Ko-tan offer than to give place beside him to the Dor-ul-Otho?
And so he invited Tarzan to ascend the pyramid and take his
place upon the stone bench that topped it. As they reached
the step below the sacred pinnacle Ko-tan continued as though
to mount to his throne, but Tarzan laid a detaining hand upon
his arm.

"None may sit upon a level with the gods," he admonished,
stepping confidently up and seating himself upon the throne.
The abashed Ko-tan showed his embarrassment, an embarrassment he feared to voice lest he incur the wrath of the king of
kings.

"But," added Tarzan, "a god may honor his faithful servant

by inviting him to a place at his side. Come, Ko-tan; thus would I honor you in the name of Jad-ben-Otho."

The ape-man's policy had for its basis an attempt not only to arouse the fearful respect of Ko-tan but to do it without making of him an enemy at heart, for he did not know how strong a hold the religion of the Ho-don had upon them, for since the time that he had prevented Ta-den and Om-at from quarreling over a religious difference the subject had been utterly taboo among them. He was therefore quick to note the evident though wordless resentment of Ko-tan at the suggestion that he entirely relinquish his throne to his guest. On the whole, however, the effect had been satisfactory as he could see from the renewed evidence of awe upon the faces of the warriors.

At Tarzan's direction the business of the court continued where it had been interrupted by his advent. It consisted principally in the settling of disputes between warriors. There was present one who stood upon the step just below the throne and which Tarzan was to learn was the place reserved for the higher chiefs of the allied tribes which made up Ko-tan's kingdom. The one who attracted Tarzan's attention was a stalwart warrior of powerful physique and massive, lion-like features. He was addressing Ko-tan on a question that is as old as government and that will continue in unabated importance until man ceases to exist. It had to do with a boundary dispute with one of his neighbors.

The matter itself held little or no interest for Tarzan, but he was impressed by the appearance of the speaker and when Ko-tan addressed him as Ja-don the ape-man's interest was permanently crystallized, for Ja-don was the father of Ta-den. That the knowledge would benefit him in any way seemed rather a remote possibility since he could not reveal to Ja-don

his friendly relations with his son without admitting the falsity
of his claims to godship.

When the affairs of the audience were concluded Ko-tan
suggested that the son of Jad-ben-Otho might wish to visit the
temple in which were performed the religious rites coincident
to the worship of the Great God. And so the ape-man was
conducted by the king himself, followed by the warriors of his
court, through the corridors of the palace toward the northern
end of the group of buildings within the royal enclosure.

The temple itself was really a part of the palace and similar
in architecture. There were several ceremonial places of varying sizes, the purposes of which Tarzan could only conjecture.
Each had an altar in the west end and another in the east and
were oval in shape, their longest diameter lying due east and
west. Each was excavated from the summit of a small hillock
and all were without roofs. The western altars invariably
were a single block of stone the top of which was hollowed
into an oblong basin. Those at the eastern ends were similar
blocks of stone with flat tops and these latter, unlike those at
the opposite ends of the ovals were invariably stained or
painted a reddish brown, nor did Tarzan need to examine them
closely to be assured of what his keen nostrils already had told
him—that the brown stains were dried and drying human
blood.

Below these temple courts were corridors and apartments
reaching far into the bowels of the hills, dim, gloomy passages
that Tarzan glimpsed as he was led from place to place on his
tour of inspection of the temple. A messenger had been dispatched by Ko-tan to announce the coming visit of the son
of Jad-ben-Otho with the result that they were accompanied
through the temple by a considerable procession of priests
whose distinguishing mark of profession seemed to consist in

grotesque headdresses; sometimes hideous faces carved from wood and entirely concealing the countenances of their wearers; or again, the head of a wild beast cunningly fitted over the head of a man. The high priest alone wore no such headdress. He was an old man with close-set, cunning eyes and a cruel, thin-lipped mouth.

At first sight of him Tarzan realized that here lay the greatest danger to his ruse, for he saw at a glance that the man was antagonistic toward him and his pretensions, and he knew too that doubtless of all the people of Pal-ul-don the high priest was most likely to harbor the truest estimate of Jad-ben-Otho, and, therefore, would look with suspicion on one who claimed to be the son of a fabulous god.

No matter what suspicion lurked within his crafty mind, Lu-don, the high priest of A-lur, did not openly question Tarzan's right to the title of Dor-ul-Otho, and it may be that he was restrained by the same doubts which had originally restrained Ko-tan and his warriors—the doubt that is at the bottom of the minds of all blasphemers even and which is based upon the fear that after all there may be a god. So, for the time being at least Lu-don played safe. Yet Tarzan knew as well as though the man had spoken aloud his inmost thoughts that it was in the heart of the high priest to tear the veil from his imposture.

At the entrance to the temple Ko-tan had relinquished the guidance of the guest to Lu-don and now the latter led Tarzan through those portions of the temple that he wished him to see. He showed him the great room where the votive offerings were kept, gifts from the barbaric chiefs of Pal-ul-don and from their followers. These things ranged in value from presents of dried fruits to massive vessels of beaten gold, so that in the great main storeroom and its connecting chambers and corridors was an accumulation of wealth that amazed even the eyes of the owner of the secret of the treasure vaults of Opar.

Moving to and fro throughout the temple were sleek black Waz-don slaves, fruits of the Ho-don raids upon the villages of their less civilized neighbors. As they passed the barred entrance to a dim corridor, Tarzan saw within a great company of pithecanthropi of all ages and of both sexes, Ho-don as well as Waz-don, the majority of them squatted upon the stone floor in attitudes of utter dejection while some paced back and forth, their features stamped with the despair of utter hopelessness.

"And who are these who lie here thus unhappily?" he asked of Lu-don. It was the first question that he had put to the high priest since entering the temple, and instantly he regretted that he had asked it, for Lu-don turned upon him a face upon which the expression of suspicion was but thinly veiled.

"Who should know better than the son of Jad-ben-Otho?" he retorted.

"The questions of Dor-ul-Otho are not with impunity answered with other questions," said the ape-man quietly, "and it may interest Lu-don, the high priest, to know that the blood of a false priest upon the altar of his temple is not displeasing in the eyes of Jad-ben-Otho."

Lu-don paled as he answered Tarzan's question. "They are the offerings whose blood must refresh the eastern altars as the sun returns to your father at the day's end."

"And who told you," asked Tarzan, "that Jad-ben-Otho was pleased that his people were slain upon his altars? What if you were mistaken?"

"Then countless thousands have died in vain," replied Lu-don.

Ko-tan and the surrounding warriors and priests were listening attentively to the dialogue. Some of the poor victims behind the barred gateway had heard and rising, pressed close to the barrier through which one was conducted just before sunset each day, never to return.

"Liberate them!" cried Tarzan with a wave of his hand toward the imprisoned victims of a cruel superstition, "for I can tell you in the name of Jad-ben-Otho that you are mistaken."

CHAPTER X

The Forbidden Garden

Lu-DON paled. "It is sacrilege," he cried; "for countless ages have the priests of the Great God offered each night a life to the spirit of Jad-ben-Otho as it returned below the western horizon to its master, and never has the Great God given sign that he was displeased."

"Stop!" commanded Tarzan. "It is the blindness of the priesthood that has failed to read the messages of their god. Your warriors die beneath the knives and clubs of the Waz-don; your hunters are taken by *ja* and *jato;* no day goes by but witnesses the deaths of few or many in the villages of the Ho-don, and one death each day of those that die are the toll which Jad-ben-Otho has exacted for the lives you take upon the eastern altar. What greater sign of his displeasure could you require, O stupid priest?"

Lu-don was silent. There was raging within him a great conflict between his fear that this indeed might be the son of god and his hope that it was not, but at last his fear won and he bowed his head. "The son of Jad-ben-Otho has spoken," he said, and turning to one of the lesser priests: "Remove the bars and return these people from whence they came."

He thus addressed did as he was bid and as the bars came down the prisoners, now all fully aware of the miracle that had saved them, crowded forward and throwing themselves upon their knees before Tarzan raised their voices in thanksgiving.

117

Ko-tan was almost as staggered as the high priest by this ruthless overturning of an age-old religious rite. "But what," he cried, "may we do that will be pleasing in the eyes of Jad-ben-Otho?" turning a look of puzzled apprehension toward the ape-man.

"If you seek to please your god," he replied, "place upon your altars such gifts of food and apparel as are most welcome in the city of your people. These things will Jad-ben-Otho bless, when you may distribute them among those of the city who need them most. With such things are your storerooms filled as I have seen with mine own eyes, and other gifts will be brought when the priests tell the people that in this way they find favor before their god," and Tarzan turned and signified that he would leave the temple.

As they were leaving the precincts devoted to the worship of their deity, the ape-man noticed a small but rather ornate building that stood entirely detached from the others as though it had been cut from a little pinnacle of limestone which had stood out from its fellows. As his interested glance passed over it he noticed that its door and windows were barred.

"To what purpose is that building dedicated?" he asked of Lu-don. "Who do you keep imprisoned there?"

"It is nothing," replied the high priest nervously, "there is no one there. The place is vacant. Once it was used but not now for many years," and he moved on toward the gateway which led back into the palace. Here he and the priests halted while Tarzan with Ko-tan and his warriors passed out from the sacred precincts of the temple grounds.

The one question which Tarzan would have asked he had feared to ask for he knew that in the hearts of many lay a suspicion as to his genuineness, but he determined that before he slept he would put the question to Ko-tan, either directly or indirectly—as to whether there was, or had been recently within the city of A-lur a female of the same race as his.

As their evening meal was being served to them in the banquet hall of Ko-tan's palace by a part of the army of black slaves upon whose shoulders fell the burden of all the heavy and menial tasks of the city, Tarzan noticed that there came to the eyes of one of the slaves what was apparently an expression of startled recognition, as he looked upon the ape-man for the first time in the banquet hall of Ko-tan. And again later he saw the fellow whisper to another slave and nod his head in his direction. The ape-man did not recall ever having seen this Waz-don before and he was at a loss to account for an explanation of the fellow's interest in him, and presently the incident was all but forgotten.

Ko-tan was surprised and inwardly disgusted to discover that his godly guest had no desire to gorge himself upon rich foods and that he would not even so much as taste the villainous brew of the Ho-don. To Tarzan the banquet was a dismal and tiresome affair, since so great was the interest of the guests in gorging themselves with food and drink that they had no time for conversation, the only vocal sounds being confined to a continuous grunting which, together with their table manners reminded Tarzan of a visit he had once made to the famous Berkshire herd of His Grace, the Duke of Westminster at Woodhouse, Chester.

One by one the diners succumbed to the stupifying effects of the liquor with the result that the grunting gave place to snores, so presently Tarzan and the slaves were the only conscious creatures in the banquet hall.

Rising, the ape-man turned to a tall black who stood behind him. "I would sleep," he said, "show me to my apartment."

As the fellow conducted him from the chamber the slave who had shown surprise earlier in the evening at sight of him, spoke again at length to one of his fellows. The latter cast a half-frightened look in the direction of the departing ape-man. "If you are right," he said, "they should reward us with our liberty, but if you are wrong, O Jad-ben-Otho, what will be our fate?"

"But I am not wrong!" cried the other.

"Then there is but one to tell this to, for I have heard that he looked sour when this Dor-ul-Otho was brought to the temple and that while the so-called son of Jad-ben-Otho was there he gave this one every cause to fear and hate him. I mean Lu-don, the high priest."

"You know him?" asked the other slave.

"I have worked in the temple," replied his companion.

"Then go to him at once and tell him, but be sure to exact the promise of our freedom for the proof."

And so a black Waz-don came to the temple gate and asked to see Lu-don, the high priest, on a matter of great importance, and though the hour was late Lu-don saw him, and when he had heard his story he promised him and his friend not only their freedom but many gifts if they could prove the correctness of their claims.

And as the slave talked with the high priest in the temple at A-lur the figure of a man groped its way around the shoulder of Pastar-ul-ved and the moonlight glistened from the shiny barrel of an Enfield that was strapped to the naked back, and brass cartridges shed tiny rays of reflected light from their polished cases where they hung in the bandoliers across the broad brown shoulders and the lean waist.

Tarzan's guide conducted him to a chamber overlooking the blue lake where he found a bed similar to that which he had seen in the villages of the Waz-don, merely a raised dais of stone upon which was piled great quantities of furry pelts.

And so he lay down to sleep, the question that he most wished to put still unasked and unanswered.

With the coming of a new day he was awake and wandering about the palace and the palace grounds before there was sign of any of the inmates of the palace other than slaves, or at least he saw no others at first, though presently he stumbled upon an enclosure which lay almost within the center of the palace grounds surrounded by a wall that piqued the ape-man's curiosity, since he had determined to investigate as fully as possible every part of the palace and its environs.

This place, whatever it might be, was apparently without doors or windows but that it was at least partially roofless was evidenced by the sight of the waving branches of a tree which spread above the top of the wall near him. Finding no other method of access, the ape-man uncoiled his rope and throwing it over the branch of the tree where it projected beyond the wall, was soon climbing with the ease of a monkey to the summit.

There he found that the wall surrounded an enclosed garden in which grew trees and shrubs and flowers in riotous profusion. Without waiting to ascertain whether the garden was empty or contained Ho-don, Waz-don, or wild beasts, Tarzan dropped lightly to the sward on the inside and without further loss of time commenced a systematic investigation of the enclosure.

His curiosity was aroused by the very evident fact that the place was not for general use, even by those who had free access to other parts of the palace grounds and so there was added to its natural beauties an absence of mortals which rendered its exploration all the more alluring to Tarzan since it suggested that in such a place might he hope to come upon the object of his long and difficult search.

In the garden were tiny artificial streams and little pools of water, flanked by flowering bushes, as though it all had been designed by the cunning hand of some master gardener, so faithfully did it carry out the beauties and contours of nature upon a miniature scale.

The interior surface of the wall was fashioned to represent the white cliffs of Pal-ul-don, broken occasionally by small replicas of the verdure-filled gorges of the original.

Filled with admiration and thoroughly enjoying each new surprise which the scene offered, Tarzan moved slowly around the garden, and as always he moved silently. Passing through a miniature forest he came presently upon a tiny area of flower-studded sward and at the same time beheld before him the first Ho-don female he had seen since entering the palace. A young and beautiful woman stood in the center of the little open space, stroking the head of a bird which she held against her golden breastplate with one hand. Her profile was presented to the ape-man and he saw that by the standards of any land she would have been accounted more than lovely.

Seated in the grass at her feet, with her back toward him, was a female Waz-don slave. Seeing that she he sought was not there and apprehensive that an alarm be raised were he discovered by the two women, Tarzan moved back to hide himself in the foliage, but before he had succeeded the Ho-don girl turned quickly toward him as though apprised of his presence by that unnamed sense, the manifestations of which are more or less familiar to us all.

At sight of him her eyes registered only her surprise though there was no expression of terror reflected in them, nor did she scream or even raise her well-modulated voice as she addressed him.

"Who are you," she asked, "who enters thus boldly the Forbidden Garden?"

At sound of her mistress' voice the slave maiden turned

quickly, rising to her feet. "Tarzan-jad-guru!" she exclaimed in tones of mingled astonishment and relief.

"You know him?" cried her mistress turning toward the slave and affording Tarzan an opportunity to raise a cautioning finger to his lips lest Pan-at-lee further betray him, for it was Pan-at-lee indeed who stood before him, no less a source of surprise to him than had his presence been to her.

Thus questioned by her mistress and simultaneously admonished to silence by Tarzan, Pan-at-lee was momentarily silenced and then haltingly she groped for a way to extricate herself from her dilemma. "I thought—" she faltered, "but no, I am mistaken—I thought that he was one whom I had seen before near the Kor-ul-gryf."

The Ho-don looked first at one and then at the other an expression of doubt and questioning in her eyes. "But you have not answered me," she continued presently; "who are you?"

"You have not heard then," asked Tarzan, "of the visitor who arrived at your king's court yesterday?"

"You mean," she exclaimed, "that you are the Dor-ul-Otho?" And now the erstwhile doubting eyes reflected naught but awe.

"I am he," replied Tarzan; "and you?"

"I am O-lo-a, daughter of Ko-tan, the king," she replied.

So this was O-lo-a, for love of whom Ta-den had chosen exile rather than priesthood. Tarzan had approached more closely the dainty barbarian princess. "Daughter of Ko-tan," he said, "Jad-ben-Otho is pleased with you and as a mark of his favor he has preserved for you through many dangers him whom you love."

"I do not understand," replied the girl but the flush that mounted to her cheek belied her words. "Bu-lat is a guest in the palace of Ko-tan, my father. I do not know that he has faced any danger. It is to Bu-lat that I am betrothed."

"But it is not Bu-lat whom you love," said Tarzan.

Again the flush and the girl half turned her face away. "Have I then displeased the Great God?" she asked.

"No," replied Tarzan; "as I told you he is well satisfied and for your sake he has saved Ta-den for you."

"Jad-ben-Otho knows all," whispered the girl, "and his son shares his great knowledge."

"No," Tarzan hastened to correct her lest a reputation for omniscience might prove embarrassing. "I know only what Jad-ben-Otho wishes me to know."

"But tell me," she said, "I shall be reunited with Ta-den? Surely the son of god can read the future."

The ape-man was glad that he had left himself an avenue of escape. "I know nothing of the future," he replied, "other than what Jad-ben-Otho tells me. But I think you need have no fear for the future if you remain faithful to Ta-den and Ta-den's friends."

"You have seen him?" asked O-lo-a. "Tell me, where is he?"

"Yes," replied Tarzan, "I have seen him. He was with Om-at, the *gund* of Kor-ul-ja."

"A prisoner of the Waz-don?" interrupted the girl.

"Not a prisoner but an honored guest," replied the ape-man.

"Wait," he exclaimed, raising his face toward the heavens; "do not speak. I am receiving a message from Jad-ben-Otho, my father."

The two women dropped to their knees, covering their faces with their hands, stricken with awe at the thought of the awful nearness of the Great God. Presently Tarzan touched O-lo-a on the shoulder.

"Rise," he said. "Jad-ben-Otho has spoken. He has told me that this slave girl is from the tribe of Kor-ul-ja, where Ta-den is, and that she is betrothed to Om-at, their chief. Her name is Pan-at-lee."

O-lo-a turned questioningly toward Pan-at-lee. The latter nodded, her simple mind unable to determine whether or not she and her mistress were the victims of a colossal hoax. "It is even as he says," she whispered.

O-lo-a fell upon her knees and touched her forehead to Tarzan's feet. "Great is the honor that Jad-ben-Otho has done his poor servant," she cried. "Carry to him my poor thanks for the happiness that he has brought to O-lo-a."

"It would please my father," said Tarzan, "if you were to cause Pan-at-lee to be returned in safety to the village of her people."

"What cares Jad-ben-Otho for such as she?" asked O-lo-a, a slight trace of hauteur in her tone.

"There is but one god," replied Tarzan, "and he is the god of the Waz-don as well as of the Ho-don; of the birds and the beasts and the flowers and of everything that grows upon the earth or beneath the waters. If Pan-at-lee does right she is greater in the eyes of Jad-ben-Otho than would be the daughter of Ko-tan should she do wrong."

It was evident that O-lo-a did not quite understand this interpretation of divine favor, so contrary was it to the teachings of the priesthood of her people. In one respect only did Tarzan's teachings coincide with her belief—that there was but one god. For the rest she had always been taught that he was solely the god of the Ho-don in every sense, other than that other creatures were created by Jad-ben-Otho to serve some useful purpose for the benefit of the Ho-don race. And now to be told by the son of god that she stood no higher in divine esteem than the black handmaiden at her side was indeed a shock to her pride, her vanity, and her faith. But who could question the word of Dor-ul-Otho, especially when she had with her own eyes seen him in actual communion with god in heaven?

"The will of Jad-ben-Otho be done," said O-lo-a meekly, "if it lies within my power. But it would be best, O Dor-ul-Otho, to communicate your father's wish directly to the king."

"Then keep her with you," said Tarzan, "and see that no harm befalls her."

O-lo-a looked ruefully at Pan-at-lee. "She was brought to me but yesterday," she said, "and never have I had slave woman who pleased me better. I shall hate to part with her."

"But there are others," said Tarzan.

"Yes," replied O-lo-a, "there are others, but there is only **one** Pan-at-lee."

"Many slaves are brought to the city?" asked Tarzan.

"Yes," she replied.

"And many strangers come from other lands?" he asked.

She shook her head negatively. "Only the Ho-don from **the** other side of the Valley of Jad-ben-Otho," she replied, "and they are not strangers."

"Am I then the first stranger to enter the gates of A-lur?" he asked.

"Can it be," she parried, "that the son of Jad-ben-Otho need question a poor ignorant mortal like O-lo-a?"

"As I told you before," replied Tarzan, "Jad-ben-Otho alone is all-knowing."

"Then if he wished you to know this thing," retorted O-lo-a quickly, "you would know it."

Inwardly the ape-man smiled that this little heathen's astuteness should beat him at his own game, yet in a measure her evasion of the question might be an answer to it. "There have been other strangers here then recently?" he persisted.

"I cannot tell you what I do not know," she replied. "Always is the palace of Ko-tan filled with rumors, but how much fact and how much fancy how may a woman of the palace know?"

"There has been such a rumor then?" he asked.

"It was only rumor that reached the Forbidden Garden," she replied.

"It described, perhaps, a woman of another race?" As he put the question and awaited her answer he thought that his heart ceased to beat, so grave to him was the issue at stake.

The girl hesitated before replying, and then: "No," she said, "I cannot speak of this thing, for if it be of sufficient importance to elicit the interest of the gods then indeed would I be subject to the wrath of my father should I discuss it."

"In the name of Jad-ben-Otho I command you to speak," said Tarzan. "In the name of Jad-ben-Otho in whose hands lies the fate of Ta-den!"

The girl paled. "Have mercy!" she cried, "and for the sake of Ta-den I will tell you all that I know."

"Tell what?" demanded a stern voice from the shrubbery behind them. The three turned to see the figure of Ko-tan emerging from the foliage. An angry scowl distorted his kingly features but at sight of Tarzan it gave place to an expression of surprise not unmixed with fear. "Dor-ul-Otho!" he exclaimed, "I did not know that it was you," and then, raising his head and squaring his shoulders he said, "but there are places where even the son of the Great God may not walk and this, the Forbidden Garden of Ko-tan, is one."

It was a challenge but despite the king's bold front there was a note of apology in it, indicating that in his superstitious mind there flourished the inherent fear of man for his Maker. "Come, Dor-ul-Otho," he continued, "I do not know all this foolish child has said to you but whatever you would know Ko-tan, the king, will tell you. O-lo-a, go to your quarters immediately," and he pointed with stern finger toward the opposite end of the garden.

The princess, followed by Pan-at-lee, turned at once and left them.

"We will go this way," said Ko-tan and preceding, led Tarzan in another direction. Close to that part of the wall which they approached Tarzan perceived a grotto in the miniature cliff into the interior of which Ko-tan led him, and down a rocky stairway to a gloomy corridor the opposite end of which opened into the palace proper. Two armed warriors stood at this entrance to the Forbidden Garden, evidencing how jealously were the sacred precincts of the place guarded.

In silence Ko-tan led the way back to his own quarters in the palace. A large chamber just outside the room toward which Ko-tan was leading his guest was filled with chiefs and warriors awaiting the pleasure of their ruler. As the two entered, an aisle was formed for them the length of the chamber, down which they passed in silence.

Close to the farther door and half hidden by the warriors who stood before him was Lu-don, the high priest. Tarzan glimpsed him but briefly but in that short period he was aware of a cunning and malevolent expression upon the cruel countenance that he was subconsciously aware boded him no good, and then with Ko-tan he passed into the adjoining room and the hangings dropped.

At the same moment the hideous headdress of an under priest appeared in the entrance of the outer chamber. Its owner, pausing for a moment, glanced quickly around the interior and then having located him whom he sought moved rapidly in the direction of Lu-don. There was a whispered conversation which was terminated by the high priest.

"Return immediately to the quarters of the princess," he said, "and see that the slave is sent to me at the temple at once."

The under priest turned and departed upon his mission while Lu-don also left the apartment and directed his footsteps toward the sacred enclosure over which he ruled.

A half-hour later a warrior was ushered into the presence of Ko-tan. "Lu-don, the high priest, desires the presence of Ko-tan, the king, in the temple," he announced, "and it is his wish that he come alone."

Ko-tan nodded to indicate that he accepted the command which even the king must obey. "I will return presently, Dor-ul-Otho," he said to Tarzan, "and in the meantime my warriors and my slaves are yours to command."

CHAPTER XI

The Sentence of Death

B UT it was an hour before the king re-entered the apartment and in the meantime the ape-man had occupied himself in examining the carvings upon the walls and the numerous speci- mens of the handicraft of Pal-ul-donian artisans which com- bined to impart an atmosphere of richness and luxury to the apartment.

The limestone of the country, close-grained and of marble whiteness yet worked with comparative ease with crude im- plements, had been wrought by cunning craftsmen into bowls and urns and vases of considerable grace and beauty. Into the carved designs of many of these virgin gold had been hammered, presenting the effect of a rich and magnificent cloisonné. A barbarian himself the art of barbarians had al- ways appealed to the ape-man to whom they represented a natural expression of man's love of the beautiful to even a greater extent than the studied and artificial efforts of civiliza- tion. Here was the real art of old masters, the other the cheap imitation of the chromo.

It was while he was thus pleasurably engaged that Ko-tan returned. As Tarzan, attracted by the movement of the hang- ings through which the king entered, turned and faced him he was almost shocked by the remarkable alteration of the king's appearance. His face was livid; his hands trembled as with palsy, and his eyes were wide as with fright. His appearance was one apparently of a combination of consuming anger and withering fear. Tarzan looked at him questioningly.

"You have had bad news, Ko-tan?" he asked.

The king mumbled an unintelligible reply. Behind there thronged into the apartment so great a number of warriors that they choked the entrance-way. The king looked apprehensively to right and left. He cast terrified glances at the ape-man and then raising his face and turning his eyes upward he cried: "Jad-ben-Otho be my witness that I do not this thing of my own accord." There was a moment's silence which was again broken by Ko-tan. "Seize him," he cried to the warriors about him, "for Lu-don, the high priest, swears that he is an impostor."

To have offered armed resistance to this great concourse of warriors in the very heart of the palace of their king would have been worse than fatal. Already Tarzan had come far by his wits and now that within a few hours he had had his hopes and his suspicions partially verified by the vague admissions of O-lo-a he was impressed with the necessity of inviting no mortal risk that he could avoid.

"Stop!" he cried, raising his palm against them. "What is the meaning of this?"

"Lu-don claims he has proof that you are not the son of Jad-ben-Otho," replied Ko-tan. "He demands that you be brought to the throneroom to face your accusers. If you are what you claim to be none knows better than you that you need have no fear in acquiescing to his demands, but remember always that in such matters the high priest commands the king and that I am only the bearer of these commands, not their author."

Tarzan saw that Ko-tan was not entirely convinced of his duplicity as was evidenced by his palpable design to play safe.

"Let not your warriors seize me," he said to Ko-tan, "lest Jad-ben-Otho, mistaking their intention, strike them dead."

The effect of his words was immediate upon the men in the
front rank of those who faced him, each seeming suddenly to
acquire a new modesty that compelled him to self-effacement
behind those directly in his rear—a modesty that became
rapidly contagious.

The ape-man smiled. "Fear not," he said, "I will go willingly
to the audience chamber to face the blasphemers who accuse
me."

Arrived at the great throneroom a new complication arose.
Ko-tan would not acknowledge the right of Lu-don to occupy
the apex of the pyramid and Lu-don would not consent to
occupying an inferior position while Tarzan, to remain consistent with his high claims, insisted that no one should stand
above him, but only to the ape-man was the humor of the situation apparent.

To relieve the situation Ja-don suggested that all three of
them occupy the throne, but this suggestion was repudiated by
Ko-tan who argued that no mortal other than a king of Pal-ul-
don had ever sat upon the high eminence, and that furthermore
there was not room for three there.

"But who," said Tarzan, "is my accuser and who is my
judge?"

"Lu-don is your accuser," explained Ko-tan.

"And Lu-don is your judge," cried the high priest.

"I am to be judged by him who accuses me then," said Tarzan. "It were better to dispense then with any formalities and
ask Lu-don to sentence me." His tone was ironical and his
sneering face, looking straight into that of the high priest, but
caused the latter's hatred to rise to still greater proportions.

It was evident that Ko-tan and his warriors saw the justice
of Tarzan's implied objection to this unfair method of dispens-
ing justice. "Only Ko-tan can judge in the throneroom of his
palace," said Ja-don, "let him hear Lu-don's charges and the

testimony of his witnesses, and then let Ko-tan's judgment be final."

Ko-tan, however, was not particularly enthusiastic over the prospect of sitting in trial upon one who might after all very possibly be the son of his god, and so he temporized, seeking for an avenue of escape. "It is purely a religious matter," he said, "and it is traditional that the kings of Pal-ul-don interfere not in questions of the church."

"Then let the trial be held in the temple," cried one of the chiefs, for the warriors were as anxious as their king to be relieved of all responsibility in the matter. This suggestion was more than satisfactory to the high priest who inwardly condemned himself for not having thought of it before.

"It is true," he said, "this man's sin is against the temple. Let him be dragged thither then for trial."

"The son of Jad-ben-Otho will be dragged nowhere," cried Tarzan. "But when this trial is over it is possible that the corpse of Lu-don, the high priest, will be dragged from the temple of the god he would desecrate. Think well, then, Lu-don before you commit this folly."

His words, intended to frighten the high priest from his position failed utterly in consummating their purpose. Lu-don showed no terror at the suggestion the ape-man's words implied.

"Here is one," thought Tarzan, "who, knowing more of his religion than any of his fellows, realizes fully the falsity of my claims as he does the falsity of the faith he preaches."

He realized, however, that his only hope lay in seeming indifference to the charges. Ko-tan and the warriors were still under the spell of their belief in him and upon this fact must he depend in the final act of the drama that Lu-don was staging for his rescue from the jealous priest whom he knew had already passed sentence upon him in his own heart.

With a shrug he descended the steps of the pyramid. "It matters not to Dor-ul-Otho," he said, "where Lu-don enrages his god, for Jad-ben-Otho can reach as easily into the chambers of the temple as into the throneroom of Ko-tan."

Immeasurably relieved by this easy solution of their problem the king and the warriors thronged from the throneroom toward the temple grounds, their faith in Tarzan increased by his apparent indifference to the charges against him. Lu-don led them to the largest of the altar courts.

Taking his place behind the western altar he motioned Ko-tan to a place upon the platform at the left hand of the altar and directed Tarzan to a similar place at the right.

As Tarzan ascended the platform his eyes narrowed angrily at the sight which met them. The basin hollowed in the top of the altar was filled with water in which floated the naked corpse of a new-born babe. "What means this?" he cried angrily, turning upon Lu-don.

The latter smiled malevolently. "That you do not know," he replied, "is but added evidence of the falsity of your claim. He who poses as the son of god did not know that as the last rays of the setting sun flood the eastern altar of the temple the lifeblood of an adult reddens the white stone for the edification of Jad-ben-Otho; and that when the sun rises again from the body of its maker it looks first upon this western altar and rejoices in the death of a new-born babe each day, the ghost of which accompanies it across the heavens by day as the ghost of the adult returns with it to Jad-ben-Otho at night.

"Even the little children of the Ho-don know these things, while he who claims to be the son of Jad-ben-Otho knows them not; and if this proof be not enough, there is more. Come, Waz-don," he cried, pointing to a tall slave who stood with a group of other blacks and priests on the temple floor at the left of the altar.

The fellow came forward fearfully. "Tell us what you know of this creature," cried Lu-don, pointing to Tarzan.

"I have seen him before," said the Waz-don. "I am of the tribe of Kor-ul-lul, and one day recently a party of which I was one encountered a few of the warriors of the Kor-ul-ja upon the ridge which separates our villages. Among the enemy was this strange creature whom they called Tarzan-jad-guru; and terrible indeed was he for he fought with the strength of many men so that it required twenty of us to subdue him. But he did not fight as a god fights, and when a club struck him upon the head he sank unconscious as might an ordinary mortal.

"We carried him with us to our village as a prisoner but he escaped after cutting off the head of the warrior we left to guard him and carrying it down into the gorge and tying it to the branch of a tree upon the opposite side."

"The word of a slave against that of a god!" cried Ja-don, who had shown previously a friendly interest in the pseudo godling.

"It is only a step in the progress toward truth," interjected Lu-don. "Possibly the evidence of the only princess of the house of Ko-tan will have greater weight with the great chief from the north, though the father of a son who fled the holy offer of the priesthood may not receive with willing ears any testimony against another blasphemer."

Ja-don's hand leaped to his knife, but the warriors next him laid detaining fingers upon his arms. "You are in the temple of Jad-ben-Otho, Ja-don," they cautioned and the great chief was forced to swallow Lu-don's affront though it left in his heart bitter hatred of the high priest.

And now Ko-tan turned toward Lu-don. "What knoweth my daughter of this matter?" he asked. "You would not bring a princess of my house to testify thus publicly?"

"No," replied Lu-don, "not in person, but I have here one

who will testify for her." He beckoned to an under priest. "Fetch the slave of the princess," he said.

His grotesque headdress adding a touch of the hideous to the scene, the priest stepped forward dragging the reluctant Pan-at-lee by the wrist.

"The Princess O-lo-a was alone in the Forbidden Garden with but this one slave," explained the priest, "when there suddenly appeared from the foliage nearby this creature who claims to be the Dor-ul-Otho. When the slave saw him the princess says that she cried aloud in startled recognition and called the creature by name—Tarzan-jad-guru—the same name that the slave from Kor-ul-lul gave him. This woman is not from Kor-ul-lul but from Kor-ul-ja, the very tribe with which the Kor-ul-lul says the creature was associating when he first saw him. And further the princess said that when this woman, whose name is Pan-at-lee, was brought to her yesterday she told a strange story of having been rescued from a Tor-o-don in the Kor-ul-gryf by a creature such as this, whom she spoke of then as Tarzan-jad-guru; and of how the two were pursued in the bottom of the gorge by two monster *gryfs,* and of how the man led them away while Pan-at-lee escaped, only to be taken prisoner in the Kor-ul-lul as she was seeking to return to her own tribe.

"Is it not plain now," cried Lu-don, "that this creature is no god. Did he tell you that he was the son of god?" he almost shouted, turning suddenly upon Pan-at-lee.

The girl shrank back terrified. "Answer me, slave!" cried the high priest.

"He seemed more than mortal," parried Pan-at-lee.

"Did he tell you that he was the son of god? Answer my question," insisted Lu-don.

"No," she admitted in a low voice, casting an appealing look of forgiveness at Tarzan who returned a smile of encouragement and friendship.

"That is no proof that he is not the son of god," cried Ja-don. "Dost think Jad-ben-Otho goes about crying 'I am god! I am god!' Hast ever heard him Lu-don? No, you have not. Why should his son do that which the father does not do?"

"Enough," cried Lu-don. "The evidence is clear. The creature is an impostor and I, the head priest of Jad-ben-Otho in the city of A-lur, do condemn him to die." There was a moment's silence during which Lu-don evidently paused for the dramatic effect of his climax. "And if I am wrong may Jad-ben-Otho pierce my heart with his lightnings as I stand here before you all."

The lapping of the wavelets of the lake against the foot of the palace wall was distinctly audible in the utter and almost breathless silence which ensued. Lu-don stood with his face turned toward the heavens and his arms outstretched in the attitude of one who bares his breast to the dagger of an executioner. The warriors and the priests and the slaves gathered in the sacred court awaited the consuming vengeance of their god.

It was Tarzan who broke the silence. "Your god ignores you Lu-don," he taunted, with a sneer that he meant to still further anger the high priest, "he ignores you and I can prove it before the eyes of your priests and your people."

"Prove it, blasphemer! How can you prove it?"

"You have called me a blasphemer," replied Tarzan, "you have proved to your own satisfaction that I am an impostor, that I, an ordinary mortal, have posed as the son of god. Demand then that Jad-ben-Otho uphold his godship and the dignity of his priesthood by directing his consuming fires through my own bosom."

Again there ensued a brief silence while the onlookers waited for Lu-don to thus consummate the destruction of this presumptuous impostor.

"You dare not," taunted Tarzan, "for you know that I would be struck dead no quicker than were you."

"You lie," cried Lu-don, "and I would do it had I not but just received a message from Jad-ben-Otho directing that your fate be different."

A chorus of admiring and reverential "Ahs" arose from the priesthood. Ko-tan and his warriors were in a state of mental confusion. Secretly they hated and feared Lu-don, but so ingrained was their sense of reverence for the office of the high priest that none dared raise a voice against him.

None? Well, there was Ja-don, fearless old Lion-man of the north. "The proposition was a fair one," he cried. "Invoke the lightnings of Jad-ben-Otho upon this man if you would ever convince us of his guilt."

"Enough of this," snapped Lu-don. "Since when was Ja-don created high priest? Seize the prisoner," he cried to the priests and warriors, "and on the morrow he shall die in the manner that Jad-ben-Otho has willed."

There was no immediate movement on the part of any of the warriors to obey the high priest's command, but the lesser priests on the other hand, imbued with the courage of fanaticism leaped eagerly forward like a flock of hideous harpies to seize upon their prey.

The game was up. That Tarzan knew. No longer could cunning and diplomacy usurp the functions of the weapons of defense he best loved. And so the first hideous priest who leaped to the platform was confronted by no suave ambassador from heaven, but rather a grim and ferocious beast whose temper savored more of hell.

The altar stood close to the western wall of the enclosure.

There was just room between the two for the high priest to stand during the performance of the sacrificial ceremonies and only Lu-don stood there now behind Tarzan, while before him were perhaps two hundred warriors and priests.

The presumptuous one who would have had the glory of first laying arresting hands upon the blasphemous impersonator rushed forward with outstretched hand to seize the ape-man. Instead it was he who was seized; seized by steel fingers that snapped him up as though he had been a dummy of straw, grasped him by one leg and the harness at his back and raised him with giant arms high above the altar. Close at his heels were others ready to seize the ape-man and drag him down, and beyond the altar was Lu-don with drawn knife advancing toward him.

There was no instant to waste, nor was it the way of the ape-man to fritter away precious moments in the uncertainty of belated decision. Before Lu-don or any other could guess what was in the mind of the condemned, Tarzan with all the force of his great muscles dashed the screaming hierophant in the face of the high priest, and, as though the two actions were one, so quickly did he move, he had leaped to the top of the altar and from there to a handhold upon the summit of the temple wall. As he gained a footing there he turned and looked down upon those beneath. For a moment he stood in silence and then he spoke.

"Who dare believe," he cried, "that Jad-ben-Otho would forsake his son?" and then he dropped from their sight upon the other side.

There were two at least left within the enclosure whose hearts leaped with involuntary elation at the success of the ape-man's maneuver, and one of them smiled openly. This was Ja-don, and the other, Pan-at-lee.

The brains of the priest that Tarzan had thrown at the

head of Lu-don had been dashed out against the temple wall while the high priest himself had escaped with only a few bruises, sustained in his fall to the hard pavement. Quickly scrambling to his feet he looked around in fear, in terror and finally in bewilderment, for he had not been a witness to the ape-man's escape. "Seize him," he cried; "seize the blasphemer," and he continued to look around in search of his victim with such a ridiculous expression of bewilderment that more than a single warrior was compelled to hide his smiles beneath his palm.

The priests were rushing around wildly, exhorting the warriors to pursue the fugitive but these awaited now stolidly the command of their king or high priest. Ko-tan, more or less secretly pleased by the discomfiture of Lu-don, waited for that worthy to give the necessary directions which he presently did when one of his acolytes excitedly explained to him the manner of Tarzan's escape.

Instantly the necessary orders were issued and priests and warriors sought the temple exit in pursuit of the ape-man. His departing words, hurled at them from the summit of the temple wall, had had little effect in impressing the majority that his claims had not been disproven by Lu-don, but in the hearts of the warriors was admiration for a brave man and in many the same unholy gratification that had risen in that of their ruler at the discomfiture of Lu-don.

A careful search of the temple grounds revealed no trace of the quarry. The secret recesses of the subterranean chambers, familiar only to the priesthood, were examined by these while the warriors scattered through the palace and the palace grounds without the temple. Swift runners were dispatched to the city to arouse the people there that all might be upon the lookout for Tarzan the Terrible. The story of his imposture and of his escape, and the tales that the Waz-don slaves had brought into the city concerning him were soon spread throughout A-lur, nor did they lose aught in the spreading, so that before an hour had passed the women and children were hiding behind barred doorways while the warriors crept apprehensively through the streets expecting momentarily to be pounced upon by a ferocious demon who, bare-handed, did victorious battle with huge *gryfs* and whose lightest pastime consisted in tearing strong men limb from limb.

CHAPTER XII

The Giant Stranger

A ND while the warriors and the priests of A-lur searched the the temple and the palace and the city for the vanished ape-man there entered the head of Kor-ul-ja down the precipitous trail from the mountains, a naked stranger bearing an Enfield upon his back. Silently he moved downward toward the bottom of the gorge and there where the ancient trail unfolded more levelly before him he swung along with easy strides, though always with the utmost alertness against possible dangers. A gentle breeze came down from the mountains behind him so that only his ears and his eyes were of value in detecting the presence of danger ahead. Generally the trail followed along the banks of the winding brooklet at the bottom of the gorge, but in some places where the waters tumbled over a precipitous ledge the trail made a detour along the side of the gorge, and again it wound in and out among rocky outcroppings, and presently where it rounded sharply the projecting shoulder of a cliff the stranger came suddenly face to face with one who was ascending the gorge.

Separated by a hundred paces the two halted simultaneously. Before him the stranger saw a tall white warrior, naked but for a loin cloth, cross belts, and a girdle. The man was armed with a heavy, knotted club and a short knife, the latter hanging in its sheath at his left hip from the end of one of his cross belts, the opposite belt supporting a leathern pouch at his right side. It was Ta-den hunting alone in the gorge of his friend, the chief of Kor-ul-ja. He contemplated the stranger with surprise but no wonder, since he recognized in him a member of the race with which his experience of Tarzan the Terrible had made him familiar and also, thanks to his friendship for the ape-man, he looked upon the newcomer without hostility.

The latter was the first to make outward sign of his intentions, raising his palm toward Ta-den in that gesture which has been a symbol of peace from pole to pole since man ceased to walk upon his knuckles. Simultaneously he advanced a few paces and halted.

Ta-den, assuming that one so like Tarzan the Terrible must be a fellow-tribesman of his lost friend, was more than glad to accept this overture of peace, the sign of which he returned in kind as he ascended the trail to where the other stood. "Who are you?" he asked, but the newcomer only shook his head to indicate that he did not understand.

By signs he tried to carry to the Ho-don the fact that he was following a trail that had led him over a period of many days from some place beyond the mountains and Ta-den was convinced that the newcomer sought Tarzan-jad-guru. He wished, however, that he might discover whether as friend or foe.

The stranger perceived the Ho-don's prehensile thumbs and great toes and his long tail with an astonishment which he sought to conceal, but greater than all was the sense of relief that the first inhabitant of this strange country whom he had met had proven friendly, so greatly would he have been handicapped by the necessity for forcing his way through a hostile land.

Ta-den, who had been hunting for some of the smaller mammals, the meat of which is especially relished by the Ho-don, forgot his intended sport in the greater interest of his new discovery. He would take the stranger to Om-at and possibly together the two would find some way of discovering the true intentions of the newcomer. And so again through signs he apprised the other that he would accompany him and together they descended toward the cliffs of Om-at's people.

As they approached these they came upon the women and children working under guard of the old men and the youths —gathering the wild fruits and herbs which constitute a part of their diet, as well as tending the small acres of growing crops which they cultivate. The fields lay in small level patches that had been cleared of trees and brush. Their farm implements consisted of metal-shod poles which bore a closer resemblance to spears than to tools of peaceful agriculture. Supplementing these were others with flattened blades that were neither hoes nor spades, but instead possessed the appearance of an unhappy attempt to combine the two implements in one.

At first sight of these people the stranger halted and unslung his bow for these creatures were black as night, their bodies entirely covered with hair. But Ta-den, interpreting the doubt in the other's mind, reassured him with a gesture and a smile. The Waz-don, however, gathered around excitedly jabbering questions in a language which the stranger discovered his guide understood though it was entirely unintelligible to the former. They made no attempt to molest him and he was now sure that he had fallen among a peaceful and friendly people.

It was but a short distance now to the caves and when they reached these Ta-den led the way aloft upon the wooden pegs, assured that this creature whom he had discovered would have no more difficulty in following him than had Tarzan the Terrible. Nor was he mistaken for the other mounted with ease until presently the two stood within the recess before the cave of Om-at, the chief.

The latter was not there and it was mid-afternoon before he returned, but in the meantime many warriors came to look upon the visitor and in each instance the latter was more thoroughly impressed with the friendly and peaceable spirit of his hosts, little guessing that he was being entertained by a ferocious and warlike tribe who never before the coming of Ta-den and Tarzan had suffered a stranger among them.

At last Om-at returned and the guest sensed intuitively that he was in the presence of a great man among these people, possibly a chief or king, for not only did the attitude of the other black warriors indicate this but it was written also in the mein and bearing of the splendid creature who stood looking at him while Ta-den explained the circumstances of their meeting. "And I believe, Om-at," concluded the Ho-don, "that he seeks Tarzan the Terrible."

At the sound of that name, the first intelligible word that had fallen upon the ears of the stranger since he had come among them, his face lightened. "Tarzan!" he cried, "Tarzan of the Apes!" and by signs he tried to tell them that it was he whom he sought.

They understood, and also they guessed from the expression of his face that he sought Tarzan from motives of affection rather than the reverse, but of this Om-at wished to make sure. He pointed to the stranger's knife, and repeating Tarzan's name, seized Ta-den and pretended to stab him, immediately turning questioningly toward the stranger.

The latter shook his head vehemently and then first placing a hand above his heart he raised his palm in the symbol of peace.

"He is a friend of Tarzan-jad-guru," exclaimed Ta-den.

"Either a friend or a great liar," replied Om-at.

"Tarzan," continued the stranger, "you know him? He lives? O God, if I could only speak your language." And again reverting to sign language he sought to ascertain where Tarzan was. He would pronounce the name and point in different directions, in the cave, down into the gorge, back toward the mountains, or out upon the valley below, and each time he would raise his brows questioningly and voice the universal "eh?" of interrogation which they could not fail to understand. But always Om-at shook his head and spread his palms in a gesture which indicated that while he understood the question he was ignorant as to the whereabouts of the ape-man, and then the black chief attempted as best he might to explain to the stranger what he knew of the whereabouts of Tarzan.

He called the newcomer Jar-don, which in the language of Pal-ul-don means "stranger," and he pointed to the sun and said *as*. This he repeated several times and then he held up one hand with the fingers outspread and touching them one by one, including the thumb, repeated the word *adenen* until the stranger understood that he meant five. Again he pointed to the sun and describing an arc with his forefinger starting at the eastern horizon and terminating at the western, he repeated again the words *as adenen*. It was plain to the stranger that the words meant that the sun had crossed the heavens five times. In other words, five days had passed. Om-at then pointed to the cave where they stood, pronouncing Tarzan's name and imitating a walking man with the first and second fingers of his right hand upon the floor of the recess, sought to show that Tarzan had walked out of the cave and climbed upward on the pegs five days before, but this was as far as the sign language would permit him to go.

This far the stranger followed him and, indicating that he understood he pointed to himself and then indicating the pegs leading above announced that he would follow Tarzan.

"Let us go with him," said Om-at, "for as yet we have not punished the Kor-ul-lul for killing our friend and ally."

"Persuade him to wait until morning," said Ta-den, "that you may take with you many warriors and make a great raid upon the Kor-ul-lul, and this time, Om-at, do not kill your prisoners. Take as many as you can alive and from some of them we may learn the fate of Tarzan-jad-guru."

"Great is the wisdom of the Ho-don," replied Om-at. "It shall be as you say, and having made prisoners of all the Kor-ul-lul we shall make them tell us what we wish to know. And then we shall march them to the rim of Kor-ul-gryf and push them over the edge of the cliff."

Ta-den smiled. He knew that they would not take prisoner all the Kor-ul-lul warriors—that they would be fortunate if they took one and it was also possible that they might even be driven back in defeat, but he knew too that Om-at would not hesitate to carry out his threat if he had the opportunity, so implacable was the hatred of these neighbors for each other.

It was not difficult to explain Om-at's plan to the stranger or to win his consent since he was aware, when the great black had made it plain that they would be accompanied by many warriors, that their venture would probably lead them into a hostile country and every safeguard that he could employ he was glad to avail himself of, since the furtherance of his quest was the paramount issue.

He slept that night upon a pile of furs in one of the compartments of Om-at's ancestral cave, and early the next day following the morning meal they sallied forth, a hundred savage warriors swarming up the face of the sheer cliff and out upon the summit of the ridge, the main body preceded by two warriors whose duties coincided with those of the point of modern military maneuvers, safeguarding the column against the danger of too sudden contact with the enemy.

Across the ridge they went and down into the Kor-ul-lul and there almost immediately they came upon a lone and unarmed Waz-don who was making his way fearfully up the gorge toward the village of his tribe. Him they took prisoner which, strangely, only added to his terror since from the moment that he had seen them and realized that escape was impossible, he had expected to be slain immediately.

"Take him back to Kor-ul-ja," said Om-at, to one of his warriors, "and hold him there unharmed until I return."

And so the puzzled Kor-ul-lul was led away while the savage company moved stealthily from tree to tree in its closer advance upon the village. Fortune smiled upon Om-at in that it gave him quickly what he sought—a battle royal, for they had not yet come in sight of the caves of the Kor-ul-lul when they encountered a considerable band of warriors headed down the gorge upon some expedition.

Like shadows the Kor-ul-ja melted into the concealment of the foliage upon either side of the trail. Ignorant of impending danger, safe in the knowledge that they trod their own domain where each rock and stone was as familiar as the features of their mates, the Kor-ul-lul walked innocently into the ambush. Suddenly the quiet of that seeming peace was shattered by a savage cry and a hurled club felled a Kor-ul-lul.

The cry was a signal for a savage chorus from a hundred Kor-ul-ja throats with which were soon mingled the war cries of their enemies. The air was filled with flying clubs and then as the two forces mingled, the battle resolved itself into a number of individual encounters as each warrior singled out a foe and closed upon him. Knives gleamed and flashed in the mottling sunlight that filtered through the foliage of the trees above. Sleek black coats were streaked with crimson stains.

In the thick of the fight the smooth brown skin of the stranger

mingled with the black bodies of friend and foe. Only his
keen eyes and his quick wit had shown him how to differentiate
between Kor-ul-lul and Kor-ul-ja since with the single exception of apparel they were identical, but at the first rush of the
enemy he had noticed that their loin cloths were not of the
leopard-marked hides such as were worn by his allies.

Om-at, after dispatching his first antagonist, glanced at Jar-
don. "He fights with the ferocity of *jato*," mused the chief.
"Powerful indeed must be the tribe from which he and Tarzan-
jad-guru come," and then his whole attention was occupied by
a new assailant.

The fighters surged to and fro through the forest until those
who survived were spent with exhaustion. All but the stranger
who seemed not to know the sense of fatigue. He fought on
when each new antagonist would have gladly quit, and when
there were no more Kor-ul-lul who were not engaged, he
leaped upon those who stood pantingly facing the exhausted
Kor-ul-ja.

And always he carried upon his back the peculiar thing
which Om-at had thought was some manner of strange weapon
but the purpose of which he could not now account for in
view of the fact that Jar-don never used it, and that for the
most part it seemed but a nuisance and needless incumbrance
since it banged and smashed against its owner as he leaped,
catlike, hither and thither in the course of his victorious duels.
The bow and arrows he had tossed aside at the beginning of
the fight but the Enfield he would not discard, for where he
went he meant that it should go until its mission had been fulfilled.

Presently the Kor-ul-ja, seemingly shamed by the example of
Jar-don closed once more with the enemy, but the latter, moved
no doubt to terror by the presence of the stranger, a tireless

demon who appeared invulnerable to their attacks, lost heart and sought to flee. And then it was that at Om-at's command his warriors surrounded a half-dozen of the most exhausted and made them prisoners.

It was a tired, bloody, and elated company that returned victorious to the Kor-ul-ja. Twenty of their number were carried back and six of these were dead men. It was the most glorious and successful raid that the Kor-ul-ja had made upon the Kor-ul-lul in the memory of man, and it marked Om-at as the greatest of chiefs, but that fierce warrior knew that advantage had lain upon his side largely because of the presence of his strange ally. Nor did he hesitate to give credit where credit belonged, with the result that Jar-don and his exploits were upon the tongue of every member of the tribe of Kor-ul-ja and great was the fame of the race that could produce two such as he and Tarzan-jad-guru.

And in the gorge of Kor-ul-lul beyond the ridge the survivors spoke in bated breath of this second demon that had joined forces with their ancient enemy.

Returned to his cave Om-at caused the Kor-ul-lul prisoners to be brought into his presence singly, and each he questioned as to the fate of Tarzan. Without exception they told him the same story—that Tarzan had been taken prisoner by them five days before but that he had slain the warrior left to guard him and escaped, carrying the head of the unfortunate sentry to the opposite side of Kor-ul-lul where he had left it suspended by its hair from the branch of a tree. But what had become of him after, they did not know; not one of them, until the last prisoner was examined, he whom they had taken first—the unarmed Kor-ul-lul making his way from the direction of the Valley of Jad-ben-Otho toward the caves of his people.

This one, when he discovered the purpose of their questioning, bartered with them for the lives and liberty of himself and his fellows. "I can tell you much of this terrible man of whom you ask, Kor-ul-ja," he said. "I saw him yesterday and I know where he is, and if you will promise to let me and my fellows return in safety to the caves of our ancestors I will tell you all, and truthfully, that which I know."

"You will tell us anyway," replied Om-at, "or we shall kill you."

"You will kill me anyway," retorted the prisoner, "unless you make me this promise; so if I am to be killed the thing I know shall go with me."

"He is right, Om-at," said Ta-den; "promise him that they shall have their liberty."

"Very well," said Om-at. "Speak Kor-ul-lul, and when you have told me all, you and your fellows may return unharmed to your tribe."

"It was thus," commenced the prisoner. "Three days since I was hunting with a party of my fellows near the mouth of Kor-ul-lul not far from where you captured me this morning, when we were surprised and set upon by a large number of Ho-don who took us prisoners and carried us to A-lur where a few were chosen to be slaves and the rest were cast into a chamber beneath the temple where are held for sacrifice the victims that are offered by the Ho-don to Jad-ben-Otho upon the sacrificial altars of the temple at A-lur.

"It seemed then that indeed was my fate sealed and that lucky were those who had been selected for slaves among the Ho-don, for they at least might hope to escape—those in the chamber with me must be without hope.

"But yesterday a strange thing happened. There came to the temple, accompanied by all the priests and by the king and many of his warriors, one whom all did great reverence, and when he came to the barred gateway leading to the chamber in which we wretched ones awaited our fate, I saw to my surprise that it was none other than that terrible man who had so recently been a prisoner in the village of Kor-ul-lul—he whom you call Tarzan-jad-guru but whom they addressed as Dor-ul-Otho. And he looked upon us and questioned the high priest and when he was told of the purpose for which we were imprisoned there he grew angry and cried that it was not the will of Jad-ben-Otho that his people be thus sacrificed, and he commanded the high priest to liberate us, and this was done.

"The Ho-don prisoners were permitted to return to their homes and we were led beyond the City of A-lur and set upon our way toward Kor-ul-lul. There were three of us, but many are the dangers that lie between A-lur and Kor-ul-lul and we were only three and unarmed. Therefore none of us reached the village of our people and only one of us lives. I have spoken."

"That is all you know concerning Tarzan-jad-guru?" asked Om-at.

"That is all I know," replied the prisoner, "other than that he whom they call Lu-don, the high priest at A-lur, was very angry, and that one of the two priests who guided us out of the city said to the other that the stranger was not Dor-ul-Otho at all; that Lu-don had said so and that he had also said that he would expose him and that he should be punished with death for his presumption. That is all they said within my hearing.

"And now, chief of Kor-ul-ja, let us depart."

Om-at nodded. "Go your way," he said, "and Ab-on, send warriors to guard them until they are safely within the Kor-ul-lul.

"Jar-don," he said beckoning to the stranger, "come with me," and rising he led the way toward the summit of the cliff, and when they stood upon the ridge Om-at pointed down into the valley toward the City of A-lur gleaming in the light of the western sun.

"There is Tarzan-jad-guru," he said, and Jar-don understood.

CHAPTER XIII

The Masquerader

As TARZAN dropped to the ground beyond the temple wall there was in his mind no intention to escape from the City of A-lur until he had satisfied himself that his mate was not a prisoner there, but how, in this strange city in which every man's hand must be now against him, he was to live and prosecute his search was far from clear to him.

There was only one place of which he knew that he might find even temporary sanctuary and that was the Forbidden Garden of the king. There was thick shrubbery in which a man might hide, and water and fruits. A cunning jungle creature, if he could reach the spot unsuspected, might remain concealed there for a considerable time, but how he was to traverse the distance between the temple grounds and the garden unseen was a question the seriousness of which he fully appreciated.

"Mighty is Tarzan," he soliloquized, "in his native jungle, but in the cities of man he is little better than they."

Depending upon his keen observation and sense of location he felt safe in assuming that he could reach the palace grounds by means of the subterranean corridors and chambers of the temple through which he had been conducted the day before, nor any slightest detail of which had escaped his keen eyes. That would be better, he reasoned, than crossing the open grounds above where his pursuers would naturally immediately follow him from the temple and quickly discover him.

And so a dozen paces from the temple wall he disappeared from sight of any chance observer above, down one of the stone stairways that led to the apartments beneath. The way that he had been conducted the previous day had followed the windings and turnings of numerous corridors and apartments, but Tarzan, sure of himself in such matters, retraced the route accurately without hesitation.

He had little fear of immediate apprehension here since he believed that all the priests of the temple had assembled in the court above to witness his trial and his humiliation and his death, and with this idea firmly implanted in his mind he rounded the turn of the corridor and came face to face with an under priest, his grotesque headdress concealing whatever emotion the sight of Tarzan may have aroused.

However, Tarzan had one advantage over the masked votary of Jad-ben-Otho in that the moment he saw the priest he knew his intention concerning him, and therefore was not compelled to delay action. And so it was that before the priest could determine on any suitable line of conduct in the premises a long, keen knife had been slipped into his heart.

As the body lunged toward the floor Tarzan caught it and snatched the headdress from its shoulders, for the first sight of the creature had suggested to his ever-alert mind a bold scheme for deceiving his enemies.

The headdress saved from such possible damage as it must have sustained had it fallen to the floor with the body of its owner, Tarzan relinquished his hold upon the corpse, set the headdress carefully upon the floor and stooping down severed the tail of the Ho-don close to its root. Near by at his right was a small chamber from which the priest had evidently just emerged and into this Tarzan dragged the corpse, the headdress, and the tail.

Quickly cutting a thin strip of hide from the loin cloth of the priest, Tarzan tied it securely about the upper end of the severed member and then tucking the tail under his loin cloth behind him, secured it in place as best he could. Then he fitted the headdress over his shoulders and stepped from the apartment, to all appearances a priest of the temple of Jad-ben-Otho unless one examined too closely his thumbs and his great toes.

He had noticed that among both the Ho-don and the Waz-don it was not at all unusual that the end of the tail be carried in one hand, and so he caught his own tail up thus lest the lifeless appearance of it dragging along behind him should arouse suspicion.

Passing along the corridor and through the various chambers he emerged at last into the palace grounds beyond the temple. The pursuit had not yet reached this point though he was conscious of a commotion not far behind him. He met now both warriors and slaves but none gave him more than a passing glance, a priest being too common a sight about the palace.

And so, passing the guards unchallenged, he came at last to the inner entrance to the Forbidden Garden and there he paused and scanned quickly that portion of the beautiful spot that lay before his eyes. To his relief it seemed unoccupied and congratulating himself upon the ease with which he had so far outwitted the high powers of A-lur he moved rapidly to the opposite end of the enclosure. Here he found a patch of flowering shrubbery that might safely have concealed a dozen men.

Crawling well within he removed the uncomfortable head-dress and sat down to await whatever eventualities fate might have in store for him the while he formulated plans for the future. The one night that he had spent in A-lur had kept him up to a late hour, apprising him of the fact that while there were few abroad in the temple grounds at night, there were yet enough to make it possible for him to fare forth under cover of his disguise without attracting the unpleasant attention of the guards, and, too, he had noticed that the priesthood constituted a privileged class that seemed to come and go at will and unchallenged throughout the palace as well as the temple. Altogether then, he decided, night furnished the most propitious hours for his investigation—by day he could lie up in the shrubbery of the Forbidden Garden, reasonably free from detection. From beyond the garden he heard the voices of men calling to one another both far and near, and he guessed that diligent was the search that was being prosecuted for him.

The idle moments afforded him an opportunity to evolve a more satisfactory scheme for attaching his stolen caudal appendage. He arranged it in such a way that it might be quickly assumed or discarded, and this done he fell to examining the weird mask that had so effectively hidden his features.

The thing had been very cunningly wrought from a single block of wood, very probably a section of a tree, upon which the features had been carved and afterward the interior hollowed out until only a comparatively thin shell remained. Two semicircular notches had been rounded out from opposite sides of the lower edge. These fitted snugly over his shoulders, aprons of wood extending downward a few inches upon his chest and back. From these aprons hung long tassels or switches of hair tapering from the outer edges toward the center which reached below the bottom of his torso. It required but the most cursory examination to indicate to the ape-man that these ornaments consisted of human scalps, taken, doubtless, from the heads of the sacrifices upon the eastern altars. The headdress itself had been carved to depict in for-

mal design a hideous face that suggested both man and *gryf.*
There were the three white horns, the yellow face with the
blue bands encircling the eyes and the red hood which took
the form of the posterior and anterior aprons.

As Tarzan sat within the concealing foliage of the shrubbery
meditating upon the hideous priest-mask which he held in his
hands he became aware that he was not alone in the garden.
He sensed another presence and presently his trained ears
detected the slow approach of naked feet across the sward.
At first he suspected that it might be one stealthily searching
the Forbidden Garden for him but a little later the figure came
within the limited area of his vision which was circumscribed
by stems and foliage and flowers. He saw then that it was
the Princess O-lo-a and that she was alone and walking with
bowed head as though in meditation—sorrowful meditation for
there were traces of tears upon her lids.

Shortly after his ears warned him that others had entered
the garden—men they were and their footsteps proclaimed
that they walked neither slowly nor meditatively. They came
directly toward the princess and when Tarzan could see them
he discovered that both were priests.

"O-lo-a, Princess of Pal-ul-don," said one, addressing her,
"the stranger who told us that he was the son of Jad-ben-Otho
has but just fled from the wrath of Lu-don, the high priest,
who exposed him and all his wicked blasphemy. The temple,
and the palace, and the city are being searched and we have
been sent to search the Forbidden Garden, since Ko-tan, the
king, said that only this morning he found him here, though
how he passed the guards he could not guess."

"He is not here," said O-lo-a. "I have been in the garden
for some time and have seen nor heard no other than myself.
However, search it if you will."

"No," said the priest who had before spoken, "it is not necessary since he could not have entered without your knowledge and the connivance of the guards, and even had he, the priest who preceded us must have seen him."

"What priest?" asked O-lo-a.

"One passed the guards shortly before us," explained the man.

"I did not see him," said O-lo-a.

"Doubtless he left by another exit," remarked the second priest.

"Yes, doubtless," acquiesced O-lo-a, "but it is strange that I did not see him." The two priests made their obeisance and turned to depart.

"Stupid as Buto, the rhinoceros," soliloquized Tarzan, who considered Buto a very stupid creature indeed. "It should be easy to outwit such as these."

The priests had scarce departed when there came the sound of feet running rapidly across the garden in the direction of the princess to an accompaniment of rapid breathing as of one almost spent, either from fatigue or excitement.

"Pan-at-lee," exclaimed O-lo-a, "what has happened? You look as terrified as the doe for which you were named!"

"O Princess of Pal-ul-don," cried Pan-at-lee, "they would have killed him in the temple. They would have killed the wondrous stranger who claimed to be the Dor-ul-Otho."

"But he escaped," said O-lo-a. "You were there. Tell me about it."

"The head priest would have had him seized and slain, but when they rushed upon him he hurled one in the face of Lu-don with the same ease that you might cast your breastplates at me; and then he leaped upon the altar and from there to the top of the temple wall and disappeared below. They are searching for him, but, O Princess, I pray that they do not find him."

"And why do you pray that?" asked O-lo-a. "Has not one who has so blasphemed earned death?"

"Ah, but you do not know him," replied Pan-at-lee.

"And you do, then?" retorted O-lo-a quickly. "This morning you betrayed yourself and then attempted to deceive me. The slaves of O-lo-a do not such things with impunity. He is then the same Tarzan-jad-guru of whom you told me? Speak woman and speak only the truth."

Pan-at-lee drew herself up very erect, her little chin held high, for was not she too among her own people already as good as a princess? "Pan-at-lee, the Kor-ul-ja does not lie," she said, "to protect herself."

"Then tell me what you know of this Tarzan-jad-guru," insisted O-lo-a.

"I know that he is a wondrous man and very brave," said Pan-at-lee, "and that he saved me from the Tor-o-don and the *gryf* as I told you, and that he is indeed the same who came into the garden this morning; and even now I do not know that he is not the son of Jad-ben-Otho for his courage and his strength are more than those of mortal man, as are also his kindness and his honor: for when he might have harmed me he protected me, and when he might have saved himself he thought only of me. And all this he did because of his friendship for Om-at, who is *gund* of Kor-ul-ja and with whom I should have mated had the Ho-don not captured me."

"He was indeed a wonderful man to look upon," mused O-lo-a, "and he was not as are other men, not alone in the conformation of his hands and feet or the fact that he was tailless, but there was that about him which made him seem different in ways more important than these."

"And," supplemented Pan-at-lee, her savage little heart loyal to the man who had befriended her and hoping to win for him the consideration of the princess even though it might not avail him; "and," she said, "did he not know all about Ta-den and even his whereabouts. Tell me, O Princess, could mortal know such things as these?"

"Perhaps he saw Ta-den," suggested O-lo-a.

"But how would he know that you loved Ta-den," parried Pan-at-lee. "I tell you, my Princess, that if he is not a god he is at least more than Ho-don or Waz-don. He followed me from the cave of Es-sat in Kor-ul-ja across Kor-ul-lul and two wide ridges to the very cave in Kor-ul-gryf where I hid, though many hours had passed since I had come that way and my bare feet left no impress upon the ground. What mortal man could do such things as these? And where in all Pal-ul-don would virgin maid find friend and protector in a strange male other than he?"

"Perhaps Lu-don may be mistaken—perhaps he is a god," said O-lo-a, influenced by her slave's enthusiastic championing of the stranger.

"But whether god or man he is too wonderful to die," cried Pan-at-lee. "Would that I might save him. If he lived he might even find a way to give you your Ta-den, Princess."

"Ah, if he only could," sighed O-lo-a, "but alas it is too late for tomorrow I am to be given to Bu-lot."

"He who came to your quarters yesterday with your father?" asked Pan-at-lee.

"Yes; the one with the awful round face and the big belly," exclaimed the Princess disgustedly. "He is so lazy he will neither hunt nor fight. To eat and to drink is all that Bu-lot is fit for, and he thinks of naught else except these things and his slave women. But come, Pan-at-lee, gather for me some of these beautiful blossoms. I would have them spread around my couch tonight that I may carry away with me in the morning the memory of the fragrance that I love best and which I know that I shall not find in the village of Mo-sar, the father of Bu-lot. I will help you, Pan-at-lee, and we will gather armfuls of them, for I love to gather them as I love nothing else —they were Ta-den's favorite flowers."

The two approached the flowering shrubbery where Tarzan hid, but as the blooms grew plentifully upon every bush the ape-man guessed there would be no necessity for them to enter the patch far enough to discover him. With little exclamations of pleasure as they found particularly large or perfect blooms the two moved from place to place upon the outskirts of Tarzan's retreat.

"Oh, look, Pan-at-lee," cried O-lo-a presently; "there is the king of them all. Never did I see so wonderful a flower—No! I will get it myself—it is so large and wonderful no other hand shall touch it," and the princess wound in among the bushes toward the point where the great flower bloomed upon a bush above the ape-man's head.

So sudden and unexpected her approach that there was no opportunity to escape and Tarzan sat silently trusting that fate might be kind to him and lead Ko-tan's daughter away before her eyes dropped from the high-growing bloom to him. But as the girl cut the long stem with her knife she looked down straight into the smiling face of Tarzan-jad-guru.

With a stifled scream she drew back and the ape-man rose and faced her.

"Have no fear, Princess," he assured her. "It is the friend of Ta-den who salutes you," raising her fingers to his lips.

Pan-at-lee came now excitedly forward. "O Jad-ben-Otho, it is he!"

"And now that you have found me," queried Tarzan, "will you give me up to Lu-don, the high priest?"

Pan-at-lee threw herself upon her knees at O-lo-a's feet. "Princess! Princess!" she beseeched, "do not discover him to his enemies."

"But Ko-tan, my father," whispered O-lo-a fearfully, "if he knew of my perfidy his rage would be beyond naming. Even though I am a princess Lu-don might demand that I be sacrificed to appease the wrath of Jad-ben-Otho, and between the two of them I should be lost."

"But they need never know," cried Pan-at-lee, "that you have seen him unless you tell them yourself for as Jad-ben-Otho is my witness I will never betray you."

"Oh, tell me, stranger," implored O-lo-a, "are you indeed a god?"

"Jad-ben-Otho is not more so," replied Tarzan truthfully.

"But why do you seek to escape then from the hands of mortals if you are a god?" she asked.

"When gods mingle with mortals," replied Tarzan, "they are no less vulnerable than mortals. Even Jad-ben-Otho, should he appear before you in the flesh, might be slain."

"You have seen Ta-den and spoken with him?" she asked with apparent irrelevancy.

"Yes, I have seen him and spoken with him," replied the ape-man. "For the duration of a moon I was with him constantly."

"And—" she hesitated—"he—" she cast her eyes toward the ground and a flush mantled her cheek—"he still loves me?" and Tarzan knew that she had been won over.

"Yes," he said, "Ta-den speaks only of O-lo-a and he waits and hopes for the day when he can claim her."

"But tomorrow they give me to Bu-lot," she said sadly.

"May it be always tomorrow," replied Tarzan, "for tomorrow never comes."

"Ah, but this unhappiness will come, and for all the tomorrows of my life I must pine in misery for the Ta-den who will never be mine."

"But for Lu-don I might have helped you," said the ape-man. "And who knows that I may not help you yet?"

"Ah, if you only could, Dor-ul-Otho," cried the girl, "and I know that you would if it were possible for Pan-at-lee has told me how brave you are, and at the same time how kind."

"Only Jad-ben-Otho knows what the future may bring," said Tarzan. "And now you two go your way lest someone should discover you and become suspicious."

"We will go," said O-lo-a, "but Pan-at-lee will return with food. I hope that you escape and that Jad-ben-Otho is pleased with what I have done." She turned and walked away and Pan-at-lee followed while the ape-man again resumed his hiding.

At dusk Pan-at-lee came with food and having her alone Tarzan put the question that he had been anxious to put since his conversation earlier in the day with O-lo-a.

"Tell me," he said, "what you know of the rumors of which O-lo-a spoke of the mysterious stranger which is supposed to be hidden in A-lur. Have you too heard of this during the short time that you have been here?"

"Yes," said Pan-at-lee, "I have heard it spoken of among the other slaves. It is something of which all whisper among themselves but of which none dares to speak aloud. They say that there is a strange she hidden in the temple and that Lu-don wants her for a priestess and that Ko-tan wants her for a wife and that neither as yet dares take her for fear of the other."

"Do you know where she is hidden in the temple?" asked Tarzan.

"No," said Pan-at-lee. "How should I know? I do not

even know that it is more than a story and I but tell you that which I have heard others say."

"There was only one," asked Tarzan, "whom they spoke of?"

"No, they speak of another who came with her but none seems to know what became of this one."

Tarzan nodded. "Thank you Pan-at-lee," he said. "You may have helped me more than either of us guess."

"I hope that I have helped you," said the girl as she turned back toward the palace.

"And I hope so too," exclaimed Tarzan emphatically.

CHAPTER XIV

The Temple of the Gryf

WHEN night had fallen Tarzan donned the mask and the dead tail of the priest he had slain in the vaults beneath the temple. He judged that it would not do to attempt again to pass the guard, especially so late at night as it would be likely to arouse comment and suspicion, and so he swung into the tree that overhung the garden wall and from its branches dropped to the ground beyond.

Avoiding too grave risk of apprehension the ape-man passed through the grounds to the court of the palace, approaching the temple from the side opposite to that at which he had left it at the time of his escape. He came thus it is true through a portion of the grounds with which he was unfamiliar but he preferred this to the danger of following the beaten track between the palace apartments and those of the temple. Having a definite goal in mind and endowed as he was with an almost miraculous sense of location he moved with great assurance through the shadows of the temple yard.

Taking advantage of the denser shadows close to the walls and of what shrubs and trees there were he came without mishap at last to the ornate building concerning the purpose of which he had asked Lu-don only to be put off with the assertion that it was forgotten—nothing strange in itself but given possible importance by the apparent hesitancy of the priest to discuss its use and the impression the ape-man had gained at the time that Lu-don lied.

And now he stood at last alone before the structure which was three stories in height and detached from all the other temple buildings. It had a single barred entrance which was carved from the living rock in representation of the head of a *gryf*, whose wide-open mouth constituted the doorway. The head, hood, and front paws of the creature were depicted as though it lay crouching with its lower jaw on the ground between its outspread paws. Small oval windows, which were likewise barred, flanked the doorway.

Seeing that the coast was clear, Tarzan stepped into the darkened entrance where he tried the bars only to discover that they were ingeniously locked in place by some device with which he was unfamiliar and that they also were probably too strong to be broken even if he could have risked the noise which would have resulted. Nothing was visible within the darkened interior and so, momentarily baffled, he sought the windows. Here also the bars refused to yield up their secret, but again Tarzan was not dismayed since he had counted upon nothing different.

If the bars would not yield to his cunning they would yield to his giant strength if there proved no other means of ingress, but first he would assure himself that this latter was the case. Moving entirely around the building he examined it carefully. There were other windows but they were similarly barred. He stopped often to look and listen but he saw no one and the sounds that he heard were too far away to cause him any apprehension.

He glanced above him at the wall of the building. Like so many of the other walls of the city, palace, and temple, it was ornately carved and there were too the peculiar ledges that ran sometimes in a horizontal plane and again were tilted at an angle, giving ofttimes an impression of irregularity and even crookedness to the buildings. It was not a difficult wall to climb, at least not difficult for the ape-man.

But he found the bulky and awkward headdress a considerable handicap and so he laid it aside upon the ground at the foot of the wall. Nimbly he ascended to find the windows of the second floor not only barred but curtained within. He did not delay long at the second floor since he had in mind an idea that he would find the easiest entrance through the roof which he had noticed was roughly dome shaped like the throneroom of Ko-tan. Here there were apertures. He had seen them from the ground, and if the construction of the interior resembled even slightly that of the throneroom, bars would not be necessary upon these apertures, since no one could reach them from the floor of the room.

There was but a single question: would they be large enough to admit the broad shoulders of the ape-man.

He paused again at the third floor, and here, in spite of the hangings, he saw that the interior was lighted and simultaneously there came to his nostrils from within a scent that stripped from him temporarily any remnant of civilization that might have remained and left him a fierce and terrible bull of the jungles of Kerchak. So sudden and complete was the metamorphosis that there almost broke from the savage lips the hideous challenge of his kind, but the cunning brute-mind saved him this blunder.

And now he heard voices within—the voice of Lu-don he could have sworn, demanding. And haughty and disdainful came the answering words though utter hopelessness spoke in the tones of this other voice which brought Tarzan to the pinnacle of frenzy.

The dome with its possible apertures was forgotten. Every

consideration of stealth and quiet was cast aside as the ape-
man drew back his mighty fist and struck a single terrific blow
upon the bars of the small window before him, a blow that
sent the bars and the casing that held them clattering to the
floor of the apartment within.

Instantly Tarzan dove headforemost through the aperture
carrying the hangings of antelope hide with him to the floor
below. Leaping to his feet he tore the entangling pelt from
about his head only to find himself in utter darkness and in
silence. He called aloud a name that had not passed his lips
for many weary months. "Jane, Jane," he cried, "where are
you?" But there was only silence in reply.

Again and again he called, groping with outstretched hands
through the Stygian blackness of the room, his nostrils assailed
and his brain tantalized by the delicate effluvia that had first
assured him that his mate had been within this very room.
And he had heard her dear voice combatting the base demands
of the vile priest. Ah, if he had but acted with greater caution!
If he had but continued to move with quiet and stealth he
might even at this moment be holding her in his arms while
the body of Lu-don, beneath his foot, spoke eloquently of
vengeance achieved. But there was no time now for idle self-
reproaches.

He stumbled blindly forward, groping for he knew not what
till suddenly the floor beneath him tilted and he shot downward into a darkness even more utter than that above. He
felt his body strike a smooth surface and he realized that he
was hurtling downward as through a polished chute while
from above there came the mocking tones of a taunting laugh
and the voice of Lu-don screamed after him: "Return to thy
father, O Dor-ul-Otho!"

The ape-man came to a sudden and painful stop upon a
rocky floor. Directly before him was an oval window crossed

by many bars, and beyond he saw the moonlight playing on the waters of the blue lake below. Simultaneously he was conscious of a familiar odor in the air of the chamber, which a quick glance revealed in the semidarkness as of considerable proportion.

It was the faint, but unmistakable odor of the *gryf*, and now Tarzan stood silently listening. At first he detected no sounds other than those of the city that came to him through the window overlooking the lake; but presently, faintly, as though from a distance he heard the shuffling of padded feet along a stone pavement, and as he listened he was aware that the sound approached.

Nearer and nearer it came, and now even the breathing of the beast was audible. Evidently attracted by the noise of his descent into its cavernous retreat it was approaching to investigate. He could not see it but he knew that it was not far distant, and then, deafeningly there reverberated through those gloomy corridors the mad bellow of the *gryf*.

Aware of the poor eyesight of the beast, and his own eyes now grown accustomed to the darkness of the cavern, the apeman sought to elude the infuriated charge which he well knew no living creature could withstand. Neither did he dare risk the chance of experimenting upon this strange *gryf* with the tactics of the Tor-o-don that he had found so efficacious upon that other occasion when his life and liberty had been the stakes for which he cast. In many respects the conditions were dissimilar. Before, in broad daylight, he had been able to approach the *gryf* under normal conditions in its natural state, and the *gryf* itself was one that he had seen subjected to the authority of man, or at least of a manlike creature; but here he was confronted by an imprisoned beast in the full swing of a furious charge and he had every reason to suspect that this *gryf* might never have felt the restraining influence of authority, confined as it was in this gloomy pit to serve likely but the single purpose that Tarzan had already seen so graphically portrayed in his own experience of the past few moments.

To elude the creature, then, upon the possibility of discovering some loophole of escape from his predicament seemed to the ape-man the wisest course to pursue. Too much was at stake to risk an encounter that might be avoided—an encounter the outcome of which there was every reason to apprehend would seal the fate of the mate that he had just found, only to lose again so harrowingly. Yet high as his disappointment and chagrin ran, hopeless as his present estate now appeared, there tingled in the veins of the savage lord a warm glow of thanksgiving and elation. She lived! After all these weary months of hopelessness and fear he had found her. She lived!

To the opposite side of the chamber, silently as the wraith of a disembodied soul, the swift jungle creature moved from the path of the charging Titan that, guided solely in the semidarkness by its keen ears, bore down upon the spot toward which Tarzan's noisy entrance into its lair had attracted it. Along the further wall the ape-man hurried. Before him now appeared the black opening of the corridor from which the beast had emerged into the larger chamber. Without hesitation Tarzan plunged into it. Even here his eyes, long accustomed to darkness that would have seemed total to you or to me, saw dimly the floor and the walls within a radius of a few feet—enough at least to prevent him plunging into any unguessed abyss, or dashing himself upon solid rock at a sudden turning.

The corridor was both wide and lofty, which indeed it must be to accommodate the colossal proportions of the creature whose habitat it was, and so Tarzan encountered no difficulty in moving with reasonable speed along its winding trail. He was aware as he proceeded that the trend of the passage was downward, though not steeply, but it seemed interminable and he wondered to what distant subterranean lair it might lead. There was a feeling that perhaps after all he might better have remained in the larger chamber and risked all on the chance of subduing the *gryf* where there was at least sufficient room and light to lend to the experiment some slight chance of success. To be overtaken here in the narrow confines of the black corridor where he was assured the *gryf* could not see him at all would spell almost certain death and now he heard the thing approaching from behind. Its thunderous bellows fairly shook the cliff from which the cavernous chambers were excavated. To halt and meet this monstrous incarnation of fury with a futile *whee-oo!* seemed to Tarzan the height of insanity and so he continued along the corridor, increasing his pace as he realized that the *gryf* was overhauling him.

Presently the darkness lessened and at the final turning of the passage he saw before him an area of moonlight. With renewed hope he sprang rapidly forward and emerged from the mouth of the corridor to find himself in a large circular enclosure the towering white walls of which rose high upon every side—smooth perpendicular walls upon the sheer face of which was no slightest foothold. To his left lay a pool of water, one side of which lapped the foot of the wall at this point. It was, doubtless, the wallow and the drinking pool of the *gryf*.

And now the creature emerged from the corridor and Tarzan retreated to the edge of the pool to make his last stand. There was no staff with which to enforce the authority of his voice, but yet he made his stand for there seemed naught else to do. Just beyond the entrance to the corridor the *gryf* paused, turning its weak eyes in all directions as though searching for its prey. This then seemed the psychological moment for his attempt and raising his voice in peremptory command the ape-man voiced the weird *whee-oo!* of the Tor-o-don. Its effect upon the *gryf* was instantaneous and complete—with a terrific bellow it lowered its three horns and dashed madly in the direction of the sound.

To right nor to left was any avenue of escape, for behind him lay the placid waters of the pool, while down upon him from before thundered annihilation. The mighty body seemed already to tower above him as the ape-man turned and dove into the dark waters.

Dead in her breast lay hope. Battling for life during harrowing months of imprisonment and danger and hardship it had fitfully flickered and flamed only to sink after each renewal to smaller proportions than before and now it had died out entirely leaving only cold, charred embers that Jane Clayton knew would never again be rekindled. Hope was dead as she faced Lu-don, the high priest, in her prison quarters in the Temple of the *Gryf* at A-lur. Both time and hardship had failed to leave their impress upon her physical beauty—the contours of her perfect form, the glory of her radiant loveliness had defied them, yet to these very attributes she owed the danger which now confronted her, for Lu-don desired her. From the lesser priests she had been safe, but from Lu-don, she was not safe, for Lu-don was not as they, since the high priestship of Pal-ul-don may descend from father to son.

Ko-tan, the king, had wanted her and all that had so far saved her from either was the fear of each for the other, but at last Lu-don had cast aside discretion and had come in the silent watches of the night to claim her. Haughtily had she repulsed him, seeking ever to gain time, though what time might bring her of relief or renewed hope she could not even

remotely conjecture. A leer of lust and greed shone hungrily
upon his cruel countenance as he advanced across the room to
seize her. She did not shrink nor cower, but stood there very
erect, her chin up, her level gaze freighted with the loathing
and contempt she felt for him. He read her expression and
while it angered him, it but increased his desire for possession.
Here indeed was a queen, perhaps a goddess; fit mate for the
high priest.

"You shall not!" she said as he would have touched her.
"One of us shall die before ever your purpose is accomplished."

He was close beside her now. His laugh grated upon her
ears. "Love does not kill," he replied mockingly.

He reached for her arm and at the same instant something
clashed against the bars of one of the windows, crashing them
inward to the floor, to be followed almost simultaneously by
a human figure which dove headforemost into the room, its
head enveloped in the skin window hangings which it carried
with it in its impetuous entry.

Jane Clayton saw surprise and something of terror too leap
to the countenance of the high priest and then she saw him
spring forward and jerk upon a leather thong that depended
from the ceiling of the apartment. Instantly there dropped
from above a cunningly contrived partition that fell between
them and the intruder, effectively barring him from them and
at the same time leaving him to grope upon its opposite side
in darkness, since the only cresset the room contained was
upon their side of the partition.

Faintly from beyond the wall Jane heard a voice calling, but
whose it was and what the words she could not distinguish.
Then she saw Lu-don jerk upon another thong and wait in
evident expectancy of some consequent happening. He did
not have long to wait. She saw the thong move suddenly as
though jerked from above and then Lu-don smiled and with

another signal put in motion whatever machinery it was that raised the partition again to its place in the ceiling.

Advancing into that portion of the room that the partition had shut off from them, the high priest knelt upon the floor, and down tilting a section of it, revealed the dark mouth of a shaft leading below. Laughing loudly he shouted into the hole: "Return to thy father, O Dor-ul-Otho!"

Making fast the catch that prevented the trapdoor from opening beneath the feet of the unwary until such time as Lu-don chose the high priest rose again to his feet.

"Now, Beautiful One!" he cried, and then; "Ja-don! what do you here?"

Jane Clayton turned to follow the direction of Lu-don's eyes and there she saw framed in the entrance-way to the apartment the mighty figure of a warrior, upon whose massive features sat an expression of stern and uncompromising authority.

"I come from Ko-tan, the king," replied Ja-don, "to remove the beautiful stranger to the Forbidden Garden."

"The king defies me, the high priest of Jad-ben-Otho?" cried Lu-don.

"It is the king's command—I have spoken," snapped Ja-don, in whose manner was no sign of either fear or respect for the priest.

Lu-don well knew why the king had chosen this messenger whose heresy was notorious, but whose power had as yet protected him from the machinations of the priest. Lu-don cast a surreptitious glance at the thongs hanging from the ceiling. Why not? If he could but maneuver to entice Ja-don to the opposite side of the chamber!

"Come," he said in a conciliatory tone, "let us discuss the matter," and moved toward the spot where he would have Ja-don follow him.

"There is nothing to discuss," replied Ja-don, yet he followed the priest, fearing treachery.

Jane watched them. In the face and figure of the warrior she found reflected those admirable traits of courage and honor that the profession of arms best develops. In the hypocritical priest there was no redeeming quality. Of the two then she might best choose the warrior. With him there was a chance —with Lu-don, none. Even the very process of exchange from one prison to another might offer some possibility of escape. She weighed all these things and decided, for Lu-don's quick glance at the thongs had not gone unnoticed nor uninterpreted by her.

"Warrior," she said, addressing Ja-don, "if you would live enter not that portion of the room."

Lu-don cast an angry glance upon her. "Silence, slave!" he cried.

"And where lies the danger?" Ja-don asked of Jane, ignoring Lu-don.

The woman pointed to the thongs. "Look," she said, and before the high priest could prevent she had seized that which controlled the partition which shot downward separating Lu-don from the warrior and herself.

Ja-don looked inquiringly at her. "He would have tricked me neatly but for you," he said; "kept me imprisoned there while he secreted you elsewhere in the mazes of his temple."

"He would have done more than that," replied Jane, as she pulled upon the other thong. "This releases the fastenings of a trapdoor in the floor beyond the partition. When you stepped on that you would have been precipitated into a pit beneath the temple. Lu-don has threatened me with this fate often. I do not know that he speaks the truth, but he says that a demon of the temple is imprisoned there—a huge *gryf*."

"There is a *gryf* within the temple," said Ja-don. "What with it and the sacrifices, the priests keep us busy supplying them with prisoners, though the victims are sometimes those for whom Lu-don has conceived hatred among our own people. He has had his eyes upon me for a long time. This would have been his chance but for you. Tell me, woman, why you warned me. Are we not all equally your jailers and your enemies?"

"None could be more horrible than Lu-don," she replied; "and you have the appearance of a brave and honorable warrior. I could not hope, for hope has died and yet there is the possibility that among so many fighting men, even though they be of another race than mine, there is one who would accord honorable treatment to a stranger within his gates—even though she be a woman."

Ja-don looked at her for a long minute. "Ko-tan would make you his queen," he said. "That he told me himself and surely that were honorable treatment from one who might make you a slave."

"Why, then, would he make me queen?" she asked.

Ja-don came closer as though in fear his words might be overheard. "He believes, although he did not tell me so in fact, that you are of the race of gods. And why not? Jad-ben-Otho is tailless, therefore it is not strange that Ko-tan should suspect that only the gods are thus. His queen is dead leaving only a single daughter. He craves a son and what more desirable than that he should found a line of rulers for Pal-uldon descended from the gods?"

"But I am already wed," cried Jane. "I cannot wed another. I do not want him or his throne."

"Ko-tan is king," replied Ja-don simply as though that explained and simplified everything.

"You will not save me then?" she asked.

"If you were in Ja-lur," he replied, "I might protect you, even against the king."

"What and where is Ja-lur?" she asked, grasping at any straw.

"It is the city where I rule," he answered. "I am chief there and of all the valley beyond."

"Where is it?" she insisted, and "is it far?"

"No," he replied, smiling, "it is not far, but do not think of that—you could never reach it. There are too many to pursue and capture you. If you wish to know, however, it lies up the river that empties into Jad-ben-lul whose waters kiss the walls of A-lur—up the western fork it lies with water upon three sides. Impregnable city of Pal-ul-don—alone of all the cities it has never been entered by a foeman since it was built there while Jad-ben-Otho was a boy."

"And there I would be safe?" she asked.

"Perhaps," he replied.

Ah, dead Hope; upon what slender provocation would you seek to glow again! She sighed and shook her head, realizing the inutility of Hope—yet the tempting bait dangled before her mind's eye—Ja-lur!

"You are wise," commented Ja-don interpreting her sigh. "Come now, we will go to the quarters of the princess beside the Forbidden Garden. There you will remain with O-lo-a, the king's daughter. It will be better than this prison you have occupied."

"And Ko-tan?" she asked, a shudder passing through her slender frame.

"There are ceremonies," explained Ja-don, "that may occupy several days before you become queen, and one of them may be difficult of arrangement." He laughed, then.

"What?" she asked.

"Only the high priest may perform the marriage ceremony for a king," he explained.

"Delay!" she murmured; "blessed delay!" Tenacious indeed of life is Hope even though it be reduced to cold and lifeless char—a veritable phoenix.

CHAPTER XV

"The King Is Dead!"

As THEY conversed Ja-don had led her down the stone stairway that leads from the upper floors of the Temple of the *Gryf* to the chambers and the corridors that honeycomb the rocky hills from which the temple and the palace are hewn and now they passed from one to the other through a doorway upon one side of which two priests stood guard and upon the other two warriors. The former would have halted Ja-don when they saw who it was that accompanied him for well known throughout the temple was the quarrel between king and high priest for possession of this beautiful stranger.

"Only by order of Lu-don may she pass," said one, placing himself directly in front of Jane Clayton, barring her progress. Through the hollow eyes of the hideous mask the woman could see those of the priest beneath gleaming with the fires of fanaticism. Ja-don placed an arm about her shoulders and laid his hand upon his knife.

"She passes by order of Ko-tan, the king," he said, "and by virtue of the fact that Ja-don, the chief, is her guide. Stand aside!"

The two warriors upon the palace side pressed forward. "We are here, *gund* of Ja-lur," said one, addressing Ja-don, "to receive and obey your commands."

The second priest now interposed. "Let them pass," he admonished his companion. "We have received no direct com-

180

mands from Lu-don to the contrary and it is a law of the temple and the palace that chiefs and priests may come and go without interference."

"But I know Lu-don's wishes," insisted the other.

"He told you then that Ja-don must not pass with the stranger?"

"No—but——"

"Then let them pass, for they are three to two and will pass anyway—we have done our best."

Grumbling, the priest stepped aside. "Lu-don will exact an accounting," he cried angrily.

Ja-don turned upon him. "And get it when and where he will," he snapped.

They came at last to the quarters of the Princess O-lo-a where, in the main entrance-way, loitered a small guard of palace warriors and several stalwart black eunuchs belonging to the princess, or her women. To one of the latter Ja-don relinquished his charge.

"Take her to the princess," he commanded, "and see that she does not escape."

Through a number of corridors and apartments lighted by stone cressets the eunuch led Lady Greystoke halting at last before a doorway concealed by hangings of *jato* skin, where the guide beat with his staff upon the wall beside the door.

"O-lo-a, Princess of Pal-ul-don," he called, "here is the stranger woman, the prisoner from the temple."

"Bid her enter," Jane heard a sweet voice from within command.

The eunuch drew aside the hangings and Lady Greystoke stepped within. Before her was a low-ceiled room of moderate size. In each of the four corners a kneeling figure of stone seemed to be bearing its portion of the weight of the ceiling upon its shoulders. These figures were evidently intended to represent Waz-don slaves and were not without bold artistic beauty. The ceiling itself was slightly arched to a central dome which was pierced to admit light by day, and air. Upon one side of the room were many windows, the other three walls being blank except for a doorway in each. The princess lay upon a pile of furs which were arranged over a low stone dais in one corner of the apartment and was alone except for a single Waz-don slave girl who sat upon the edge of the dais near her feet.

As Jane entered O-lo-a beckoned her to approach and when she stood beside the couch the girl half rose upon an elbow and surveyed her critically.

"How beautiful you are," she said simply.

Jane smiled, sadly; for she had found that beauty may be a curse.

"That is indeed a compliment," she replied quickly, "from one so radiant as the Princess O-lo-a."

"Ah!" exclaimed the princess delightedly; "you speak my language! I was told that you were of another race and from some far land of which we of Pal-ul-don have never heard."

"Lu-don saw to it that the priests instructed me," explained Jane; "but I am from a far country, Princess; one to which I long to return—and I am very unhappy."

"But Ko-tan, my father, would make you his queen," cried the girl; "that should make you very happy."

"But it does not," replied the prisoner; "I love another to whom I am already wed. Ah, Princess, if you had known what it was to love and to be forced into marriage with another you would sympathize with me."

The Princess O-lo-a was silent for a long moment. "I know," she said at last, "and I am very sorry for you; but if the king's daughter cannot save herself from such a fate who may save a slave woman? for such in fact you are."

The drinking in the great banquet hall of the palace of Ko-tan, king of Pal-ul-don had commenced earlier this night than was usual, for the king was celebrating the morrow's betrothal of his only daughter to Bu-lot, son of Mo-sar, the chief, whose great-grandfather had been king of Pal-ul-don and who thought that he should be king, and Mo-sar was drunk and so was Bu-lot, his son. For that matter nearly all of the warriors, including the king himself, were drunk. In the heart of Ko-tan was no love either for Mo-sar, or Bu-lot, nor did either of these love the king. Ko-tan was giving his daughter to Bu-lot in the hope that the alliance would prevent Mo-sar from insisting upon his claims to the throne, for, next to Ja-don, Mo-sar was the most powerful of the chiefs and while Ko-tan looked with fear upon Ja-don, too, he had no fear that the old Lion-man would attempt to seize the throne, though which way he would throw his influence and his warriors in the event that Mo-sar declare war upon Ko-tan, the king could not guess.

Primitive people who are also warlike are seldom inclined toward either tact or diplomacy even when sober; but drunk they know not the words, if aroused. It was really Bu-lot who started it.

"This," he said, "I drink to O-lo-a," and he emptied his tankard at a single gulp. "And this," seizing a full one from a neighbor, "to her son and mine who will bring back the throne of Pal-ul-don to its rightful owners!"

"The king is not yet dead!" cried Ko-tan, rising to his feet; "nor is Bu-lot yet married to his daughter—and there is yet time to save Pal-ul-don from the spawn of the rabbit breed."

The king's angry tone and his insulting reference to Bu-lot's well-known cowardice brought a sudden, sobering silence upon the roistering company. Every eye turned upon Bu-lot and Mo-sar, who sat together directly opposite the king. The first was very drunk though suddenly he seemed quite sober. He was so drunk that for an instant he forgot to be a coward, since his reasoning powers were so effectually paralyzed by the fumes of liquor that he could not intelligently weigh the consequences of his acts. It is reasonably conceivable that a drunk and angry rabbit might commit a rash deed. Upon no other hypothesis is the thing that Bu-lot now did explicable. He rose suddenly from the seat to which he had sunk after delivering his toast and seizing the knife from the sheath of the warrior upon his right hurled it with terrific force at Ko-tan. Skilled in the art of throwing both their knives and their clubs are the warriors of Pal-ul-don and at this short distance and coming as it did without warning there was no defense and but one possible result—Ko-tan, the king, lunged forward across the table, the blade buried in his heart.

A brief silence followed the assassin's cowardly act. White with terror, now, Bu-lot fell slowly back toward the doorway at his rear, when suddenly angry warriors leaped with drawn knives to prevent his escape and to avenge their king. But Mo-sar now took his stand beside his son.

"Ko-tan is dead!" he cried. "Mo-sar is king! Let the loyal warriors of Pal-ul-don protect their ruler!"

Mo-sar commanded a goodly following and these quickly surrounded him and Bu-lot, but there were many knives against them and now Ja-don pressed forward through those who confronted the pretender.

"Take them both!" he shouted. "The warriors of Pal-ul-don will choose their own king after the assassin of Ko-tan has paid the penalty of his treachery."

Directed now by a leader whom they both respected and admired those who had been loyal to Ko-tan rushed forward upon the faction that had surrounded Mo-sar. Fierce and terrible was the fighting, devoid, apparently, of all else than the ferocious lust to kill and while it was at its height Mo-sar and Bu-lot slipped unnoticed from the banquet hall.

To that part of the palace assigned to them during their visit to A-lur they hastened. Here were their servants and the lesser warriors of their party who had not been bidden to the feast of Ko-tan. These were directed quickly to gather together their belongings for immediate departure. When all was ready, and it did not take long, since the warriors of Pal-ul-don require but little impedimenta on the march, they moved toward the palace gate.

Suddenly Mo-sar approached his son. "The princess," he whispered. "We must not leave the city without her—she is half the battle for the throne."

Bu-lot, now entirely sober, demurred. He had had enough of fighting and of risk. "Let us get out of A-lur quickly," he urged, "or we shall have the whole city upon us. She would not come without a struggle and that would delay us too long."

"There is plenty of time," insisted Mo-sar. "They are still fighting in the *pal-e-don-so*. It will be long before they miss us and, with Ko-tan dead, long before any will think to look to the safety of the princess. Our time is now—it was made for us by Jad-ben-Otho. Come!"

Reluctantly Bu-lot followed his father, who first instructed the warriors to await them just inside the gateway of the palace. Rapidly the two approached the quarters of the princess. Within the entrance-way only a handful of warriors were on guard. The eunuchs had retired.

"There is fighting in the *pal-e-don-so*," Mo-sar announced in feigned excitement as they entered the presence of the guards. "The king desires you to come at once and has sent us to guard the apartments of the princess. Make haste!" he commanded as the men hesitated.

The warriors knew him and that on the morrow the princess was to be betrothed to Bu-lot, his son. If there was trouble what more natural than that Mo-sar and Bu-lot should be intrusted with the safety of the princess. And then, too, was not Mo-sar a powerful chief to whose orders disobedience might prove a dangerous thing? They were but common fighting men disciplined in the rough school of tribal warfare, but they had learned to obey a superior and so they departed for the banquet hall—the *place-where-men-eat.*

Barely waiting until they had disappeared Mo-sar crossed to the hangings at the opposite end of the entrance-hall and followed by Bu-lot made his way toward the sleeping apartment of O-lo-a and a moment later, without warning, the two men burst in upon the three occupants of the room. At sight of them O-lo-a sprang to her feet.

"What is the meaning of this?" she demanded angrily.

Mo-sar advanced and halted before her. Into his cunning mind had entered a plan to trick her. If it succeeded it would prove easier than taking her by force, and then his eyes fell upon Jane Clayton and he almost gasped in astonishment and admiration, but he caught himself and returned to the business of the moment.

"O-lo-a," he cried, "when you know the urgency of our mission you will forgive us. We have sad news for you. There has been an uprising in the palace and Ko-tan, the king, has been slain. The rebels are drunk with liquor and now on their way here. We must get you out of A-lur at once—there is not a moment to lose. Come, and quickly!"

"My father dead?" cried O-lo-a, and suddenly her eyes went wide. "Then my place is here with my people," she cried. "If Ko-tan is dead I am queen until the warriors choose a new ruler—that is the law of Pal-ul-don. And if I am queen none can make me wed whom I do not wish to wed—and Jad-ben-

Otho knows I never wished to wed thy cowardly son. Go!"
She pointed a slim forefinger imperiously toward the doorway.

Mo-sar saw that neither trickery nor persuasion would avail
now and every precious minute counted. He looked again at
the beautiful woman who stood beside O-lo-a. He had never
before seen her but he well knew from palace gossip that she
could be no other than the godlike stranger whom Ko-tan had
planned to make his queen.

"Bu-lot," he cried to his son, "take you your own woman
and I will take—mine!" and with that he sprang suddenly
forward and seizing Jane about the waist lifted her in his
arms, so that before O-lo-a or Pan-at-lee might even guess his
purpose he had disappeared through the hangings near the
foot of the dais and was gone with the stranger woman struggling and fighting in his grasp.

And then Bu-lot sought to seize O-lo-a, but O-lo-a had her
Pan-at-lee—fierce little tiger-girl of the savage Kor-ul-ja—Pan-
at-lee whose name belied her—and Bu-lot found that with the
two of them his hands were full. When he would have lifted
O-lo-a and borne her away Pan-at-lee seized him around the
legs and strove to drag him down. Viciously he kicked her,
but she would not desist, and finally, realizing that he might
not only lose his princess but be so delayed as to invite capture if he did not rid himself of this clawing, scratching she-
jato, he hurled O-lo-a to the floor and seizing Pan-at-lee by
the hair drew his knife and——

The curtains behind him suddenly parted. In two swift
bounds a lithe figure crossed the room and before ever the
knife of Bu-lot reached its goal his wrist was seized from behind and a terrific blow crashing to the base of his brain
dropped him, lifeless, to the floor. Bu-lot, coward, traitor, and
assassin, died without knowing who struck him down.

As Tarzan of the Apes leaped into the pool in the *gryf* pit of the temple at A-lur one might have accounted for his act on the hypothesis that it was the last blind urge of self-preservation to delay, even for a moment, the inevitable tragedy in which each some day must play the leading rôle upon his little stage; but no—those cool, gray eyes had caught the sole possibility for escape that the surroundings and the circumstances offered—a tiny, moonlit patch of water glimmering through a small aperture in the cliff at the surface of the pool upon it farther side. With swift, bold strokes he swam for speed alone knowing that the water would in no way deter his pursuer. Nor did it. Tarzan heard the great splash as the huge creature plunged into the pool behind him; he heard the churning waters as it forged rapidly onward in his wake. He was nearing the opening—would it be large enough to permit the passage of his body? That portion of it which showed above the surface of the water most certainly would not. His life, then, depended upon how much of the aperture was submerged. And now it was directly before him and the *gryf* directly behind. There was no alternative—there was no other hope. The ape-man threw all the resources of his great strength into the last few strokes, extended his hands before him as a cutwater, submerged to the water's level and shot forward toward the hole.

Frothing with rage was the baffled Lu-don as he realized how neatly the stranger she had turned his own tables upon him. He could of course escape the Temple of the *Gryf* in which her quick wit had temporarily imprisoned him; but during the delay, however brief, Ja-don would find time to steal her from the temple and deliver her to Ko-tan. But he would have her yet—that the high priest swore in the names of Jad-ben-Otho and all the demons of his faith. He hated

Ko-tan. Secretly he had espoused the cause of Mo-sar, in whom he would have a willing tool. Perhaps, then, this would give him the opportunity he had long awaited—a pretext for inciting the revolt that would dethrone Ko-tan and place Mo-sar in power—with Lu-don the real ruler of Pal-ul-don. He licked his thin lips as he sought the window through which Tarzan had entered and now Lu-don's only avenue of escape. Cautiously he made his way across the floor, feeling before him with his hands, and when they discovered that the trap was set for him an ugly snarl broke from the priest's lips. "The she-devil!" he muttered; "but she shall pay, she shall pay—ah, Jad-ben-Otho; how she shall pay for the trick she has played upon Lu-don!"

He crawled through the window and climbed easily downward to the ground. Should he pursue Ja-don and the woman, chancing an encounter with the fierce chief, or bide his time until treachery and intrigue should accomplish his design? He chose the latter solution, as might have been expected of such as he.

Going to his quarters he summoned several of his priests—those who were most in his confidence and who shared his ambitions for absolute power of the temple over the palace—all men who hated Ko-tan.

"The time has come," he told them, "when the authority of the temple must be placed definitely above that of the palace. Ko-tan must make way for Mo-sar, for Ko-tan has defied your high priest. Go then, Pan-sat, and summon Mo-sar secretly to the temple, and you others go to the city and prepare the faithful warriors that they may be in readiness when the time comes."

For another hour they discussed the details of the coup d'état that was to overthrow the government of Pal-ul-don. One knew a slave who, as the signal sounded from the temple gong, would thrust a knife into the heart of Ko-tan, for the price of liberty. Another held personal knowledge of an officer of the palace that he could use to compel the latter to admit a number of Lu-don's warriors to various parts of the palace. With Mo-sar as the cat's-paw, the plan seemed scarce possible of failure and so they separated, going upon their immediate errands to palace and to city.

As Pan-sat entered the palace grounds he was aware of a sudden commotion in the direction of the *pal-e-don-so* and a few minutes later Lu-don was surprised to see him return to the apartments of the high priest, breathless and excited.

"What now, Pan-sat?" cried Lu-don. "Are you pursued by demons?"

"O master, our time has come and gone while we sat here planning. Ko-tan is already dead and Mo-sar fled. His friends are fighting with the warriors of the palace but they have no head, while Ja-don leads the others. I could learn but little from frightened slaves who had fled at the outburst of the quarrel. One told me that Bu-lot had slain the king and that he had seen Mo-sar and the assassin hurrying from the palace."

"Ja-don," muttered the high priest. "The fools will make him king if we do not act and act quickly. Get into the city, Pan-sat—let your feet fly and raise the cry tha Ja-don has killed the king and is seeking to wrest the throne from O-lo-a. Spread the word as you know best how to spread it that Ja-don has threatened to destroy the priests and hurl the altars of the temple into Jad-ben-lul. Rouse the warriors of the city and urge them to attack at once. Lead them into the temple by the secret way that only the priests know and from here we may spew them out upon the palace before they learn the truth. Go, Pan-sat, immediately—delay not an instant."

"But stay," he called as the under priest turned to leave the apartment; "saw or heard you anything of the strange white woman that Ja-don stole from the Temple of the *Gryf* where we have had her imprisoned?"

"Only that Ja-don took her into the palace where he threatened the priests with violence if they did not permit him to pass," replied Pan-sat. "This they told me, but where within the palace she is hidden I know not."

"Ko-tan ordered her to the Forbidden Garden," said Lu-don, "doubtless we shall find her there. And now, Pan-sat, be upon your errand."

In a corridor by Lu-don's chamber a hideously masked priest leaned close to the curtained aperture that led within. Were he listening he must have heard all that passed between Pan-sat and the high priest, and that he had listened was evidenced by his hasty withdrawal to the shadows of a nearby passage as the lesser priest moved across the chamber toward the doorway. Pan-sat went his way in ignorance of the near presence that he almost brushed against as he hurried toward the secret passage that leads from the temple of Jad-ben-Otho, far beneath the palace, to the city beyond, nor did he sense the silent creature following in his footsteps.

CHAPTER XVI

The Secret Way

I T W A S a baffled *gryf* that bellowed in angry rage as Tarzan's
sleek brown body cutting the moonlit waters shot through
the aperture in the wall of the *gryf* pool and out into the lake
beyond. The ape-man smiled as he thought of the comparative
ease with which he had defeated the purpose of the high priest
but his face clouded again at the ensuing remembrance of the
grave danger that threatened his mate. His sole object now
must be to return as quickly as he might to the chamber where
he had last seen her on the third floor of the Temple of the
Gryf, but how he was to find his way again into the temple
grounds was a question not easy of solution.

In the moonlight he could see the sheer cliff rising from the
water for a great distance along the shore—far beyond the precincts of the temple and the palace—towering high above him,
a seemingly impregnable barrier against his return. Swimming close in, he skirted the wall searching diligently for some
foothold, however slight, upon its smooth, forbidding surface.
Above him and quite out of reach were numerous apertures,
but there were no means at hand by which he could reach
them. Presently, however, his hopes were raised by the sight
of an opening level with the surface of the water. It lay just
ahead and a few strokes brought him to it—cautious strokes
that brought forth no sound from the yielding waters. At the
nearer side of the opening he stopped and reconnoitered.

There was no one in sight. Carefully he raised his body to the threshold of the entrance-way, his smooth brown hide glistening in the moonlight as it shed the water in tiny sparkling rivulets.

Before him stretched a gloomy corridor, unlighted save for the faint illumination of the diffused moonlight that penetrated it for but a short distance from the opening. Moving as rapidly as reasonable caution warranted, Tarzan followed the corridor into the bowels of the cave. There was an abrupt turn and then a flight of steps at the top of which lay another corridor running parallel with the face of the cliff. This passage was dimly lighted by flickering cressets set in niches in the walls at considerable distances apart. A quick survey showed the ape-man numerous openings upon each side of the corridor and his quick ears caught sounds that indicated that there were other beings not far distant—priests, he concluded, in some of the apartments letting upon the passageway.

To pass undetected through this hive of enemies appeared quite beyond the range of possibility. He must again seek disguise and knowing from experience how best to secure such he crept stealthily along the corridor toward the nearest doorway. Like Numa, the lion, stalking a wary prey he crept with quivering nostrils to the hangings that shut off his view from the interior of the apartment beyond. A moment later his head disappeared within; then his shoulders, and his lithe body, and the hangings dropped quietly into place again. A moment later there filtered to the vacant corridor without a brief, gasping gurgle and again silence. A minute passed; a second, and a third, and then the hangings were thrust aside and a grimly masked priest of the temple of Jad-ben-Otho strode into the passageway.

With bold steps he moved along and was about to turn into a diverging gallery when his attention was aroused by voices coming from a room upon his left. Instantly the figure halted and crossing the corridor stood with an ear close to the skins that concealed the occupants of the room from him, and him from them. Presently he leaped back into the concealing shadows of the diverging gallery and immediately thereafter the hangings by which he had been listening parted and a priest emerged to turn quickly down the main corridor. The eavesdropper waited until the other had gained a little distance and then stepping from his place of concealment followed silently behind.

The way led along the corridor which ran parallel with the face of the cliff for some little distance and then Pan-sat, taking a cresset from one of the wall niches, turned abruptly into a small apartment at his left. The tracker followed cautiously in time to see the rays of the flickering light dimly visible from an aperture in the floor before him. Here he found a series of steps, similar to those used by the Waz-don in scaling the cliff to their caves, leading to a lower level.

First satisfying himself that his guide was continuing upon his way unsuspecting, the other descended after him and continued his stealthy stalking. The passageway was now both narrow and low, giving but bare headroom to a tall man, and it was broken often by flights of steps leading always downward. The steps in each unit seldom numbered more than six and sometimes there was only one or two but in the aggregate the tracker imagined that they had descended between fifty and seventy-five feet from the level of the upper corridor when the passageway terminated in a small apartment at one side of which was a little pile of rubble.

Setting his cresset upon the ground, Pan-sat commenced hurriedly to toss the bits of broken stone aside, presently revealing a small aperture at the base of the wall upon the opposite side of which there appeared to be a further accumulation

of rubble. This he also removed until he had a hole of sufficient size to permit the passage of his body, and leaving the
cresset still burning upon the floor the priest crawled through
the opening he had made and disappeared from the sight of
the watcher hiding in the shadows of the narrow passageway
behind him.

No sooner, however, was he safely gone than the other followed, finding himself, after passing through the hole, on a
little ledge about halfway between the surface of the lake and
the top of the cliff above. The ledge inclined steeply upward,
ending at the rear of a building which stood upon the edge of
the cliff and which the second priest entered just in time to
see Pan-sat pass out into the city beyond.

As the latter turned a nearby corner the other emerged from
the doorway and quickly surveyed his surroundings. He was
satisfied the priest who had led him hither had served his
purpose in so far as the tracker was concerned. Above him,
and perhaps a hundred yards away, the white walls of the
palace gleamed against the northern sky. The time that it had
taken him to acquire definite knowledge concerning the secret
passageway between the temple and the city he did not count
as lost, though he begrudged every instant that kept him from
the prosecution of his main objective. It had seemed to him,
however, necessary to the success of a bold plan that he had
formulated upon overhearing the conversation between Lu-don
and Pan-sat as he stood without the hangings of the apartment
of the high priest.

Alone against a nation of suspicious and half-savage enemies
he could scarce hope for a successful outcome to the one great
issue upon which hung the life and happiness of the creature
he loved best. For her sake he must win allies and it was for
this purpose that he had sacrificed these precious moments,
but now he lost no further time in seeking to regain entrance

to the palace grounds that he might search out whatever new prison they had found in which to incarcerate his lost love.

He found no difficulty in passing the guards at the entrance to the palace for, as he had guessed, his priestly disguise disarmed all suspicion. As he approached the warriors he kept his hands behind him and trusted to fate that the sickly light of the single torch which stood beside the doorway would not reveal his un-Pal-ul-donian feet. As a matter of fact so accustomed were they to the comings and goings of the priesthood that they paid scant attention to him and he passed on into the palace grounds without even a moment's delay.

His goal now was the Forbidden Garden and this he had little difficulty in reaching though he elected to enter it over the wall rather than to chance arousing any suspicion on the part of the guards at the inner entrance, since he could imagine no reason why a priest should seek entrance there thus late at night.

He found the garden deserted, nor any sign of her he sought. That she had been brought hither he had learned from the conversation he had overheard between Lu-don and Pan-sat, and he was sure that there had been no time or opportunity for the high priest to remove her from the palace grounds. The garden he knew to be devoted exclusively to the uses of the princess and her women and it was only reasonable to assume therefore that if Jane had been brought to the garden it could only have been upon an order from Ko-tan. This being the case the natural assumption would follow that he would find her in some other portion of O-lo-a's quarters.

Just where these lay he could only conjecture, but it seemed reasonable to believe that they must be adjacent to the garden, so once more he scaled the wall and passing around its end directed his steps toward an entrance-way which he judged must lead to that portion of the palace nearest the Forbidden Garden.

To his surprise he found the place unguarded and then there fell upon his ear from an interior apartment the sound of voices raised in anger and excitement. Guided by the sound he quickly traversed several corridors and chambers until he stood before the hangings which separated him from the chamber from which issued the sounds of altercation. Raising the skins slightly he looked within. There were two women battling with a Ho-don warrior. One was the daughter of Ko-tan and the other Pan-at-lee, the Kor-ul-ja.

At the moment that Tarzan lifted the hangings, the warrior threw O-lo-a viciously to the ground and seizing Pan-at-lee by the hair drew his knife and raised it above her head. Casting the encumbering headdress of the dead priest from his shoulders the ape-man leaped across the intervening space and seizing the brute from behind struck him a single terrible blow.

As the man fell forward dead, the two women recognized Tarzan simultaneously. Pan-at-lee fell upon her knees and would have bowed her head upon his feet had he not, with an impatient gesture, commanded her to rise. He had no time to listen to their protestations of gratitude or answer the numerous questions which he knew would soon be flowing from those two feminine tongues.

"Tell me," he cried, "where is the woman of my own race whom Ja-don brought here from the temple?"

"She is but this moment gone," cried O-lo-a. "Mo-sar, the father of this thing here," and she indicated the body of Bu-lot with a scornful finger, "seized her and carried her away."

"Which way?" he cried. "Tell me quickly, in what direction he took her."

"That way," cried Pan-at-lee, pointing to the doorway

through which Mo-sar had passed. "They would have taken the princess and the stranger woman to Tu-lur, Mo-sar's city by the Dark Lake."

"I go to find her," he said to Pan-at-lee, "she is my mate. And if I survive I shall find means to liberate you too and return you to Om-at."

Before the girl could reply he had disappeared behind the hangings of the door near the foot of the dais. The corridor through which he ran was illy lighted and like nearly all its kind in the Ho-don city wound in and out and up and down, but at last it terminated at a sudden turn which brought him into a courtyard filled with warriors, a portion of the palace guard that had just been summoned by one of the lesser palace chiefs to join the warriors of Ko-tan in the battle that was raging in the banquet hall.

At sight of Tarzan, who in his haste had forgotten to recover his disguising headdress, a great shout arose. "Blasphemer!" "Defiler of the temple!" burst hoarsely from savage throats, and mingling with these were a few who cried, "Dor-ul-Otho!" evidencing the fact that there were among them still some who clung to their belief in his divinity.

To cross the courtyard armed only with a knife, in the face of this great throng of savage fighting men seemed even to the giant ape-man a thing impossible of achievement. He must use his wits now and quickly too, for they were closing upon him. He might have turned and fled back through the corridor but flight now even in the face of dire necessity would but delay him in his pursuit of Mo-sar and his mate.

"Stop!" he cried, raising his palm against them. "I am the Dor-ul-Otho and I come to you with a word from Ja-don, who it is my father's will shall be your king now that Ko-tan is slain. Lu-don, the high priest, has planned to seize the palace and destroy the loyal warriors that Mo-sar may be made king

—Mo-sar who will be the tool and creature of Lu-don. Follow me. There is no time to lose if you would prevent the traitors whom Lu-don has organized in the city from entering the palace by a secret way and overpowering Ja-don and the faithful band within."

For a moment they hesitated. At last one spoke. "What guarantee have we," he demanded, "that it is not you who would betray us and by leading us now away from the fighting in the banquet hall cause those who fight at Ja-don's side to be defeated?"

"My life will be your guarantee," replied Tarzan. "If you find that I have not spoken the truth you are sufficient in numbers to execute whatever penalty you choose. But come, there is not time to lose. Already are the lesser priests gathering their warriors in the city below," and without waiting for any further parley he strode directly toward them in the direction of the gate upon the opposite side of the courtyard which led toward the principal entrance to the palace ground.

Slower in wit than he, they were swept away by his greater initiative and that compelling power which is inherent to all natural leaders. And so they followed him, the giant ape-man with a dead tail dragging the ground behind him—a demi-god where another would have been ridiculous. Out into the city he led them and down toward the unpretentious building that hid Lu-don's secret passageway from the city to the temple, and as they rounded the last turn they saw before them a gathering of warriors which was being rapidly augmented from all directions as the traitors of A-lur mobilized at the call of the priesthood.

"You spoke the truth, stranger," said the chief who marched at Tarzan's side, "for there are the warriors with the priests among them, even as you told us."

"And now," replied the ape-man, "that I have fulfilled my

promise I will go my way after Mo-sar, who has done me a great wrong. Tell Ja-don that Jad-ben-Otho is upon his side, nor do you forget to tell him also that it was the Dor-ul-Otho who thwarted Lu-don's plan to seize the palace."

"I will not forget," replied the chief. "Go your way. We are enough to overpower the traitors."

"Tell me," asked Tarzan, "how I may know this city of Tu-lur?"

"It lies upon the south shore of the second lake below A-lur," replied the chief, "the lake that is called Jad-in-lul."

They were now approaching the band of traitors, who evidently thought that this was another contingent of their own party since they made no effort either toward defense or retreat. Suddenly the chief raised his voice in a savage war cry that was immediately taken up by his followers, and simultaneously, as though the cry were a command, the entire party broke into a mad charge upon the surprised rebels.

Satisfied with the outcome of his suddenly conceived plan and sure that it would work to the disadvantage of Lu-don, Tarzan turned into a side street and pointed his steps toward the outskirts of the city in search of the trail that led southward toward Tu-lur.

CHAPTER XVII

By Jad-bal-lul

A s MO-SAR carried Jane Clayton from the palace of Ko-tan, the king, the woman struggled incessantly to regain her freedom. He tried to compel her to walk, but despite his threats and his abuse she would not voluntarily take a single step in the direction in which he wished her to go. Instead she threw herself to the ground each time he sought to place her upon her feet, and so of necessity he was compelled to carry her though at last he tied her hands and gagged her to save himself from further lacerations, for the beauty and slenderness of the woman belied her strength and courage. When he came at last to where his men had gathered he was glad indeed to turn her over to a couple of stalwart warriors, but these too were forced to carry her since Mo-sar's fear of the vengeance of Ko-tan's retainers would brook no delays.

And thus they came down out of the hills from which A-lur is carved, to the meadows that skirt the lower end of Jad-ben-lul, with Jane Clayton carried between two of Mo-sar's men. At the edge of the lake lay a fleet of strong canoes, hollowed from the trunks of trees, their bows and sterns carved in the semblance of grotesque beasts or birds and vividly colored by some master in that primitive school of art, which fortunately is not without its devotees today.

Into the stern of one of these canoes the warriors tossed their captive at a sign from Mo-sar, who came and stood beside her as the warriors were finding their places in the canoes and selecting their paddles.

"Come, Beautiful One," he said, "let us be friends and you shall not be harmed. You will find Mo-sar a kind master if you do his bidding," and thinking to make a good impression on her he removed the gag from her mouth and the thongs from her wrists, knowing well that she could not escape surrounded as she was by his warriors, and presently, when they were out on the lake, she would be as safely imprisoned as though he held her behind bars.

And so the fleet moved off to the accompaniment of the gentle splashing of a hundred paddles, to follow the windings of the rivers and lakes through which the waters of the Valley of Jad-ben-Otho empty into the great morass to the south. The warriors, resting upon one knee, faced the bow and in the last canoe Mo-sar tiring of his fruitless attempts to win responses from his sullen captive, squatted in the bottom of the canoe with his back toward her and resting his head upon the gunwale sought sleep.

Thus they moved in silence between the verdure-clad banks of the little river through which the waters of Jad-ben-lul emptied—now in the moonlight, now in dense shadow where great trees overhung the stream, and at last out upon the waters of another lake, the black shores of which seemed far away under the weird influence of a moonlight night.

Jane Clayton sat alert in the stern of the last canoe. For months she had been under constant surveillance, the prisoner first of one ruthless race and now the prisoner of another. Since the long-gone day that Hauptmann Fritz Schneider and his band of native German troops had treacherously wrought the Kaiser's work of rapine and destruction on the Greystoke bungalow and carried her away to captivity she had not drawn a free breath. That she had survived unharmed the countless dangers through which she had passed she attributed solely to the beneficence of a kind and watchful Providence.

At first she had been held on the orders of the German High Command with a view of her ultimate value as a hostage and during these months she had been subjected to neither hardship nor oppression, but when the Germans had become hard pressed toward the close of their unsuccessful campaign in East Africa it had been determined to take her further into the interior and now there was an element of revenge in their motives, since it must have been apparent that she could no longer be of any possible military value.

Bitter indeed were the Germans against that half-savage mate of hers who had cunningly annoyed and harassed them with a fiendishness of persistence and ingenuity that had resulted in a noticeable loss in morale in the sector he had chosen for his operations. They had to charge against him the lives of certain officers that he had deliberately taken with his own hands, and one entire section of trench that had made possible a disastrous turning movement by the British. Tarzan had out-generaled them at every point. He had met cunning with cunning and cruelty with cruelties until they feared and loathed his very name. The cunning trick that they had played upon him in destroying his home, murdering his retainers, and covering the abduction of his wife in such a way as to lead him to believe that she had been killed, they had regretted a thousand times, for a thousandfold had they paid the price for their senseless ruthlessness, and now, unable to wreak their vengeance directly upon him, they had conceived the idea of inflicting further suffering upon his mate.

In sending her into the interior to avoid the path of the victorious British, they had chosen as her escort Lieutenant Erich Obergatz who had been second in command of Schneider's company, and who alone of its officers had escaped the consuming vengeance of the ape-man. For a long time Obergatz had held her in a native village, the chief of which was still under the domination of his fear of the ruthless German oppressors. While here only hardships and discomforts assailed her, Obergatz himself being held in leash by the orders of his distant superior but as time went on the life in the village grew to be a veritable hell of cruelties and oppressions practised by the arrogant Prussian upon the villagers and the members of his native command—for time hung heavily upon the hands of the lieutenant and with idleness combining with the personal discomforts he was compelled to endure, his none too agreeable temper found an outlet first in petty interference with the chiefs and later in the practise of absolute cruelties upon them.

What the self-sufficient German could not see was plain to Jane Clayton—that the sympathies of Obergatz' native soldiers lay with the villagers and that all were so heartily sickened by his abuse that it needed now but the slightest spark to detonate the mine of revenge and hatred that the pig-headed Hun had been assiduously fabricating beneath his own person.

And at last it came, but from an unexpected source in the form of a German native deserter from the theater of war. Footsore, weary, and spent, he dragged himself into the village late one afternoon, and before Obergatz was even aware of his presence the whole village knew that the power of Germany in Africa was at an end. It did not take long for the lieutenant's native soldiers to realize that the authority that held them in service no longer existed and that with it had gone the power to pay them their miserable wage. Or at least, so they reasoned. To them Obergatz no longer represented aught else than a powerless and hated foreigner, and short indeed would have been his shrift had not a native woman who had conceived a doglike affection for Jane Clayton hurried to her with word of the murderous plan, for the fate of the innocent white woman lay in the balance beside that of the guilty Teuton.

"Already they are quarreling as to which one shall possess you," she told Jane.

"When will they come for us?" asked Jane. "Did you hear them say?"

"Tonight," replied the woman, "for even now that he has none to fight for him they still fear the white man. And so they will come at night and kill him while he sleeps."

Jane thanked the woman and sent her away lest the suspicion of her fellows be aroused against her when they discovered that the two whites had learned of their intentions. The woman went at once to the hut occupied by Obergatz. She had never gone there before and the German looked up in surprise as he saw who his visitor was.

Briefly she told him what she had heard. At first he was inclined to bluster arrogantly, with a great display of bravado but she silenced him peremptorily.

"Such talk is useless," she said shortly. "You have brought upon yourself the just hatred of these people. Regardless of the truth or falsity of the report which has been brought to them, they believe in it and there is nothing now between you and your Maker other than flight. We shall both be dead before morning if we are unable to escape from the village unseen. If you go to them now with your silly protestations of authority you will be dead a little sooner, that is all."

"You think it is as bad as that?" he said, a noticeable alteration in his tone and manner.

"It is precisely as I have told you," she replied. "They will come tonight and kill you while you sleep. Find me pistols and a rifle and ammunition and we will pretend that we go into the jungle to hunt. That you have done often. Perhaps it will arouse suspicion that I accompany you but that we must chance. And be sure my dear Herr Lieutenant to bluster and curse and abuse your servants unless they note a change in your manner and realizing your fear know that you suspect their intention. If all goes well then we can go out into the jungle to hunt and we need not return.

"But first and now you must swear never to harm me, or otherwise it would be better that I called the chief and turned you over to him and then put a bullet into my own head, for unless you swear as I have asked I were no better alone in the jungle with you than here at the mercies of these degraded blacks."

"I swear," he replied solemnly, "in the names of my God and my Kaiser that no harm shall befall you at my hands, Lady Greystoke."

"Very well," she said, "we will make this pact to assist each other to return to civilization, but let it be understood that there is and never can be any semblance even of respect for you upon my part. I am drowning and you are the straw. Carry that always in your mind, German."

If Obergatz had held any doubt as to the sincerity of her word it would have been wholly dissipated by the scathing contempt of her tone. And so Obergatz, without further parley, got pistols and an extra rifle for Jane, as well as bandoleers of cartridges. In his usual arrogant and disagreeable manner he called his servants, telling them that he and the white *kali* were going out into the brush to hunt. The beaters would go north as far as the little hill and then circle back to the east and in toward the village. The gun carriers he directed to take the extra pieces and precede himself and Jane slowly toward the east, waiting for them at the ford about half a mile distant. The blacks responded with greater alacrity than usual and it was noticeable to both Jane and Obergatz that they left the village whispering and laughing.

"The swine think it is a great joke," growled Obergatz, "that the afternoon before I die I go out and hunt meat for them."

As soon as the gun bearers disappeared in the jungle beyond the village the two Europeans followed along the same trail, nor was there any attempt upon the part of Obergatz' native soldiers, or the warriors of the chief to detain them, for they too doubtless were more than willing that the whites should bring them in one more mess of meat before they killed them.

A quarter of a mile from the village, Obergatz turned toward the south from the trail that led to the ford and hurrying onward the two put as great a distance as possible between them and the village before night fell. They knew from the habits of their erstwhile hosts that there was little danger of pursuit by night since the villagers held Numa, the lion, in too great respect to venture needlessly beyond their stockade during the hours that the king of beasts was prone to choose for hunting.

And thus began a seemingly endless sequence of frightful days and horror-laden nights as the two fought their way toward the south in the face of almost inconceivable hardships, privations, and dangers. The east coast was nearer but Obergatz positively refused to chance throwing himself into the hands of the British by returning to the territory which they now controlled, insisting instead upon attempting to make his way through an unknown wilderness to South Africa where, among the Boers, he was convinced he would find willing sympathizers who would find some way to return him in safety to Germany, and the woman was perforce compelled to accompany him.

And so they had crossed the great thorny, waterless steppe

and come at last to the edge of the morass before Pal-ul-don. They had reached this point just before the rainy season when the waters of the morass were at their lowest ebb. At this time a hard crust is baked upon the dried surface of the marsh and there is only the open water at the center to materially impede progress. It is a condition that exists perhaps not more than a few weeks, or even days at the termination of long periods of drought, and so the two crossed the otherwise almost impassable barrier without realizing its latent terrors. Even the open water in the center chanced to be deserted at the time by its frightful denizens which the drought and the receding waters had driven southward toward the mouth of Pal-ul-don's largest river which carries the waters out of the Valley of Jad-ben-Otho.

Their wanderings carried them across the mountains and into the Valley of Jad-ben-Otho at the source of one of the larger streams which bears the mountain waters down into the valley to empty them into the main river just below The Great Lake on whose northern shore lies A-lur. As they had come down out of the mountains they had been surprised by a party of Ho-don hunters. Obergatz had escaped while Jane had been taken prisoner and brought to A-lur. She had neither seen nor heard aught of the German since that time and she did not know whether he had perished in this strange land, or succeeded in successfully eluding its savage denizens and making his way at last into South Africa.

For her part, she had been incarcerated alternately in the palace and the temple as either Ko-tan or Lu-don succeeded in wresting her temporarily from the other by various strokes of cunning and intrigue. And now at last she was in the power of a new captor, one whom she knew from the gossip of the temple and the palace to be cruel and degraded. And she was in the stern of the last canoe, and every enemy back was toward her, while almost at her feet Mo-sar's loud snores gave ample evidence of his unconsciousness to his immediate surroundings.

The dark shore loomed closer to the south as Jane Clayton, Lady Greystoke, slid quietly over the stern of the canoe into the chill waters of the lake. She scarcely moved other than to keep her nostrils above the surface while the canoe was yet discernible in the last rays of the declining moon. Then she struck out toward the southern shore.

Alone, unarmed, all but naked, in a country overrun by savage beasts and hostile men, she yet felt for the first time in many months a sensation of elation and relief. She was free! What if the next moment brought death, she knew again, at least a brief instant of absolute freedom. Her blood tingled to the almost forgotten sensation and it was with difficulty that she restrained a glad triumphant cry as she clambered from the quiet waters and stood upon the silent beach.

Before her loomed a forest, darkly, and from its depths came those nameless sounds that are a part of the night life of the jungle—the rustling of leaves in the wind, the rubbing together of contiguous branches, the scurrying of a rodent, all magnified by the darkness to sinister and awe-inspiring proportions; the hoot of an owl, the distant scream of a great cat, the barking of wild dogs, attested the presence of the myriad life she could not see—the savage life, the free life of which she was now a part. And then there came to her, possibly for the first time since the giant ape-man had come into her life, a fuller realization of what the jungle meant to him, for though alone and unprotected from its hideous dangers she yet felt its lure upon her and an exaltation that she had not dared hope to feel again.

Ah, if that mighty mate of hers were but by her side! What utter joy and bliss would be hers! She longed for no more than this. The parade of cities, the comforts and luxuries of civilization held forth no allure half as insistent as the glorious freedom of the jungle.

A lion moaned in the blackness to her right, eliciting delicious thrills that crept along her spine. The hair at the back of her head seemed to stand erect—yet she was unafraid. The muscles bequeathed her by some primordial ancestor reacted instinctively to the presence of an ancient enemy—that was all. The woman moved slowly and deliberately toward the wood. Again the lion moaned; this time nearer. She sought a low-hanging branch and finding it swung easily into the friendly shelter of the tree. The long and perilous journey with Obergatz had trained her muscles and her nerves to such unaccustomed habits. She found a safe resting place such as Tarzan had taught her was best and there she curled herself, thirty feet above the ground, for a night's rest. She was cold and uncomfortable and yet she slept, for her heart was warm with renewed hope and her tired brain had found temporary surcease from worry.

She slept until the heat of the sun, high in the heavens, awakened her. She was rested and now her body was well as her heart was warm. A sensation of ease and comfort and happiness pervaded her being. She rose upon her gently swaying couch and stretched luxuriously, her naked limbs and lithe body mottled by the sunlight filtering through the foliage above combined with the lazy gesture to impart to her appearance something of the leopard. With careful eye she scrutinized the ground below and with attentive ear she listened for any warning sound that might suggest the near presence of enemies, either man or beast. Satisfied at last that there was nothing close of which she need have fear she clambered to the ground. She wished to bathe but the lake was too exposed and just a bit too far from the safety of the trees for her to risk it until she became more familiar with her surroundings. She wandered aimlessly through the forest searching for food which she found in abundance. She ate and rested, for she had no objective as yet. Her freedom was too new to be spoiled by plannings for the future. The haunts of civilized man seemed to her now as vague and unattainable as the half-forgotten substance of a dream. If she could but live on here in peace, waiting, waiting for—*him*. It was the old hope revived. She knew that he would come some day, if he lived. She had always known that, though recently she had believed that he would come too late. If he lived! Yes, he would come if he lived, and if he did not live she were as well off here as elsewhere, for then nothing mattered, only to wait for the end as patiently as might be.

Her wanderings brought her to a crystal brook and there she drank and bathed beneath an overhanging tree that offered her quick asylum in the event of danger. It was a quiet and beautiful spot and she loved it from the first. The bottom of the brook was paved with pretty stones and bits of glassy obsidian. As she gathered a handful of the pebbles and held them up to look at them she noticed that one of her fingers was bleeding from a clean, straight cut. She fell to searching for the cause and presently discovered it in one of the fragments of volcanic glass which revealed an edge that was almost razor-like. Jane Clayton was elated. Here, God-given to her hands, was the first beginning with which she might eventually arrive at both weapons and tools—a cutting edge. Everything was possible to him who possessed it—nothing without.

She sought until she had collected many of the precious bits of stone—until the pouch that hung at her right side was almost filled. Then she climbed into the great tree to examine them at leisure. There were some that looked like knife blades, and some that could easily be fashioned into spear heads, and many smaller ones that nature seemed to have intended for the tips of savage arrows.

The spear she would essay first—that would be easiest. There was a hollow in the bole of the tree in a great crotch high above the ground. Here she cached all of her treasure except a single knifelike sliver. With this she descended to the ground and searching out a slender sapling that grew arrow-straight she hacked and sawed until she could break it off without splitting the wood. It was just the right diameter for the shaft of a spear—a hunting spear such as her beloved Waziri had liked best. How often had she watched them fashioning them, and they had taught her how to use them, too—them and the heavy war spears—laughing and clapping their hands as her proficiency increased.

She knew the arborescent grasses that yielded the longest and toughest fibers and these she sought and carried to her tree with the spear shaft that was to be. Clambering to her crotch she bent to her work, humming softly a little tune. She caught herself and smiled—it was the first time in all these bitter months that song had passed her lips or such a smile.

"I feel," she sighed, "I almost feel that John is near—my John—my Tarzan!"

She cut the spear shaft to the proper length and removed the twigs and branches and the bark, whittling and scraping at the nubs until the surface was all smooth and straight. Then she split one end and inserted a spear point, shaping the wood until it fitted perfectly. This done she laid the shaft aside and fell to splitting the thick grass stems and pounding and twisting them until she had separated and partially cleaned the fibers. These she took down to the brook and washed and brought back again and wound tightly around the cleft end of the shaft, which she had notched to receive them, and the upper part of the spear head which she had also notched slightly with a bit of stone. It was a crude spear but the best that she could attain in so short a time. Later, she promised herself, she should have others—many of them—and they would be spears of which even the greatest of the Waziri spear-men might be proud.

CHAPTER XVIII

The Lion Pit of Tu-lur

T HOUGH Tarzan searched the outskirts of the city until nearly dawn he discovered nowhere the spoor of his mate. The breeze coming down from the mountains brought to his nostrils a diversity of scents but there was not among them the slightest suggestion of her whom he sought. The natural deduction was therefore that she had been taken in some other direction. In his search he had many times crossed the fresh tracks of many men leading toward the lake and these he concluded had probably been made by Jane Clayton's abductors. It had only been to minimize the chance of error by the process of elimination that he had carefully reconnoitered every other avenue leading from A-lur toward the southeast where lay Mosar's city of Tu-lur, and now he followed the trail to the shores of Jad-ben-lul where the party had embarked upon the quiet waters in their sturdy canoes.

He found many other craft of the same description moored along the shore and one of these he commandeered for the purpose of pursuit. It was daylight when he passed through the lake which lies next below Jad-ben-lul and paddling strongly passed within sight of the very tree in which his lost mate lay sleeping.

Had the gentle wind that caressed the bosom of the lake been blowing from a southerly direction the giant ape-man and Jane Clayton would have been reunited then, but an unkind fate had willed otherwise and the opportunity passed with the passing of his canoe which presently his powerful strokes carried out of sight into the stream at the lower end of the lake.

Following the winding river which bore a considerable distance to the north before doubling back to empty into the Jad-in-lul, the ape-man missed a portage that would have saved him hours of paddling.

It was at the upper end of this portage where Mo-sar and his warriors had debarked that the chief discovered the absence of his captive. As Mo-sar had been asleep since shortly after their departure from A-lur, and as none of the warriors recalled when she had last been seen, it was impossible to conjecture with any degree of accuracy the place where she had escaped. The concensus of opinion was, however, that it had been in the narrow river connecting Jad-ben-lul with the lake next below it, which is called Jad-bal-lul, which freely translated means the lake of gold. Mo-sar had been very wroth and having himself been the only one at fault he naturally sought with great diligence to fix the blame upon another.

He would have returned in search of her had he not feared to meet a pursuing company dispatched either by Ja-don or the high priest, both of whom, he knew, had just grievances against him. He would not even spare a boatload of his warriors from his own protection to return in quest of the fugitive but hastened onward with as little delay as possible across the portage and out upon the waters of Jad-in-lul.

The morning sun was just touching the white domes of Tu-lur when Mo-sar's paddlers brought their canoes against the shore at the city's edge. Safe once more behind his own walls and protected by many warriors, the courage of the chief returned sufficiently at least to permit him to dispatch three canoes in search of Jane Clayton, and also to go as far as A-lur if possible to learn what had delayed Bu-lot, whose failure to reach the canoes with the balance of the party at the time of the flight from the northern city had in no way delayed Mo-sar's departure, his own safety being of far greater moment than that of his son.

As the three canoes reached the portage on their return journey the warriors who were dragging them from the water were suddenly startled by the appearance of two priests, carrying a light canoe in the direction of Jad-in-lul. At first they thought them the advance guard of a larger force of Lu-don's followers, although the correctness of such a theory was belied by their knowledge that priests never accepted the risks or perils of a warrior's vocation, nor even fought until driven into a corner and forced to do so. Secretly the warriors of Pal-ul-don held the emasculated priesthood in contempt and so instead of immediately taking up the offensive as they would have had the two men been warriors from A-lur instead of priests, they waited to question them.

At sight of the warriors the priests made the sign of peace and upon being asked if they were alone they answered in the affirmative.

The leader of Mo-sar's warriors permitted them to approach. "What do you here," he asked, "in the country of Mo-sar, so far from your own city?"

"We carry a message from Lu-don, the high priest, to Mo-sar," explained one.

"Is it a message of peace or of war?" asked the warrior.

"It is an offer of peace," replied the priest.

"And Lu-don is sending no warriors behind you?" queried the fighting man.

"We are alone," the priest assured him. "None in A-lur save Lu-don knows that we have come upon this errand."

"Then go your way," said the warrior.

"Who is that?" asked one of the priests suddenly, pointing

toward the upper end of the lake at the point where the river from Jad-bal-lul entered it.

All eyes turned in the direction that he had indicated to see a lone warrior paddling rapidly into Jad-in-lul, the prow of his canoe pointing toward Tu-lur. The warriors and the priests drew into the concealment of the bushes on either side of the portage.

It is the terrible man who called himself the Dor-ul-Otho," whispered one of the priests. "I would know that figure among a great multitude as far as I could see it."

"You are right, priest," cried one of the warriors who had seen Tarzan the day that he had first entered Ko-tan's palace. "It is indeed he who has been rightly called Tarzan-jad-guru."

"Hasten priests," cried the leader of the party. "You are two paddles in a light canoe. Easily can you reach Tu-lur ahead of him and warn Mo-sar of his coming, for he has but only entered the lake."

For a moment the priests demurred for they had no stomach for an encounter with this terrible man, but the warrior insisted and even went so far as to threaten them. Their canoe was taken from them and pushed into the lake and they were all but lifted bodily from their feet and put aboard it. Still protesting they were shoved out upon the water where they were immediately in full view of the lone paddler above them. Now there was no alternative. The city of Tu-lur offered the only safety and bending to their paddles the two priests sent their craft swiftly in the direction of the city.

The warriors withdrew again to the concealment of the foliage. If Tarzan had seen them and should come hither to investigate there were thirty of them against one and naturally they had no fear of the outcome, but they did not consider it necessary to go out upon the lake to meet him since they had been sent to look for the escaped prisoner and not to intercept the strange warrior, the stories of whose ferocity and prowess doubtless helped them to arrive at their decision to provoke no uncalled-for quarrel with him.

If he had seen them he gave no sign, but continued paddling steadily and strongly toward the city, nor did he increase his speed as the two priests shot out in full view. The moment the priests' canoe touched the shore by the city its occupants leaped out and hurried swiftly toward the palace gate, casting affrighted glances behind them. They sought immediate audience with Mo-sar, after warning the warriors on guard that Tarzan was approaching.

They were conducted at once to the chief, whose court was a smaller replica of that of the king at A-lur. "We come from Lu-don, the high priest," explained the spokesman. "He wishes the friendship of Mo-sar, who has always been his friend. Ja-don is gathering warriors to make himself king. Throughout the villages of the Ho-don are thousands who will obey the commands of Lu-don, the high priest. Only with Lu-don's assistance can Mo-sar become king, and the message from Lu-don is that if Mo-sar would retain the friendship of Lu-don he must return immediately the woman he took from the quarters of the Princess O-lo-a."

At this juncture a warrior entered. His excitement was evident. "The Dor-ul-Otho has come to Tu-lur and demands to see Mo-sar at once," he said.

"The Dor-ul-Otho!" exclaimed Mo-sar.

"That is the message he sent," replied the warrior, "and indeed he is not as are the people of Pal-ul-don. He is, we think, the same of whom the warriors that returned from A-lur today told us and whom some call Tarzan-jad-guru and some Dor-ul-Otho. But indeed only the son of god would dare come thus alone to a strange city, so it must be that he speaks the truth."

Mo-Sar, his heart filled with terror and indecision, turned questioningly toward the priests.

"Receive him graciously, Mo-Sar," counseled he who had spoken before, his advice prompted by the petty shrewdness of his defective brain which, under the added influence of Lu-don's tutorage leaned always toward duplicity. "Receive him graciously and when he is quite convinced of your friendship he will be off his guard, and then you may do with him as you will. But if possible, Mo-sar, and you would win the undying gratitude of Lu-don, the high priest, save him alive for my master."

Mo-sar nodded understandingly and turning to the warrior commanded that he conduct the visitor to him.

"We must not be seen by the creature," said one of the priests. "Give us your answer to Lu-don, Mo-sar, and we will go our way."

"Tell Lu-don," replied the chief, "that the woman would have been lost to him entirely had it not been for me. I sought to bring her to Tu-lur that I might save her for him from the clutches of Ja-don, but during the night she escaped. Tell Lu-don that I have sent thirty warriors to search for her. It is strange you did not see them as you came."

"We did," replied the priests, "but they told us nothing of the purpose of their journey."

"It is as I have told you," said Mo-sar, "and if they find her, assure your master that she will be kept unharmed in Tu-lur for him. Also tell him that I will send my warriors to join with his against Ja-don whenever he sends word that he wants them. Now go, for Tarzan-jad-guru will soon be here."

He signaled to a slave. "Lead the priests to the temple," he commanded, "and ask the high priest of Tu-lur to see that they are fed and permitted to return to A-lur when they will."

The two priests were conducted from the apartment by the

slave through a doorway other than that at which they had entered, and a moment later Tarzan-jad-guru strode into the presence of Mo-sar, ahead of the warrior whose duty it had been to conduct and announce him. The ape-man made no sign of greeting or of peace but strode directly toward the chief who, only by the exertion of his utmost powers of will, hid the terror that was in his heart at sight of the giant figure and the scowling face.

"I am the Dor-ul-Otho," said the ape-man in level tones that carried to the mind of Mo-sar a suggestion of cold steel; "I am Dor-ul-Otho, and I come to Tu-lur for the woman you stole from the apartments of O-lo-a, the princess."

The very boldness of Tarzan's entry into this hostile city had had the effect of giving him a great moral advantage over Mo-sar and the savage warriors who stood upon either side of the chief. Truly it seemed to them that no other than the son of Jad-ben-Otho would dare so heroic an act. Would any mortal warrior act thus boldly, and alone enter the presence of a powerful chief and, in the midst of a score of warriors, arrogantly demand an accounting? No, it was beyond reason. Mo-sar was faltering in his decision to betray the stranger by seeming friendliness. He even paled to a sudden thought— Jad-ben-Otho knew everything, even our inmost thoughts. Was it not therefore possible that this creature, if after all it should prove true that he was the Dor-ul-Otho, might even now be reading the wicked design that the priests had implanted in the brain of Mo-sar and which he had entertained so favorably? The chief squirmed and fidgeted upon the bench of hewn rock that was his throne.

"Quick," snapped the ape-man, "Where is she?"

"She is not here," cried Mo-sar.

"You lie," replied Tarzan.

"As Jad-ben-Otho is my witness, she is not in Tu-lur," insisted the chief. "You may search the palace and the temple and the entire city but you will not find her, for she is not here."

"Where is she, then?" demanded the ape-man. "You took her from the palace at A-lur. If she is not here, where is she? Tell me not that harm has befallen her," and he took a sudden threatening step toward Mo-sar, that sent the chief shrinking back in terror.

"Wait," he cried, "if you are indeed the Dor-ul-Otho you will know that I speak the truth. I took her from the palace of Ko-tan to save her for Lu-don, the high priest, lest with Ko-tan dead Ja-don seize her. But during the night she escaped from me between here and A-lur, and I have but just sent three canoes full-manned in search of her."

Something in the chief's tone and manner assured the ape-man that he spoke in part the truth, and that once again he had braved incalculable dangers and suffered loss of time futilely.

"What wanted the priests of Lu-don that preceded me here?" demanded Tarzan chancing a shrewd guess that the two he had seen paddling so frantically to avoid a meeting with him had indeed come from the high priest at A-lur.

"They came upon an errand similar to yours," replied Mo-sar; "to demand the return of the woman whom Lu-don thought I had stolen from him, thus wronging me as deeply, O Dor-ul-Otho, as have you."

"I would question the priests," said Tarzan. "Bring them hither." His peremptory and arrogant manner left Mo-sar in doubt as to whether to be more incensed, or terrified, but ever as is the way with such as he, he concluded that the first consideration was his own safety. If he could transfer the attention and the wrath of this terrible man from himself to Lu-don's priests it would more than satisfy him and if they should con-

spire to harm him, then Mo-sar would be safe in the eyes of
Jad-ben-Otho if it finally developed that the stranger was in
reality the son of god. He felt uncomfortable in Tarzan's presence and this fact rather accentuated his doubt, for thus indeed would mortal feel in the presence of a god. Now he saw
a way to escape, at least temporarily.

"I will fetch them myself, Dor-ul-Otho," he said, and turning, left the apartment. His hurried steps brought him quickly
to the temple, for the palace grounds of Tu-lur, which also included the temple as in all of the Ho-don cities, covered a
much smaller area than those of the larger city of A-lur. He
found Lu-don's messengers with the high priest of his own
temple and quickly transmitted to them the commands of the
ape-man.

"What do you intend to do with him?" asked one of the
priests.

"I have no quarrel with him," replied Mo-sar. "He came in
peace and he may depart in peace, for who knows but that he
is indeed the Dor-ul-Otho?"

"We know that he is not," replied Lu-don's emissary. "We
have every proof that he is only mortal, a strange creature
from another country. Already has Lu-don offered his life to
Jad-ben-Otho if he is wrong in his belief that this creature is
not the son of god. If the high priest of A-lur, who is the highest priest of all the high priests of Pal-ul-don is thus so sure
that the creature is an impostor as to stake his life upon his
judgment then who are we to give credence to the claims of
this stranger? No, Mo-sar, you need not fear him. He is only
a warrior who may be overcome with the same weapons that
subdue your own fighting men. Were it not for Lu-don's command that he be taken alive I would urge you to set your
warriors upon him and slay him, but the commands of Lu-don

are the commands of Jad-ben-Otho himself, and those we may not disobey."

But still the remnant of a doubt stirred within the cowardly breast of Mo-sar, urging him to let another take the initiative against the stranger.

"He is yours then," he replied, "to do with as you will. I have no quarrel with him. What you may command shall be the command of Lu-don, the high priest, and further than that I shall have nothing to do in the matter."

The priests turned to him who guided the destinies of the temple at Tu-lur. "Have you no plan?" they asked. "High indeed will he stand in the counsels of Lu-don and in the eyes of Jad-ben-Otho who finds the means to capture this impostor alive."

"There is the lion pit," whispered the high priest. "It is now vacant and what will hold *ja* and *jato* will hold this stranger if he is not the Dor-ul-Otho."

"It will hold him," said Mo-sar; "doubtless too it would hold a *gryf*, but first you would have to get the *gryf* into it."

The priests pondered this bit of wisdom thoughtfully and then one of those from A-lur spoke. "It should not be difficult," he said, "if we use the wits that Jad-ben-Otho gave us instead of the worldly muscles which were handed down to us from our fathers and our mothers and which have not even the power possessed by those of the beasts that run about on four feet."

"Lu-don matched his wits with the stranger and lost," suggested Mo-sar. "But this is your own affair. Carry it out as you see best."

"At A-lur, Ko-tan made much of this Dor-ul-Otho and the priests conducted him through the temple. It would arouse in his mind no suspicion were you to do the same, and let the high priest of Tu-lur invite him to the temple and gathering all the priests make a great show of belief in his kinship to Jad-ben-Otho. And what more natural then than that the high priest should wish to show him through the temple as did Lu-don at A-lur when Ko-tan commanded it, and if by chance he should be led through the lion pit it would be a simple matter for those who bear the torches to extinguish them suddenly and before the stranger was aware of what had happened, the stone gates could be dropped, thus safely securing him."

"But there are windows in the pit that let in light," interposed the high priest, "and even though the torches were extinguished he could still see and might escape before the stone door could be lowered."

"Send one who will cover the windows tightly with hides," said the priest from A-lur.

"The plan is a good one," said Mo-sar, seeing an opportunity for entirely eliminating himself from any suspicion of complicity, "for it will require the presence of no warriors, and thus with only priests about him his mind will entertain no suspicion of harm."

They were interrupted at this point by a messenger from the palace who brought word that the Dor-ul-Otho was becoming impatient and if the priests from A-lur were not brought to him at once he would come himself to the temple and get them. Mo-sar shook his head. He could not conceive of such brazen courage in mortal breast and glad he was that the plan evolved for Tarzan's undoing did not necessitate his active participation.

And so, while Mo-sar left for a secret corner of the palace by a roundabout way, three priests were dispatched to Tarzan and with whining words that did not entirely deceive him, they acknowledged his kinship to Jad-ben-Otho and begged him in the name of the high priest to honor the temple with a visit, when the priests from A-lur would be brought to him and would answer any questions that he put to them.

Confident that a continuation of his bravado would best serve his purpose, and also that if suspicion against him should crystallize into conviction on the part of Mo-sar and his followers that he would be no worse off in the temple than in the palace, the ape-man haughtily accepted the invitation of the high priest.

And so he came into the temple and was received in a manner befitting his high claims. He questioned the two priests of A-lur from whom he obtained only a repetition of the story that Mo-sar had told him, and then the high priest invited him to inspect the temple.

They took him first to the altar court, of which there was only one in Tu-lur. It was almost identical in every respect with those at A-lur. There was a bloody altar at the east end and the drowning basin at the west, and the grizzly fringes upon the headdresses of the priests attested the fact that the eastern altar was an active force in the rites of the temple. Through the chambers and corridors beneath they led him, and finally, with torch bearers to light their steps, into a damp and gloomy labyrinth at a low level and here in a large chamber, the air of which was still heavy with the odor of lions, the crafty priests of Tu-lur encompassed their shrewd design.

The torches were suddenly extinguished. There was a hurried confusion of bare feet moving rapidly across the stone floor. There was a loud crash as of a heavy weight of stone falling upon stone, and then surrounding the ape-man naught but the darkness and the silence of the tomb.

CHAPTER XIX

Diana of the Jungle

JANE had made her first kill and she was very proud of it. It was not a very formidable animal—only a hare; but it marked an epoch in her existence. Just as in the dim past the first hunter had shaped the destinies of mankind so it seemed that this event might shape hers in some new mold. No longer was she dependent upon the wild fruits and vegetables for sustenance. Now she might command meat, the giver of the strength and endurance she would require successfully to cope with the necessities of her primitive existence.

The next step was fire. She might learn to eat raw flesh as had her lord and master; but she shrank from that. The thought even was repulsive. She had, however, a plan for fire. She had given the matter thought, but had been too busy to put into execution so long as fire could be of no immediate use to her. Now it was different—she had something to cook and her mouth watered for the flesh of her kill. She would grill it above glowing embers. Jane hastened to her tree. Among the treasures she had gathered in the bed of the stream were several pieces of volcanic glass, clear as crystal. She sought until she had found the one in mind, which was convex. Then she hurried to the ground and gathered a little pile of powdered bark that was very dry, and some dead leaves and grasses that had lain long in the hot sun. Near at hand she arranged a supply of dead twigs and branches—small and large.

Vibrant with suppressed excitement she held the bit of glass above the tinder, moving it slowly until she had focused the sun's rays upon a tiny spot. She waited breathlessly. How slow it was! Were her high hopes to be dashed in spite of all her clever planning? No! A thin thread of smoke rose gracefully into the quiet air. Presently the tinder glowed and broke suddenly into flame. Jane clasped her hands beneath her chin with a little gurgling exclamation of delight. She had achieved fire!

She piled on twigs and then larger branches and at last dragged a small log to the flames and pushed an end of it into the fire which was crackling merrily. It was the sweetest sound that she had heard for many a month. But she could not wait for the mass of embers that would be required to cook her hare. As quickly as might be she skinned and cleaned her kill, burying the hide and entrails. That she had learned from Tarzan. It served two purposes. One was the necessity for keeping a sanitary camp and the other the obliteration of the scent that most quickly attracts the man-eaters.

Then she ran a stick through the carcass and held it above the flames. By turning it often she prevented burning and at the same time permitted the meat to cook thoroughly all the way through. When it was done she scampered high into the safety of her tree to enjoy her meal in quiet and peace. Never, thought Lady Greystoke, had aught more delicious passed her lips. She patted her spear affectionately. It had brought her this toothsome dainty and with it a feeling of greater confidence and safety than she had enjoyed since that frightful day that she and Obergatz had spent their last cartridge. She would never forget that day—it had seemed one hideous succession of frightful beast after frightful beast. They had not been long in this strange country, yet they thought that they were hardened to dangers, for daily they had had encounters with ferocious creatures; but this day—she shuddered when she thought of it. And with her last cartridge she had killed a black and yellow striped lion-thing with great saber teeth just as it was about to spring upon Obergatz who had futilely emptied his rifle into it—the last shot—his final cartridge. For another day they had carried the now useless rifles; but at last they had discarded them and thrown away the cumbersome bandoleers, as well. How they had managed to survive during the ensuing week she could never quite understand, and then the Ho-don had come upon them and captured her. Obergatz had escaped—she was living it all over again. Doubtless he was dead unless he had been able to reach this side of the valley which was quite evidently less overrun with savage beasts.

Jane's days were very full ones now, and the daylight hours seemed all too short in which to accomplish the many things she had determined upon, since she had concluded that this spot presented as ideal a place as she could find to live until she could fashion the weapons she considered necessary for the obtaining of meat and for self-defense.

She felt that she must have, in addition to a good spear, a knife, and bow and arrows. Possibly when these had been achieved she might seriously consider an attempt to fight her way to one of civilization's nearest outposts. In the meantime it was necessary to construct some sort of protective shelter in which she might feel a greater sense of security by night, for she knew that there was a possibility that any night she might receive a visit from a prowling panther, although she had as yet seen none upon this side of the valley. Aside from this danger she felt comparatively safe in her aerial retreat.

The cutting of the long poles for her home occupied all of the daylight hours that were not engaged in the search for food. These poles she carried high into her tree and with them constructed a flooring across two stout branches binding the poles together and also to the branches with fibers from the tough arboraceous grasses that grew in profusion near the stream. Similarly she built walls and a roof, the latter thatched with many layers of great leaves. The fashioning of the barred windows and the door were matters of great importance and consuming interest. The windows, there were two of them, were large and the bars permanently fixed; but the door was small, the opening just large enough to permit her to pass through easily on hands and knees, which made it easier to barricade. She lost count of the days that the house cost her; but time was a cheap commodity—she had more of it than of anything else. It meant so little to her that she had not even any desire to keep account of it. How long since she and Obergatz had fled from the wrath of the Negro villagers she did not know and she could only roughly guess at the seasons. She worked hard for two reasons; one was to hasten the completion of her little place of refuge, and the other a desire for such physical exhaustion at night that she would sleep through those dreaded hours to a new day. As a matter of fact the house was finished in less than a week—that is, it was made as safe as it ever would be, though regardless of how long she might occupy it she would keep on adding touches and refinements here and there.

Her daily life was filled with her house building and her hunting, to which was added an occasional spice of excitement contributed by roving lions. To the woodcraft that she had learned from Tarzan, that master of the art, was added a considerable store of practical experience derived from her own past adventures in the jungle and the long months with Obergatz, nor was any day now lacking in some added store of useful knowledge. To these facts was attributable her apparent immunity from harm, since they told her when *ja* was

approaching before he crept close enough for a successful
charge and, too, they kept her close to those never-failing
havens of retreat—the trees.

The nights, filled with their weird noises, were lonely and
depressing. Only her ability to sleep quickly and soundly
made them endurable. The first night that she spent in her
completed house behind barred windows and barricaded door
was one of almost undiluted peace and happiness. The night
noises seemed far removed and impersonal and the soughing
of the wind in the trees was gently soothing. Before, it had
carried a mournful note and was sinister in that it might hide
the approach of some real danger. That night she slept indeed.

She went further afield now in search of food. So far nothing
but rodents had fallen to her spear—her ambition was an antelope, since beside the flesh it would give her, and the gut for
her bow, the hide would prove invaluable during the colder
weather that she knew would accompany the rainy season.
She had caught glimpses of these wary animals and was sure
that they always crossed the stream at a certain spot above her
camp. It was to this place that she went to hunt them. With
the stealth and cunning of a panther she crept through the
forest, circling about to get up wind from the ford, pausing
often to look and listen for aught that might menace her—
herself the personification of a hunted deer. Now she moved
silently down upon the chosen spot. What luck! A beautiful
buck stood drinking in the stream. The woman wormed her
way closer. Now she lay upon her belly behind a small bush
within throwing distance of the quarry. She must rise to her
full height and throw her spear almost in the same instant and
she must throw it with great force and perfect accuracy. She
thrilled with the excitement of the minute, yet cool and steady
were her swift muscles as she rose and cast her missile. Scarce

by the width of a finger did the point strike from the spot at which it had been directed. The buck leaped high, landed upon the bank of the stream, and fell dead. Jane Clayton sprang quickly forward toward her kill.

"Bravo!" A man's voice spoke in English from the shrubbery upon the opposite side of the stream. Jane Clayton halted in her tracks—stunned, almost, by surprise. And then a strange, unkempt figure of a man stepped into view. At first she did not recognize him, but when she did, instinctively she stepped back.

"Lieutenant Obergatz!" she cried. "Can it be you?"

"It can. It is," replied the German. "I am a strange sight, no doubt; but still it is I, Erich Obergatz. And you? You have changed too, is it not?"

He was looking at her naked limbs and her golden breastplates, the loin cloth of *jato*-hide, the harness and ornaments that constitute the apparel of a Ho-don woman—the things that Lu-don had dressed her in as his passion for her grew. Not Ko-tan's daughter, even, had finer trappings.

"But why are you here?" Jane insisted. "I had thought you safely among civilized men by this time, if you still lived."

"Gott!" he exclaimed. "I do not know why I continue to live. I have prayed to die and yet I cling to life. There is no hope. We are doomed to remain in this horrible land until we die. The bog! The frightful bog! I have searched its shores for a place to cross until I have entirely circled the hideous country. Easily enough we entered; but the rains have come since and now no living man could pass that slough of slimy mud and hungry reptiles. Have I not tried it! And the beasts that roam this accursed land. They hunt me by day and by night."

"But how have you escaped them?" she asked.

"I do not know," he replied gloomily. "I have fled and fled and fled. I have remained hungry and thirsty in tree tops for

days at a time. I have fashioned weapons—clubs and spears
—and I have learned to use them. I have slain a lion with my
club. So even will a cornered rat fight. And we are no better
than rats in this land of stupendous dangers, you and I. But
tell me about yourself. If it is surprising that I live, how much
more so that you still survive."

Briefly she told him and all the while she was wondering
what she might do to rid herself of him. She could not conceive of a prolonged existence with him as her sole companion.
Better, a thousand times better, to be alone. Never had her
hatred and contempt for him lessened through the long weeks
and months of their constant companionship, and now that
he could be of no service in returning her to civilization, she
shrank from the thought of seeing him daily. And, too, she
feared him. Never had she trusted him; but now there was a
strange light in his eye that had not been there when last she
saw him. She could not interpret it—all she knew was that
it gave her a feeling of apprehension—a nameless dread.

"You lived long then in the city of A-lur?" he said, speaking
in the language of Pal-ul-don.

"You have learned this tongue?" she asked. "How?"

"I fell in with a band of half-breeds," he replied, "members
of a proscribed race that dwells in the rock-bound gut through
which the principal river of the valley empties into the morass.
They are called Waz-ho-don and their village is partly made
up of cave dwellings and partly of houses carved from the
soft rock at the foot of the cliff. They are very ignorant and
superstitious and when they first saw me and realized that I
had no tail and that my hands and feet were not like theirs
they were afraid of me. They thought that I was either god or
demon. Being in a position where I could neither escape them
nor defend myself, I made a bold front and succeeded in
impressing them to such an extent that they conducted me

to their city, which they call Bu-lur, and there they fed me and treated me with kindness. As I learned their language I sought to impress them more and more with the idea that I was a god, and I succeeded, too, until an old fellow who was something of a priest among them, or medicine-man, became jealous of my growing power. That was the beginning of the end and came near to being the end in fact. He told them that if I was a god I would not bleed if a knife was stuck into me —if I did bleed it would prove conclusively that I was not a god. Without my knowledge he arranged to stage the ordeal before the whole village upon a certain night—it was upon one of those numerous occasions when they eat and drink to Jad-ben-Otho, their pagan deity. Under the influence of their vile liquor they would be ripe for any bloodthirsty scheme the medicine-man might evolve. One of the women told me about the plan—not with any intent to warn me of danger, but prompted merely by feminine curiosity as to whether or not I would bleed if stuck with a dagger. She could not wait, it seemed, for the orderly procedure of the ordeal—she wanted to know at once, and when I caught her trying to slip a knife into my side and questioned her she explained the whole thing with the utmost naïveté. The warriors already had commenced drinking—it would have been futile to make any sort of appeal either to their intellects or their superstitions. There was but one alternative to death and that was flight. I told the woman that I was very much outraged and offended at this reflection upon my godhood and that as a mark of my disfavor I should abandon them to their fate.

" 'I shall return to heaven at once!' I exclaimed.

"She wanted to hang around and see me go, but I told her that her eyes would be blasted by the fire surrounding my departure and that she must leave at once and not return to the spot for at least an hour. I also impressed upon her the fact that should any other approach this part of the village within that time not only they, but she as well, would burst into flames and be consumed.

"She was very much impressed and lost no time in leaving, calling back as she departed that if I were indeed gone in an hour she and all the village would know that I was no less than Jad-ben-Otho himself, and so they must think me, for I can assure you that I was gone in much less than an hour, nor have I ventured close to the neighborhood of the city of Bu-lur since," and he fell to laughing in harsh, cackling notes that sent a shiver through the woman's frame.

As Obergatz talked Jane had recovered her spear from the carcass of the antelope and commenced busying herself with the removal of the hide. The man made no attempt to assist her, but stood by talking and watching her, the while he continually ran his filthy fingers through his matted hair and beard. His face and body were caked with dirt and he was naked except for a torn and greasy hide about his loins. His weapons consisted of a club and knife of Waz-don pattern, that he had stolen from the city of Bu-lur; but what more greatly concerned the woman than his filth or his armament were his cackling laughter and the strange expression in his eyes.

She went on with her work; however, removing those parts of the buck she wanted, taking only as much meat as she might consume before it spoiled, as she was not sufficiently a true jungle creature to relish it beyond that stage, and then she straightened up and faced the man.

"Lieutenant Obergatz," she said, "by a chance of accident we have met again. Certainly you would not have sought the meeting any more than I. We have nothing in common other than those sentiments which may have been engendered by my natural dislike and suspicion of you, one of the authors of all the misery and sorrow that I have endured for endless months. This little corner of the world is mine by right of discovery and occupation. Go away and leave me to enjoy here what peace I may. It is the least that you can do to amend the wrong that you have done me and mine."

The man stared at her through his fishy eyes for a moment in silence, then there broke from his lips a peal of mirthless, uncanny laughter.

"Go away! Leave you alone!" he cried. "I have found you. We are going to be good friends. There is no one else in the world but us. No one will ever know what we do or what becomes of us and now you ask me to go away and live alone in this hellish solitude." Again he laughed, though neither the muscles of his eyes or his mouth reflected any mirth—it was just a hollow sound that imitated laughter.

"Remember your promise," she said.

"Promise! Promise! What are promises? They are made to be broken—we taught the world that at Liége and Louvain. No, no! I will not go away. I shall stay and protect you."

"I do not need your protection," she insisted. "You have already seen that I can use a spear."

"Yes," he said; "but it would not be right to leave you here alone—you are but a woman. No, no; I am an officer of the Kaiser and I cannot abandon you."

Once more he laughed. "We could be very happy here together," he added.

The woman could not repress a shudder, nor, in fact, did she attempt to hide her aversion.

"You do not like me?" he asked. "Ah, well; it is too sad. But some day you will love me," and again the hideous laughter.

The woman had wrapped the pieces of the buck in the hide and this she now raised and threw across her shoulder. In her other hand she held her spear and faced the German.

"Go!" she commanded. "We have wasted enough words. This is my country and I shall defend it. If I see you about again I shall kill you. Do you understand?"

An expression of rage contorted Obergatz' features. He raised his club and started toward her.

"Stop!" she commanded, throwing her spear-hand backward for a cast. "You saw me kill this buck and you have said truthfully that no one will ever know what we do here. Put these two facts together, German, and draw your own conclusions before you take another step in my direction."

The man halted and his club-hand dropped to his side. "Come," he begged in what he intended as a conciliatory tone. "Let us be friends, Lady Greystoke. We can be of great assistance to each other and I promise not to harm you."

"Remember Liége and Louvain," she reminded him with a sneer. "I am going now—be sure that you do not follow me. As far as you can walk in a day from this spot in any direction you may consider the limits of my domain. If ever again I see you within these limits I shall kill you."

There could be no question that she meant what she said and the man seemed convinced for he but stood sullenly eyeing her as she backed from sight beyond a turn in the game trail that crossed the ford where they had met, and disappeared in the forest.

CHAPTER XX

Silently in the Night

IN A-LUR the fortunes of the city had been tossed from hand
to hand. The party of Ko-tan's loyal warriors that Tarzan
had led to the rendezvous at the entrance to the secret passage
below the palace gates had met with disaster. Their first rush
had been met with soft words from the priests. They had
been exhorted to defend the faith of their fathers from blasphemers. Ja-don was painted to them as a defiler of temples,
and the wrath of Jad-ben-Otho was prophesied for those who
embraced his cause. The priests insisted that Lu-don's only
wish was to prevent the seizure of the throne by Ja-don until
a new king could be chosen according to the laws of the Ho-
don.

The result was that many of the palace warriors joined their
fellows of the city, and when the priests saw that those whom
they could influence outnumbered those who remained loyal
to the palace, they caused the former to fall upon the latter
with the result that many were killed and only a handful succeeded in reaching the safety of the palace gates, which they
quickly barred.

The priests led their own forces through the secret passageway into the temple, while some of the loyal ones sought
out Ja-don and told him all that had happened. The fight in
the banquet hall had spread over a considerable portion of the
palace grounds and had at last resulted in the temporary de-

237

feat of those who had opposed Ja-don. This force, counseled
by under priests sent for the purpose by Lu-don, had with-
drawn within the temple grounds so that now the issue was
plainly marked as between Ja-don on the one side and Lu-don
on the other.

The former had been told of all that had occurred in the
apartments of O-lo-a to whose safety he had attended at the
first opportunity and he had also learned of Tarzan's part in
leading his men to the gathering of Lu-don's warriors.

These things had naturally increased the old warrior's former
inclinations of friendliness toward the ape-man, and now he
regretted that the other had departed from the city.

The testimony of O-lo-a and Pan-at-lee was such as to
strengthen whatever belief in the godliness of the stranger
Ja-don and others of the warriors had previously entertained,
until presently there appeared a strong tendency upon the
part of this palace faction to make the Dor-ul-Otho an issue
of their original quarrel with Lu-don. Whether this occurred
as the natural sequence to repeated narrations of the ape-man's
exploits, which lost nothing by repetition, in conjunction with
Lu-don's enmity toward him, or whether it was the shrewd design of some wily old warrior such as Ja-don, who realized the
value of adding a religious cause to their temporal one, it
were difficult to determine; but the fact remained that Ja-don's
followers developed bitter hatred for the followers of Lu-don
because of the high priest's antagonism to Tarzan.

Unfortunately however Tarzan was not there to inspire the
followers of Ja-don with the holy zeal that might have quickly
settled the dispute in the old chieftain's favor. Instead, he
was miles away and because their repeated prayers for his
presence were unanswered, the weaker spirits among them
commenced to suspect that their cause did not have divine
favor. There was also another and a potent cause for defection

from the ranks of Ja-don. It emanated from the city where the friends and relatives of the palace warriors, who were largely also the friends and relatives of Lu-don's forces, found the means, urged on by the priesthood, to circulate throughout the palace pernicious propaganda aimed at Ja-don's cause.

The result was that Lu-don's power increased while that of Ja-don waned. Then followed a sortie from the temple which resulted in the defeat of the palace forces, and though they were able to withdraw in decent order withdraw they did, leaving the palace to Lu-don, who was now virtually ruler of Pal-ul-don.

Ja-don, taking with him the princess, her women, and their slaves, including Pan-at-lee, as well as the women and children of his faithful followers, retreated not only from the palace but from the city of A-lur as well and fell back upon his own city of Ja-lur. Here he remained, recruiting his forces from the surrounding villages of the north which, being far removed from the influence of the priesthood of A-lur, were enthusiastic partizans in any cause that the old chieftain espoused, since for years he had been revered as their friend and protector.

And while these events were transpiring in the north, Tarzan-jad-guru lay in the lion pit at Tu-lur while messengers passed back and forth between Mo-sar and Lu-don as the two dickered for the throne of Pal-ul-don. Mo-sar was cunning enough to guess that should an open breach occur between himself and the high priest he might use his prisoner to his own advantage, for he had heard whisperings among even his own people that suggested that there were those who were more than a trifle inclined to belief in the divinity of the stranger and that he might indeed be the Dor-ul-Otho. Lu-don wanted Tarzan himself. He wanted to sacrifice him upon the eastern altar with his own hands before a multitude of people, since he was

not without evidence that his own standing and authority had
been lessened by the claims of the bold and heroic figure of
the stranger.

The method that the high priest of Tu-lur had employed to
trap Tarzan had left the ape-man in possession of his weapons
though there seemed little likelihood of their being of any service to him. He also had his pouch, in which were the various
odds and ends which are the natural accumulation of all receptacles from a gold meshbag to an attic. There were bits
of obsidian and choice feathers for arrows, some pieces of flint
and a couple of steel, an old knife, a heavy bone needle, and
strips of dried gut. Nothing very useful to you or me, perhaps; but nothing useless to the savage life of the ape-man.

When Tarzan realized the trick that had been so neatly
played upon him he had awaited expectantly the coming of
the lion, for though the scent of *ja* was old he was sure that
sooner or later they would let one of the beasts in upon him.
His first consideration was a thorough exploration of his prison.
He had noticed the hide-covered windows and these he immediately uncovered, letting in the light, and revealing the
fact that though the chamber was far below the level of the
temple courts it was yet many feet above the base of the hill
from which the temple was hewn. The windows were so
closely barred that he could not see over the edge of the thick
wall in which they were cut to determine what lay close in
below him. At a little distance were the blue waters of Jad-in-
lul and beyond, the verdure-clad farther shore, and beyond
that the mountains. It was a beautiful picture upon which
he looked—a picture of peace and harmony and quiet. Nor
anywhere a slightest suggestion of the savage men and beasts
that claimed this lovely landscape as their own. What a
paradise! And some day civilized man would come and—
spoil it! Ruthless axes would raze that age-old wood; black,

sticky smoke would rise from ugly chimneys against that azure sky; grimy little boats with wheels behind or upon either side would churn the mud from the bottom of Jad-in-lul, turning its blue waters to a dirty brown; hideous piers would project into the lake from squalid buildings of corrugated iron, doubtless, for of such are the pioneer cities of the world.

But would civilized man come? Tarzan hoped not. For countless generations civilization had ramped about the globe; it had dispatched its emissaries to the North Pole and the South; it had circled Pal-ul-don once, perhaps many times, but it had never touched her. God grant that it never would. Perhaps He was saving this little spot to be always just as He had made it, for the scratching of the Ho-don and the Waz-don upon His rocks had not altered the fair face of Nature.

Through the windows came sufficient light to reveal the whole interior to Tarzan. The room was fairly large and there was a door at each end—a large door for men and a smaller one for lions. Both were closed with heavy masses of stone that had been lowered in grooves running to the floor. The two windows were small and closely barred with the first iron that Tarzan had seen in Pal-ul-don. The bars were let into holes in the casing, and the whole so strongly and neatly contrived that escape seemed impossible. Yet within a few minutes of his incarceration Tarzan had commenced to undertake his escape. The old knife in his pouch was brought into requisition and slowly the ape-man began to scrape and chip away the stone from about the bars of one of the windows. It was slow work but Tarzan had the patience of absolute health.

Each day food and water were brought him and slipped quickly beneath the smaller door which was raised just sufficiently to allow the stone receptacles to pass in. The prisoner began to believe that he was being preserved for something beside lions. However that was immaterial. If they would but hold off for a few more days they might select what fate they would—he would not be there when they arrived to announce it.

And then one day came Pan-sat, Lu-don's chief tool, to the city of Tu-lur. He came ostensibly with a fair message for Mo-sar from the high priest at A-lur. Lu-don had decided that Mo-sar should be king and he invited Mo-sar to come at once to A-lur and then Pan-sat, having delivered the message, asked that he might go to the temple of Tu-lur and pray, and there he sought the high priest of Tu-lur to whom was the true message that Lu-don had sent. The two were closeted alone in a little chamber and Pan-sat whispered into the ear of the high priest.

"Mo-sar wishes to be king," he said, "and Lu-don wishes to be king. Mo-sar wishes to retain the stranger who claims to be the Dor-ul-Otho and Lu-don wishes to kill him, and now," he leaned even closer to the ear of the high priest of Tu-lur, "if you would be high priest at A-lur it is within your power."

Pan-sat ceased speaking and waited for the other's reply. The high priest was visibly affected. To be high priest at A-lur! That was almost as good as being king of all Pal-ul-don, for great were the powers of him who conducted the sacrifices upon the altars at A-lur.

"How?" whispered the high priest. "How may I become high priest at A-lur?"

Again Pan-sat leaned close: "By killing the one and bringing the other to A-lur," replied he. Then he rose and departed knowing that the other had swallowed the bait and could be depended upon to do whatever was required to win him the great prize.

Nor was Pan-sat mistaken other than in one trivial consideration. This high priest would indeed commit murder and treason to attain the high office at A-lur; but he had misunderstood which of his victims was to be killed and which to be delivered to Lu-don. Pan-sat, knowing himself all the details of the plannings of Lu-don, had made the quite natural error of assuming that the other was perfectly aware that only by publicly sacrificing the false Dor-ul-Otho could the high priest at A-lur bolster his waning power and that the assassination of Mo-sar, the pretender, would remove from Lu-don's camp the only obstacle to his combining the offices of high priest and king. The high priest at Tu-lur thought that he had been commissioned to kill Tarzan and bring Mo-sar to A-lur. He also thought that when he had done these things he would be made high priest at A-lur; but he did not know that already the priest had been selected who was to murder him within the hour that he arrived at A-lur, nor did he know that a secret grave had been prepared for him in the floor of a subterranean chamber in the very temple he dreamed of controlling.

And so when he should have been arranging the assassination of his chief he was leading a dozen heavily bribed warriors through the dark corridors beneath the temple to slay Tarzan in the lion pit. Night had fallen. A single torch guided the footsteps of the murderers as they crept stealthily upon their evil way, for they knew that they were doing the thing that their chief did not want done and their guilty consciences warned them to stealth.

In the dark of his cell the ape-man worked at his seemingly endless chipping and scraping. His keen ears detected the coming of footsteps along the corridor without—footsteps that approached the larger door. Always before had they come to the smaller door—the footsteps of a single slave who brought his food. This time there were many more than one and their coming at this time of night carried a sinister suggestion. Tarzan continued to work at his scraping and chipping. He heard them stop beyond the door. All was silence broken only by the scrape, scrape, scrape of the ape-man's tireless blade.

Those without heard it and listening sought to explain it. They whispered in low tones making their plans. Two would raise the door quickly and the others would rush in and hurl their clubs at the prisoner. They would take no chances, for the stories that had circulated in A-lur had been brought to Tu-lur—stories of the great strength and wonderful prowess of Tarzan-jad-guru that caused the sweat to stand upon the brows of the warriors, though it was cool in the damp corridor and they were twelve to one.

And then the high priest gave the signal—the door shot upward and ten warriors leaped into the chamber with poised clubs. Three of the heavy weapons flew across the room toward a darker shadow that lay in the shadow of the opposite wall, then the flare of the torch in the priest's hand lighted the interior and they saw that the thing at which they had flung their clubs was a pile of skins torn from the windows and that except for themselves the chamber was vacant.

One of them hastened to a window. All but a single bar was gone and to this was tied one end of a braided rope fashioned from strips cut from the leather window hangings.

To the ordinary dangers of Jane Clayton's existence was now added the menace of Obergatz' knowledge of her whereabouts. The lion and the panther had given her less cause for anxiety than did the return of the unscrupulous Hun, whom she had always distrusted and feared, and whose repulsiveness was now immeasurably augmented by his unkempt and filthy appearance, his strange and mirthless laughter, and his unnatural demeanor. She feared him now with a new fear as though he had suddenly become the personification of some nameless horror. The wholesome, outdoor life that she had been leading had strengthened and rebuilt her nervous system yet it seemed to her as she thought of him that if this man should ever touch her she should scream, and, possibly, even faint. Again and again during the day following their unexpected meeting the woman reproached herself for not having killed him as she would *ja* or *jato* or any other predatory beast that menaced her existence or her safety. There was no attempt at self-justification for these sinister reflections—they needed no justification. The standards by which the acts of such as you or I may be judged could not apply to hers. We have recourse to the protection of friends and relatives and the civil soldiery that upholds the majesty of the law and which may be invoked to protect the righteous weak against the unrighteous strong; but Jane Clayton comprised within herself not only the righteous weak but all the various agencies for the protection of the weak. To her, then, Lieutenant Erich Obergatz presented no different problem than did *ja*, the lion, other than that she considered the former the more dangerous animal. And so she determined that should he ignore her warning there would be no temporizing upon the occasion of their next meeting—the same swift spear that would meet *ja's* advances would meet his.

That night her snug little nest perched high in the great tree seemed less the sanctuary that it had before. What might resist the sanguinary intentions of a prowling panther would prove no great barrier to man, and influenced by this thought she slept less well than before. The slightest noise that broke the monotonous hum of the nocturnal jungle startled her into alert wakefulness to lie with straining ears in an attempt to classify the origin of the disturbance, and once she was awakened thus by a sound that seemed to come from something moving in her own tree. She listened intently—scarce breathing. Yes, there it was again. A scuffing of something soft against the hard bark of the tree. The woman reached out in the darkness and grasped her spear. Now she felt a slight sagging of one of the limbs that supported her shelter as though the thing, whatever it was, was slowly raising its weight to the branch. It came nearer. Now she thought that she could detect its breathing. It was at the door. She could hear it fumbling with the frail barrier. What could it be? It made no sound by which she might identify it. She raised herself upon her hands and knees and crept stealthily the little distance to the doorway, her spear clutched tightly in her hand. Whatever the thing was, it was evidently attempting to gain entrance without awakening her. It was just beyond the pitiful little contraption of slender boughs that she had bound together with grasses and called a door—only a few inches lay between the thing and her. Rising to her knees she reached out with her left hand and felt until she found a place where a crooked branch had left an opening a couple of inches wide near the center of the barrier. Into this she inserted the point of her spear. The thing must have heard her move within for suddenly it abandoned its efforts for stealth and tore angrily at the obstacle. At the same moment Jane thrust her spear forward with all her strength. She felt it enter flesh. There was a scream and a curse from without, followed by the crashing of a body through limbs and foliage. Her spear was almost dragged from her grasp, but she held to it until it broke free from the thing it had pierced.

It was Obergatz; the curse had told her that. From below came no further sound. Had she, then, killed him? She prayed so—with all her heart she prayed it. To be freed from the menace of this loathsome creature were relief indeed. During all the balance of the night she lay there awake, listening. Below her, she imagined, she could see the dead man with his hideous face bathed in the cold light of the moon— lying there upon his back staring up at her.

She prayed that *ja* might come and drag it away, but all during the remainder of the night she heard never another sound above the drowsy hum of the jungle. She was glad that he was dead, but she dreaded the gruesome ordeal that awaited her on the morrow, for she must bury the thing that had been Erich Obergatz and live on there above the shallow grave of the man she had slain.

She reproached herself for her weakness, repeating over and over that she had killed in self-defense, that her act was justified; but she was still a woman of today, and strong upon her were the iron mandates of the social order from which she had sprung, its interdictions and its superstitions.

At last came the tardy dawn. Slowly the sun topped the distant mountains beyond Jad-in-lul. And yet she hesitated to loosen the fastenings of her door and look out upon the thing below. But it must be done. She steeled herself and untied the rawhide thong that secured the barrier. She looked down and only the grass and the flowers looked up at her. She came from her shelter and examined the ground upon the opposite side of the tree—there was no dead man there, nor anywhere as far as she could see. Slowly she descended, keeping a wary eye and an alert ear ready for the first intimation of danger.

At the foot of the tree was a pool of blood and a little trail of crimson drops upon the grass, leading away parallel with the shore of Jad-bal-lul. Then she had not slain him! She was vaguely aware of a peculiar, double sensation of relief and regret. Now she would be always in doubt. He might return; but at least she would not have to live above his grave.

She thought some of following the bloody spoor on the chance that he might have crawled away to die later, but she gave up the idea for fear that she might find him dead nearby, or, worse yet badly wounded. What then could she do? She could not finish him with her spear—no, she knew that she could not do that, nor could she bring him back and nurse him, nor could she leave him there to die of hunger or of thirst, or to become the prey of some prowling beast. It were better then not to search for him for fear that she might find him.

That day was one of nervous starting to every sudden sound. The day before she would have said that her nerves were of iron; but not today. She knew now the shock that she had suffered and that this was the reaction. Tomorrow it might be different, but something told her that never again would her little shelter and the patch of forest and jungle that she called her own be the same. There would hang over them always the menace of this man. No longer would she pass restful nights of deep slumber. The peace of her little world was shattered forever.

That night she made her door doubly secure with additional thongs of rawhide cut from the pelt of the buck she had slain the day that she met Obergatz. She was very tired for she had lost much sleep the night before; but for a long time she lay with wide-open eyes staring into the darkness. What saw she there? Visions that brought tears to those brave and beautiful eyes—visions of a rambling bungalow that had been home to her and that was no more, destroyed by the same cruel force that haunted her even now in this remote, uncharted corner of the earth; visions of a strong man whose protecting arm would never press her close again; visions of a tall, straight son who looked at her adoringly out of brave, smiling eyes that were like his father's. Always the vision of the crude simple bungalow rather than of the stately halls that had been as much a part of her life as the other. But *he* had loved the bungalow and the broad, free acres best and so she had come to love them best, too.

At last she slept, the sleep of utter exhaustion. How long it lasted she did not know; but suddenly she was wide awake and once again she heard the scuffing of a body against the bark of her tree and again the limb bent to a heavy weight. He had returned! She went cold, trembling as with ague. Was it he, or, O God! had she killed him then and was this—? She tried to drive the horrid thought from her mind, for this way, she knew, lay madness.

And once again she crept to the door, for the thing was outside just as it had been last night. Her hands trembled as she placed the point of her weapon to the opening. She wondered if it would scream as it fell.

CHAPTER XXI

The Maniac

THE last bar that would make the opening large enough to
permit his body to pass had been removed as Tarzan
heard the warriors whispering beyond the stone door of his
prison. Long since had the rope of hide been braided. To secure one end to the remaining bar that he had left for this
purpose was the work of but a moment, and while the warriors
whispered without, the brown body of the ape-man slipped
through the small aperture and disappeared below the sill.

Tarzan's escape from the cell left him still within the walled
area that comprised the palace and temple grounds and buildings. He had reconnoitered as best he might from the window
after he had removed enough bars to permit him to pass his
head through the opening, so that he knew what lay immediately before him—a winding and usually deserted alleyway
leading in the direction of the outer gate that opened from
the palace grounds into the city.

The darkness would facilitate his escape. He might even
pass out of the palace and the city without detection. If he
could elude the guard at the palace gate the rest would be
easy. He strode along confidently, exhibiting no fear of detection, for he reasoned that thus would he disarm suspicion.
In the darkness he easily could pass for a Ho-don and in truth,
though he passed several after leaving the deserted alley, no
one accosted or detain him, and thus he came at last to the

guard of a half-dozen warriors before the palace gate. These he attempted to pass in the same unconcerned fashion and he might have succeeded had it not been for one who came running rapidly from the direction of the temple shouting: "Let no one pass the gates! The prisoner has escaped from the *pal-ul-ja!*"

Instantly a warrior barred his way and simultaneously the fellow recognized him. "*Xot tor!*" he exclaimed: "Here he is now. Fall upon him! Fall upon him! Back! Back before I kill you."

The others came forward. It cannot be said that they rushed forward. If it was their wish to fall upon him there was a noticeable lack of enthusiasm other than that which directed their efforts to persuade someone else to fall upon him. His fame as a fighter had been too long a topic of conversation for the good of the morale of Mo-sar's warriors. It were safer to stand at a distance and hurl their clubs and this they did, but the ape-man had learned something of the use of this weapon since he had arrived in Pal-ul-don. And as he learned great had grown his respect for this most primitive of arms. He had come to realize that the black savages he had known had never appreciated the possibilities of their knob sticks, nor had he, and he had discovered, too, why the Pal-ul-dons had turned their ancient spears into plowshares and pinned their faith to the heavy-ended club alone. In deadly execution it was far more effective than a spear and it answered, too, every purpose of a shield, combining the two in one and thus reducing the burden of the warrior. Thrown as they throw it, after the manner of the hammer-throwers of the Olympian games, an ordinary shield would prove more a weakness than a strength while one that would be strong enough to prove a protection would be too heavy to carry. Only another club, deftly wielded to deflect the course of an enemy missile, is in any way effective against these formidable weapons and, too, the war club of Pal-ul-don can be thrown with accuracy a far greater distance than any spear.

And now was put to the test that which Tarzan had learned from Om-at and Ta-den. His eyes and his muscles trained by a lifetime of necessity moved with the rapidity of light and his brain functioned with an uncanny celerity that suggested nothing less than prescience, and these things more than compensated for his lack of experience with the war club he handled so dexterously. Weapon after weapon he warded off and always he moved with a single idea in mind—to place himself within reach of one of his antagonists. But they were wary for they feared this strange creature to whom the superstitious fears of many of them attributed the miraculous powers of deity. They managed to keep between Tarzan and the gateway and all the time they bawled lustily for reinforcements. Should these come before he had made his escape the ape-man realized that the odds against him would be unsurmountable, and so he redoubled his efforts to carry out his design.

Following their usual tactics, two or three of the warriors were always circling behind him collecting the thrown clubs when Tarzan's attention was directed elsewhere. He himself retrieved several of them which he hurled with such deadly effect as to dispose of two of his antagonists, but now he heard the approach of hurrying warriors, the patter of their bare feet upon the stone pavement and then the savage cries which were to bolster the courage of their fellows and fill the enemy with fear.

There was no time to lose. Tarzan held a club in either hand and, swinging one he hurled it at a warrior before him and as the man dodged he rushed in and seized him, at the same time casting his second club at another of his opponents.

The Ho-don with whom he grappled reached instantly for his knife but the ape-man grasped his wrist. There was a sudden twist, the snapping of a bone and an agonized scream, then the warrior was lifted bodily from his feet and held as a shield between his fellows and the fugitive as the latter backed through the gateway. Beside Tarzan stood the single torch that lighted the entrance to the palace grounds. The warriors were advancing to the succor of their fellow when the ape-man raised his captive high above his head and flung him full in the face of the foremost attacker. The fellow went down and two directly behind him sprawled headlong over their companion as the ape-man seized the torch and cast is back into the palace grounds to be extinguished as it struck the bodies of those who led the charging reinforcements.

In the ensuing darkness Tarzan disappeared in the streets of Tu-lur beyond the palace gate. For a time he was aware of sounds of pursuit but the fact that they trailed away and died in the direction of Jad-in-lul informed him that they were searching in the wrong direction, for he had turned south out of Tu-lur purposely to throw them off his track. Beyond the outskirts of the city he turned directly toward the northwest, in which direction lay A-lur.

In his path he knew lay Jad-bal-lul, the shore of which he was compelled to skirt, and there would be a river to cross at the lower end of the great lake upon the shores of which lay A-lur. What other obstacles lay in his way he did not know but he believed that he could make better time on foot than by attempting to steal a canoe and force his way up stream with a single paddle. It was his intention to put as much distance as possible between himself and Tu-lur before he slept for he was sure that Mo-sar would not lightly accept his loss, but that with the coming of day, or possibly even before, he would dispatch warriors in search of him.

A mile or two from the city he entered a forest and here at last he felt such a measure of safety as he never knew in open spaces or in cities. The forest and the jungle were his birthright. No creature that went upon the ground upon four feet, or climbed among the trees, or crawled upon its belly had any advantage over the ape-man in his native heath. As myrrh and frankincense were the dank odors of rotting vegetation in the nostrils of the great Tarmangani. He squared his broad shoulders and lifting his head filled his lungs with the air that he loved best. The heavy fragrance of tropical blooms, the commingled odors of the myriad-scented life of the jungle went to his head with a pleasurable intoxication far more potent than aught contained in the oldest vintages of civilization.

He took to the trees now, not from necessity but from pure love of the wild freedom that had been denied him so long. Though it was dark and the forest strange yet he moved with a surety and ease that bespoke more a strange uncanny sense than wondrous skill. He heard *ja* moaning somewhere ahead and an owl hooted mournfully to the right of him—long familiar sounds that imparted to him no sense of loneliness as they might to you or to me, but on the contrary one of companionship for they betokened the presence of his fellows of the jungle, and whether friend or foe it was all the same to the ape-man.

He came at last to a little stream at a spot where the trees did not meet above it so he was forced to descend to the ground and wade through the water and upon the opposite shore he stopped as though suddenly his godlike figure had been transmuted from flesh to marble. Only his dilating nostrils bespoke his pulsing vitality. For a long moment he stood there thus and then swiftly, but with a caution and silence that were inherent in him he moved forward again, but now his whole attitude bespoke a new urge. There was a definite and masterful purpose in every movement of those steel muscles rolling softly beneath the smooth brown hide. He moved now toward a certain goal that quite evidently filled him with far greater enthusiasm than had the possible event of his return to A-lur.

And so he came at last to the foot of a great tree and there he stopped and looked up above him among the foliage where the dim outlines of a roughly rectangular bulk loomed darkly. There was a choking sensation in Tarzan's throat as he raised himself gently into the branches. It was as though his heart were swelling either to a great happiness or a great fear.

Before the rude shelter built among the branches he paused listening. From within there came to his sensitive nostrils the same delicate aroma that had arrested his eager attention at the little stream a mile away. He crouched upon the branch close to the little door.

"Jane," he called, "heart of my heart, it is I."

The only answer from within was as the sudden indrawing of a breath that was half gasp and half sigh, and the sound of a body falling to the floor. Hurriedly Tarzan sought to release the thongs which held the door but they were fastened from the inside, and at last, impatient with further delay, he seized the frail barrier in one giant hand and with a single effort tore it completely away. And then he entered to find the seemingly lifeless body of his mate stretched upon the floor.

He gathered her in his arms; her heart beat; she still breathed, and presently he realized that she had but swooned.

When Jane Clayton regained consciousness it was to find herself held tightly in two strong arms, her head pillowed upon the broad shoulder where so often before her fears had been soothed and her sorrows comforted. At first she was not sure but that it was all a dream. Timidly her hand stole to his cheek.

"John," she murmured, "tell me, is it really you?"

In reply he drew her more closely to him. "It is I," he replied. "But there is something in my throat," he said haltingly, "that makes it hard for me to speak."

She smiled and snuggled closer to him. "God has been good to us, Tarzan of the Apes," she said.

For some time neither spoke. It was enough that they were reunited and that each knew that the other was alive and safe. But at last they found their voices and when the sun rose they were still talking, so much had each to tell the other; so many questions there were to be asked and answered.

"And Jack," she asked, "where is he?"

"I do not know," replied Tarzan. "The last I heard of him he was on the Argonne Front."

"Ah, then our happiness is not quite complete," she said, a little note of sadness creeping into her voice.

"No," he replied, "but the same is true in countless other English homes today, and pride is learning to take the place of happiness in these."

She shook her head, "I want my boy," she said.

"And I too," replied Tarzan, "and we may have him yet. He was safe and unwounded the last word I had. And now," he said, "we must plan upon our return. Would you like to rebuild the bungalow and gather together the remnants of our Waziri or would you rather return to London?"

"Only to find Jack," she said. "I dream always of the bungalow and never of the city, but John, we can only dream, for Obergatz told me that he had circled this whole country and found no place where he might cross the morass."

"I am not Obergatz," Tarzan reminded her, smiling. "We will rest today and tomorrow we will set out toward the north. It is a savage country, but we have crossed it once and we can cross it again."

And so, upon the following morning, the Tarmangani and his mate went forth upon their journey across the Valley of Jad-ben-Otho, and ahead of them were fierce men and savage beasts, and the lofty mountains of Pal-ul-don; and beyond the mountains the reptiles and the morass, and beyond that the arid, thorn-covered steppe, and other savage beasts and men and weary, hostile miles of untracked wilderness between them and the charred ruins of their home.

Lieutenant Erich Obergatz crawled through the grass upon all fours, leaving a trail of blood behind him after Jane's spear had sent him crashing to the ground beneath her tree. He made no sound after the one piercing scream that had ac-knowledged the severity of his wound. He was quiet because of a great fear that had crept into his warped brain that the devil woman would pursue and slay him. And so he crawled away like some filthy beast of prey, seeking a thicket where he might lie down and hide.

He thought that he was going to die, but he did not, and with the coming of the new day he discovered that his wound was superficial. The rough obsidian-shod spear had entered the muscles of his side beneath his right arm inflicting a pain-ful, but not a fatal wound. With the realization of this fact came a renewed desire to put as much distance as possible between himself and Jane Clayton. And so he moved on, still going upon all fours because of a persistent hallucination that in this way he might escape observation. Yet though he fled his mind still revolved muddily about a central desire—while he fled from her he still planned to pursue her, and to his lust of possession was added a desire for revenge. She should pay for the suffering she had inflicted upon him. She should pay for rebuffing him, but for some reason which he did not try to explain to himself he would crawl away and hide. He would

come back though. He would come back and when he had
finished with her, he would take that smooth throat in his two
hands and crush the life from her.

He kept repeating this over and over to himself and then he
fell to laughing out loud, the cackling, hideous laughter that
had terrified Jane. Presently he realized his knees were bleeding and that they hurt him. He looked cautiously behind. No
one was in sight. He listened. He could hear no indications
of pursuit and so he rose to his feet and continued upon his
way a sorry sight—covered with filth and blood, his beard and
hair tangled and matted and filled with burrs and dried mud
and unspeakable filth. He kept no track of time. He ate
fruits and berries and tubers that he dug from the earth with
his fingers. He followed the shore of the lake and the river
that he might be near water, and when *ja* roared or moaned he
climbed a tree and hid there, shivering.

And so after a time he came up the southern shore of Jad-
ben-lul until a wide river stopped his progress. Across the
blue water a white city glimmered in the sun. He looked at it
for a long time, blinking his eyes like an owl. Slowly a recollection forced itself through his tangled brain. This was A-lur,
the City of Light. The association of ideas recalled Bu-lur
and the Waz-ho-don. They had called him Jad-ben-Otho.
He commenced to laugh aloud and stood up very straight and
strode back and forth along the shore. "I am Jad-ben-Otho,"
he cried, "I am the Great God. In A-lur is my temple and my
high priest. What is Jad-ben-Otho doing here alone in the
jungle?"

He stepped out into the water and raising his voice shrieked
loudly across toward A-lur. "I am Jad-ben-Otho!" he
screamed. "Come hither slaves and take your god to his
temple." But the distance was great and they did not hear

him and no one came, and the feeble mind was distracted by other things—a bird flying in the air, a school of minnows swimming around his feet. He lunged at them trying to catch them, and falling upon his hands and knees he crawled through the water grasping futilely at the elusive fish.

Presently it occurred to him that he was a sea lion and he forgot the fish and lay down and tried to swim by wriggling his feet in the water as though they were a tail. The hardships, the privations, the terrors, and for the past few weeks the lack of proper nourishment had reduced Erich Obergatz to little more than a gibbering idiot.

A water snake swam out upon the surface of the lake and the man pursued it, crawling upon his hands and knees. The snake swam toward the shore just within the mouth of the river where tall reeds grew thickly and Obergatz followed, making grunting noises like a pig. He lost the snake within the reeds but he came upon something else—a canoe hidden there close to the bank. He examined it with cackling laughter. There were two paddles within it which he took and threw out into the current of the river. He watched them for a while and then he sat down beside the canoe and commenced to splash his hands up and down upon the water. He liked to hear the noise and see the little splashes of spray. He rubbed his left forearm with his right palm and the dirt came off and left a white spot that drew his attention. He rubbed again upon the now thoroughly soaked blood and grime that covered his body. He was not attempting to wash himself; he was merely amused by the strange results. "I am turning white," he cried. His glance wandered from his body now that the grime and blood were all removed and caught again the white city shimmering beneath the hot sun.

"A-lur—City of Light!" he shrieked and that reminded him

again of Tu-lur and by the same process of associated ideas
that had before suggested it, he recalled that the Waz-ho-don
had thought him Jad-ben-Otho.

"I am Jad-ben-Otho!" he screamed and then his eyes fell
again upon the canoe. A new idea came and persisted. He
looked down at himself, examining his body, and seeing the
filthy loin cloth, now water soaked and more bedraggled than
before, he tore it from him and flung it into the lake. "Gods
do not wear dirty rags," he said aloud. "They do not wear
anything but wreaths and garlands of flowers and I am a god
—I am Jad-ben-Otho—and I go in state to my sacred city
of A-lur."

He ran his fingers through his matted hair and beard. The
water had softened the burrs but had not removed them. The
man shook his head. His hair and beard failed to harmonize
with his other godly attributes. He was commencing to think
more clearly now, for the great idea had taken hold of his
scattered wits and concentrated them upon a single purpose,
but he was still a maniac. The only difference being that he
was now a maniac with a fixed intent. He went out on the
shore and gathered flowers and ferns and wove them in his
beard and hair—blazing blooms of different colors—green
ferns that trailed about his ears or rose bravely upward like
the plumes in a lady's hat.

When he was satisfied that his appearance would impress
the most casual observer with his evident deity he returned to
the canoe, pushed it from shore and jumped in. The impetus
carried it into the river's current and the current bore it out
upon the lake. The naked man stood erect in the center of
the little craft, his arms folded upon his chest. He screamed
aloud his message to the city: "I am Jad-ben-Otho! Let the
high priest and the under priests attend upon me!"

As the current of the river was dissipated by the waters of

the lake the wind caught him and his craft and carried them bravely forward. Sometimes he drifted with his back toward A-lur and sometimes with his face toward it, and at intervals he shrieked his message and his commands. He was still in the middle of the lake when someone discovered him from the palace wall, and as he drew nearer, a crowd of warriors and women and children were congregated there watching him and along the temple walls were many priests and among them Lu-don, the high priest. When the boat had drifted close enough for them to distinguish the bizarre figure standing in it and for them to catch the meaning of his words Lu-don's cunning eyes narrowed. The high priest had learned of the escape of Tarzan and he feared that should he join Ja-don's forces, as seemed likely, he would attract many recruits who might still believe in him, and the Dor-ul-Otho, even if a false one, upon the side of the enemy might easily work havoc with Lu-don's plans.

The man was drifting close in. His canoe would soon be caught in the current that ran close to shore here and carried toward the river that emptied the waters of Jad-ben-lul into Jad-bal-lul. The under priests were looking toward Lu-don for instructions.

"Fetch him hither!" he commanded. "If he is Jad-ben-Otho I shall know him."

The priests hurried to the palace grounds and summoned warriors. "Go, bring the stranger to Lu-don. If he is Jad-ben-Otho we shall know him."

And so Lieutenant Erich Obergatz was brought before the high priest at A-lur. Lu-don looked closely at the naked man with the fantastic headdress.

"Where did you come from?" he asked.

"I am Jad-ben-Otho," cried the German. "I came from heaven. Where is my high priest?"

"I am the high priest," replied Lu-don.

Obergatz clapped his hands. "Have my feet bathed and food brought to me," he commanded.

Lu-don's eyes narrowed to mere slits of crafty cunning. He bowed low until his forehead touched the feet of the stranger. Before the eyes of many priests, and warriors from the palace he did it.

"Ho, slaves!" he cried, rising; "fetch water and food for the Great God," and thus the high priest acknowledged before his people the godhood of Lieutenant Erich Obergatz, nor was it long before the story ran like wildfire through the palace and out into the city and beyond that to the lesser villages all the way from A-lur to Tu-lur.

The real god had come—Jad-ben-Otho himself, and he had espoused the cause of Lu-don, the high priest. Mo-sar lost no time in placing himself at the disposal of Lu-don, nor did he mention aught about his claims to the throne. It was Mo-sar's opinion that he might consider himself fortunate were he allowed to remain in peaceful occupation of his chieftain-ship at Tu-lur, nor was Mo-sar wrong in his deductions.

But Lu-don could still use him and so he let him live and sent word to him to come to A-lur with all his warriors, for it was rumored that Ja-don was raising a great army in the north and might soon march upon the City of Light.

Obergatz thoroughly enjoyed being a god. Plenty of food and peace of mind and rest partially brought back to him the reason that had been so rapidly slipping from him; but in one respect he was madder than ever, since now no power on earth would ever be able to convince him that he was not a god. Slaves were put at his disposal and these he ordered about in godly fashion. The sane portion of his naturally cruel mind met upon common ground the mind of Lu-don, so that the two seemed always in accord. The high priest saw in the stranger a mighty force wherewith to hold forever his power over all Pal-ul-don and thus the future of Obergatz was assured so long as he cared to play god to Lu-don's high priest.

A throne was erected in the main temple court before the eastern altar where Jad-ben-Otho might sit in person and behold the sacrifices that were offered up to him there each day at sunset. So much did the cruel, half-crazed mind enjoy these spectacles that at times he even insisted upon wielding the sacrificial knife himself and upon such occasions the priests and the people fell upon their faces in awe of the dread deity.

If Obergatz taught them not to love their god more he taught them to fear him as they never had before, so that the name of Jad-ben-Otho was whispered in the city and little children were frightened into obedience by the mere mention of it. Lu-don, through his priests and slaves, circulated the information that Jad-ben-Otho had commanded all his faithful followers to flock to the standard of the high priest at A-lur and that all others were cursed, especially Ja-don and the base impostor who had posed as the Dor-ul-Otho. The curse was to take the form of early death following terrible suffering, and Lu-don caused it to be published abroad that the name of any warrior who complained of a pain should be brought to him, for such might be deemed to be under suspicion, since the first effects of the curse would result in slight pains attacking the unholy. He counseled those who felt pains to look carefully to their loyalty. The result was remarkable and immediate—half a nation without a pain, and recruits pouring into A-lur to offer their services to Lu-don while secretly hoping that the little pains they had felt in arm or leg or belly would not recur in aggravated form.

CHAPTER XXII

A Journey on a Gryf

TARZAN and Jane skirted the shore of Jad-bal-lul and crossed the river at the head of the lake. They moved in leisurely fashion with an eye to comfort and safety, for the ape-man, now that he had found his mate, was determined to court no chance that might again separate them, or delay or prevent their escape from Pal-ul-don. How they were to re-cross the morass was a matter of little concern to him as yet—it would be time enough to consider that matter when it became of more immediate moment. Their hours were filled with the happiness and content of reunion after long separation; they had much to talk of, for each had passed through many trials and vicissitudes and strange adventures, and no important hour might go unaccounted for since last they met.

It was Tarzan's intention to choose a way above A-lur and the scattered Ho-don villages below it, passing about midway between them and the mountains, thus avoiding, in so far as possible, both the Ho-don and Waz-don, for in this area lay the neutral territory that was uninhabited by either. Thus he would travel northwest until opposite the Kor-ul-ja where he planned to stop to pay his respects to Om-at and give the *gund* word of Pan-at-lee, and a plan Tarzan had for insuring her safe return to her people. It was upon the third day of their journey and they had almost reached the river that passes through A-lur when Jane suddenly clutched Tarzan's arm and pointed ahead toward the edge of a forest that they were approaching. Beneath the shadows of the trees loomed a great bulk that the ape-man instantly recognized.

"What is it?" whispered Jane.

"A *gryf*," replied the ape-man, "and we have met him in the worst place that we could possibly have found. There is not a large tree within a quarter of a mile, other than those among which he stands. Come, we shall have to go back, Jane; I cannot risk it with you along. The best we can do is to pray that he does not discover us."

"And if he does?"

"Then I shall have to risk it."

"Risk what?"

"The chance that I can subdue him as I subdued one of his fellows," replied Tarzan. "I told you—you recall?"

"Yes, but I did not picture so huge a creature. Why, John, he is as big as a battleship."

The ape-man laughed. "Not quite, though I'll admit he looks quite as formidable as one when he charges."

They were moving away slowly so as not to attract the attention of the beast.

"I believe we're going to make it," whispered the woman, her voice tense with suppressed excitement. A low rumble rolled like distant thunder from the wood. Tarzan shook his head.

" 'The big show is about to commence in the main tent,' " he quoted, grinning. He caught the woman suddenly to his breast and kissed her. "One can never tell, Jane," he said. "We'll do our best—that is all we can do. Give me your spear, and—don't run. The only hope we have lies in that little brain more than in us. If I can control it—well, let us see."

The beast had emerged from the forest and was looking about through his weak eyes, evidently in search of them.

Tarzan raised his voice in the weird notes of the Tor-o-don's cry; "Whee-oo! Whee-oo! Whee-oo!" For a moment the great beast stood motionless, his attention rivetted by the call. The ape-man advanced straight toward him, Jane Clayton at his elbow. "Whee-oo!" he cried again peremptorily. A low rumble rolled from the *gryf's* cavernous chest in answer to the call, and the beast moved slowly toward them.

"Fine!" exclaimed Tarzan. "The odds are in our favor now. You can keep your nerve?—but I do not need to ask."

"I know no fear when I am with Tarzan of the Apes," she replied softly, and he felt the pressure of her soft fingers on his arm.

And thus the two approached the giant monster of a forgotten epoch until they stood close in the shadow of a mighty shoulder. "Whee-oo!" shouted Tarzan and struck the hideous snout with the shaft of the spear. The vicious side snap that did not reach its mark—that evidently was not intended to reach its mark—was the hoped-for answer.

"Come," said Tarzan, and taking Jane by the hand he led her around behind the monster and up the broad tail to the great, horned back. "Now will we ride in the state that our forebears knew, before which the pomp of modern kings pales into cheap and tawdry insignificance. How would you like to canter through Hyde Park on a mount like this?"

"I am afraid the Bobbies would be shocked by our riding habits, John," she cried, laughingly.

Tarzan guided the *gryf* in the direction that they wished to go. Steep embankments and rivers proved no slightest obstacle to the ponderous creature.

"A prehistoric tank, this," Jane assured him, and laughing and talking they continued on their way. Once they came unexpectedly upon a dozen Ho-don warriors as the *gryf* emerged suddenly into a small clearing. The fellows were lying about in the shade of a single tree that grew alone. When they saw the beast they leaped to their feet in consternation and at their shouts the *gryf* issued his hideous, challenging bellow and charged them. The warriors fled in all directions while Tarzan belabored the beast across the snout with his spear in an effort to control him, and at last he succeeded, just as the *gryf* was almost upon one poor devil that it seemed to have singled out for its special prey. With an angry grunt the *gryf* stopped and the man, with a single backward glance that showed a face white with terror, disappeared in the jungle he had been seeking to reach.

The ape-man was elated. He had doubted that he could control the beast should it take it into its head to charge a victim and had intended abandoning it before they reached the Kor-ul-ja. Now he altered his plans—they would ride to the very village of Om-at upon the *gryf*, and the Kor-ul-ja would have food for conversation for many generations to come. Nor was it the theatric instinct of the ape-man alone that gave favor to this plan. The element of Jane's safety entered into the matter for he knew that she would be safe from man and beast alike so long as she rode upon the back of Pal-ul-don's most formidable creature.

As they proceeded slowly in the direction of the Kor-ul-ja, for the natural gait of the *gryf* is far from rapid, a handful of terrified warriors came panting into A-lur, spreading a weird story of the Dor-ul-Otho, only none dared call him the Dor-ul-Otho aloud. Instead they spoke of him as Tarzan-jad-guru and they told of meeting him mounted upon a mighty *gryf* beside the beautiful stranger woman whom Ko-tan would have made queen of Pal-ul-don. This story was brought to Lu-don who caused the warriors to be hailed to his presence, when he questioned them closely until finally he was convinced that they spoke the truth and when they had told him the direction in which the two were traveling, Lu-don guessed that they were on their way to Ja-lur to join Ja-don, a contingency that he felt must be prevented at any cost. As was his wont in the stress of emergency, he called Pan-sat into consultation and for long the two sat in close conference. When they arose a plan had been developed. Pan-sat went immediately to his own quarters where he removed the headdress and trappings of a priest to don in their stead the harness and weapons of a warrior. Then he returned to Lu-don.

"Good!" cried the latter, when he saw him. "Not even your fellow-priests or the slaves that wait upon you daily would know you now. Lose no time, Pan-sat, for all depends upon the speed with which you strike and—remember! Kill the man if you can; but in any event bring the woman to me here, alive. You understand?"

"Yes, master," replied the priest, and so it was that a lone warrior set out from A-lur and made his way northwest in the direction of Ja-lur.

The gorge next above Kor-ul-ja is uninhabited and here the wily Ja-don had chosen to mobilize his army for its descent upon A-lur. Two considerations influenced him—one being the fact that could he keep his plans a secret from the enemy he would have the advantage of delivering a surprise attack upon the forces of Lu-don from a direction that they would not expect attack, and in the meantime he would be able to keep his men from the gossip of the cities where strange tales were already circulating relative to the coming of Jad-ben-Otho in person to aid the high priest in his war against Ja-don. It took stout hearts and loyal ones to ignore the implied threats of divine vengeance that these tales suggested. Already there had been desertions and the cause of Ja-don seemed tottering to destruction.

Such was the state of affairs when a sentry posted on the knoll in the mouth of the gorge sent word that he had observed in the valley below what appeared at a distance to be nothing less than two people mounted upon the back of a *gryf*. He said that he had caught glimpses of them, as they passed open spaces, and they seemed to be traveling up the river in the direction of the Kor-ul-ja.

At first Ja-don was inclined to doubt the veracity of his informant; but, like all good generals, he could not permit even palpably false information to go uninvestigated and so he determined to visit the knoll himself and learn precisely what it was that the sentry had observed through the distorting spectacles of fear. He had scarce taken his place beside the man ere the fellow touched his arm and pointed. "They are closer now," he whispered, "you can see them plainly." And sure enough, not a quarter of a mile away Ja-don saw that which in his long experience in Pal-ul-don he had never before seen—two humans riding upon the broad back of a *gryf*.

At first he could scarce credit even this testimony of his own eyes, but soon he realized that the creatures below could be naught else than they appeared, and then he recognized the man and rose to his feet with a loud cry.

"It is he!" he shouted to those about him. "It is the Dor-ul-Otho himself."

The *gryf* and his riders heard the shout though not the words. The former bellowed terrifically and started in the direction of the knoll, and Ja-don, followed by a few of his more intrepid warriors, ran to meet him. Tarzan, loath to enter an unnecessary quarrel, tried to turn the animal, but as the beast was far from tractable it always took a few minutes to force the will of its master upon it; and so the two parties were quite close before the ape-man succeeded in stopping the mad charge of his furious mount.

Ja-don and his warriors, however, had come to the realization that this bellowing creature was bearing down upon them with evil intent and they had assumed the better part of valor and taken to trees, accordingly. It was beneath these trees that Tarzan finally stopped the *gryf*. Ja-don called down to him.

"We are friends," he cried. "I am Ja-don, Chief of Ja-lur. I and my warriors lay our foreheads upon the feet of Dor-ul-Otho and pray that he will aid us in our righteous fight with Lu-don, the high priest."

"You have not defeated him yet?" asked Tarzan. "Why I thought you would be king of Pal-ul-don long before this."

"No," replied Ja-don. "The people fear the high priest and now that he has in the temple one whom he claims to be Jad-ben-Otho many of my warriors are afraid. If they but knew that the Dor-ul-Otho had returned and that he had blessed the cause of Ja-don I am sure that victory would be ours."

Tarzan thought for a long minute and then he spoke. "Ja-don," he said, "was one of the few who believed in me and who wished to accord me fair treatment. I have a debt to pay to Ja-don and an account to settle with Lu-don, not alone on my own behalf, but principally upon that of my mate. I will go with you Ja-don to mete to Lu-don the punishment he deserves. Tell me, chief, how may the Dor-ul-Otho best serve his father's people?"

"By coming with me to Ja-lur and the villages between," replied Ja-don quickly, "that the people may see that it is indeed the Dor-ul-Otho and that he smiles upon the cause of Ja-don."

"You think that they will believe in me more now than before?" asked the ape-man.

"Who will dare doubt that he who rides upon the great *gryf* is less than a god?" returned the old chief.

"And if I go with you to the battle at A-lur," asked Tarzan, "can you assure the safety of my mate while I am gone from her?"

"She shall remain in Ja-lur with the Princess O-lo-a and my own women," replied Ja-don. "There she will be safe for there I shall leave trusted warriors to protect them. Say that you will come, O Dor-ul-Otho, and my cup of happiness will be full, for even now Ta-den, my son, marches toward A-lur with a force from the northwest and if we can attack, with the Dor-ul-Otho at our head, from the northeast our arms should be victorious."

"It shall be as you wish, Ja-don," replied the ape-man; "but first you must have meat fetched for my *gryf*."

"There are many carcasses in the camp above," replied Ja-don, "for my men have little else to do than hunt."

"Good," exclaimed Tarzan. "Have them brought at once."

And when the meat was brought and laid at a distance the ape-man slipped from the back of his fierce charger and fed him with his own hand. "See that there is always plenty of flesh for him," he said to Ja-don, for he guessed that his mastery might be short-lived should the vicious beast become over-hungry.

It was morning before they could leave for Ja-lur, but Tarzan found the *gryf* lying where he had left him the night before beside the carcasses of two antelope and a lion; but now there was nothing but the *gryf*.

"The paleontologists say that he was herbivorous," said Tarzan as he and Jane approached the beast.

The journey to Ja-lur was made through the scattered villages where Ja-don hoped to arouse a keener enthusiasm for his cause. A party of warriors preceded Tarzan that the people might properly be prepared, not only for the sight of the *gryf* but to receive the Dor-ul-Otho as became his high station. The results were all that Ja-don could have hoped and in no village through which they passed was there one who doubted the deity of the ape-man.

As they approached Ja-lur a strange warrior joined them, one whom none of Ja-don's following knew. He said he came from one of the villages to the south and that he had been treated unfairly by one of Lu-don's chiefs. For this reason he had deserted the cause of the high priest and come north in the hope of finding a home in Ja-lur. As every addition to his forces was welcome to the old chief he permitted the stranger to accompany them, and so he came into Ja-lur with them.

There arose now the question as to what was to be done with the *gryf* while they remained in the city. It was with difficulty that Tarzan had prevented the savage beast from attacking all who came near it when they had first entered the camp of Ja-don in the uninhabited gorge next to the Kor-ul-ja, but during the march to Ja-lur the creature had seemed to become accustomed to the presence of the Ho-don. The latter, however, gave him no cause for annoyance since they kept as far from him as possible and when he passed through the streets of the city he was viewed from the safety of lofty windows and roofs. However tractable he appeared to have become there would have been no enthusiastic seconding of a suggestion to turn him loose within the city. It was finally suggested that he be turned into a walled enclosure within the palace grounds and this was done, Tarzan driving him in after Jane had dismounted. More meat was thrown to him and he was left to his own devices, the awe-struck inhabitants of the palace not even venturing to climb upon the walls to look at him.

Ja-don led Tarzan and Jane to the quarters of the Princess O-lo-a who, the moment that she beheld the ape-man, threw herself to the ground and touched her forehead to his feet. Pan-at-lee was there with her and she too seemed happy to see Tarzan-jad-guru again. When they found that Jane was his mate they looked with almost equal awe upon her, since even the most sceptical of the warriors of Ja-don were now convinced that they were entertaining a god and a goddess within the city of Ja-lur, and that with the assistance of the power of these two, the cause of Ja-don would soon be victorious and the old Lion-man set upon the throne of Pal-ul-don.

From O-lo-a Tarzan learned that Ta-den had returned and that they were to be united in marriage with the weird rites of their religion and in accordance with the custom of their people as soon as Ta-den came home from the battle that was to be fought at A-lur.

The recruits were now gathering at the city and it was decided that the next day Ja-don and Tarzan would return to the main body in the hidden camp and immediately under cover of night the attack should be made in force upon Ludon's forces at A-lur. Word of this was sent to Ta-den where he awaited with his warriors upon the north side of Jad-ben-lul, only a few miles from A-lur.

In the carrying out of these plans it was necessary to leave Jane behind in Ja-don's palace at Ja-lur, but O-lo-a and her women were with her and there were many warriors to guard them, so Tarzan bid his mate good-bye with no feelings of apprehension as to her safety, and again seated upon the *gryf* made his way out of the city with Ja-don and his warriors.

At the mouth of the gorge the ape-man abandoned his huge

mount since it had served its purpose and could be of no further value to him in their attack upon A-lur, which was to be made just before dawn the following day when, as he could not have been seen by the enemy, the effect of his entry to the city upon the *gryf* would have been totally lost. A couple of sharp blows with the spear sent the big animal rumbling and growling in the direction of the Kor-ul-gryf nor was the ape-man sorry to see it depart since he had never known at what instant its short temper and insatiable appetite for flesh might turn it upon some of his companions.

Immediately upon their arrival at the gorge the march on A-lur was commenced.

CHAPTER XXIII

Taken Alive

As NIGHT fell a warrior from the palace of Ja-lur slipped into the temple grounds. He made his way to where the lesser priests were quartered. His presence aroused no suspicion as it was not unusual for warriors to have business within the temple. He came at last to a chamber where several priests were congregated after the evening meal. The rites and ceremonies of the sacrifice had been concluded and there was nothing more of a religious nature to make call upon their time until the rites at sunrise.

Now the warrior knew, as in fact nearly all Pal-ul-don knew, that there was no strong bond between the temple and the palace at Ja-lur and that Ja-don only suffered the presence of the priests and permitted their cruel and abhorrent acts because of the fact that these things had been the custom of the Ho-don of Pal-ul-don for countless ages, and rash indeed must have been the man who would have attempted to interfere with the priests or their ceremonies. That Ja-don never entered the temple was well known, and that his high priest never entered the palace, but the people came to the temple with their votive offerings and the sacrifices were made night and morning as in every other temple in Pal-ul-don.

The warrior knew these things, knew them better perhaps

than a simple warrior should have known them. And so it was here in the temple that he looked for the aid that he sought in the carrying out of whatever design he had.

As he entered the apartment where the priests were he greeted them after the manner which was customary in Pal-ul-don, but at the same time he made a sign with his finger that might have attracted little attention or scarcely been noticed at all by one who knew not its meaning. That there were those within the room who noticed it and interpreted it was quickly apparent, through the fact that two of the priests rose and came close to him as he stood just within the doorway and each of them, as he came, returned the signal that the warrior had made.

The three talked for but a moment and then the warrior turned and left the apartment. A little later one of the priests who had talked with him left also and shortly after that the other.

In the corridor they found the warrior waiting, and led him to a little chamber which opened upon a smaller corridor just beyond where it joined the larger. Here the three remained in whispered conversation for some little time and then the warrior returned to the palace and the two priests to their quarters.

The apartments of the women of the palace at Ja-lur are all upon the same side of a long, straight corridor. Each has a single door leading into the corridor and at the opposite end several windows overlooking a garden. It was in one of these rooms that Jane slept alone. At each end of the corridor was a sentinel, the main body of the guard being stationed in a room near the outer entrance to the women's quarters.

The palace slept for they kept early hours there where Ja-don ruled. The *pal-e-don-so* of the great chieftain of the north knew no such wild orgies as had resounded through the palace of the king at A-lur. Ja-lur was a quiet city by comparison with the capital, yet there was always a guard kept at every entrance to the chambers of Ja-don and his immediate family as well as at the gate leading into the temple and that which opened upon the city.

These guards, however, were small, consisting usually of not more than five or six warriors, one of whom remained awake while the others slept. Such were the conditions then when two warriors presented themselves, one at either end of the corridor, to the sentries who watched over the safety of Jane Clayton and the Princess O-lo-a, and each of the newcomers repeated to the sentinels the stereotyped words which announced that they were relieved and these others sent to watch in their stead. Never is a warrior loath to be relieved of sentry duty. Where, under different circumstances he might ask numerous questions he is now too well satisfied to escape the monotonies of that universally hated duty. And so these two men accepted their relief without question and hastened away to their pallets.

And then a third warrior entered the corridor and all of the newcomers came together before the door of the ape-man's slumbering mate. And one was the strange warrior who had met Ja-don and Tarzan outside the city of Ja-lur as they had approached it the previous day; and he was the same warrior who had entered the temple a short hour before, but the faces of his fellows were unfamiliar, even to one another, since it is seldom that a priest removes his hideous headdress in the presence even of his associates.

Silently they lifted the hangings that hid the interior of the room from the view of those who passed through the corridor, and stealthily slunk within. Upon a pile of furs in a far corner lay the sleeping form of Lady Greystoke. The bare feet of the intruders gave forth no sound as they crossed the stone floor toward her. A ray of moonlight entering through a window near her couch shone full upon her, revealing the beautiful contours of an arm and shoulder in cameo-distinctness against the dark furry pelt beneath which she slept, and the perfect profile that was turned toward the skulking three.

But neither the beauty nor the helplessness of the sleeper aroused such sentiments of passion or pity as might stir in the breasts of normal men. To the three priests she was but a lump of clay, nor could they conceive aught of that passion which had aroused men to intrigue and to murder for possession of this beautiful American girl, and which even now was influencing the destiny of undiscovered Pal-ul-don.

Upon the floor of the chamber were numerous pelts and as the leader of the trio came close to the sleeping woman he stooped and gathered up one of the smaller of these. Standing close to her head he held the rug outspread above her face. "Now," he whispered and simultaneously he threw the rug over the woman's head and his two fellows leaped upon her, seizing her arms and pinioning her body while their leader stifled her cries with the furry pelt. Quickly and silently they bound her wrists and gagged her and during the brief time that their work required there was no sound that might have been heard by occupants of the adjoining apartments.

Jerking her roughly to her feet they forced her toward a window but she refused to walk, throwing herself instead upon the floor. They were very angry and would have resorted to cruelties to compel her obedience but dared not, since the wrath of Lu-don might fall heavily upon whoever mutilated his fair prize.

And so they were forced to lift and carry her bodily. Nor was the task any sinecure since the captive kicked and struggled as best she might, making their labor as arduous as possible. But finally they succeeded in getting her through the window and into the garden beyond where one of the two priests from the Ja-lur temple directed their steps toward a small barred gateway in the south wall of the enclosure.

Immediately beyond this a flight of stone stairs led downward toward the river and at the foot of the stairs were moored several canoes. Pan-sat had indeed been fortunate in enlisting aid from those who knew the temple and the palace so well, or otherwise he might never have escaped from Ja-lur with his captive. Placing the woman in the bottom of a light canoe Pan-sat entered it and took up the paddle. His companions unfastened the moorings and shoved the little craft out into the current of the stream. Their traitorous work completed they turned and retraced their steps toward the temple, while Pan-sat, paddling strongly with the current, moved rapidly down the river that would carry him to the Jad-ben-lul and A-lur.

The moon had set and the eastern horizon still gave no hint of approaching day as a long file of warriors wound stealthily through the darkness into the city of A-lur. Their plans were all laid and there seemed no likelihood of their miscarriage. A messenger had been dispatched to Ta-den whose forces lay northwest of the city. Tarzan, with a small contingent, was to enter the temple through the secret passageway, the location of which he alone knew, while Ja-don, with the greater proportion of the warriors, was to attack the palace gates.

The ape-man, leading his little band, moved stealthily through the winding alleys of A-lur, arriving undetected at the building which hid the entrance to the secret passageway. This spot being best protected by the fact that its existence was unknown to others than the priests, was unguarded. To facilitate the passage of his little company through the narrow winding, uneven tunnel, Tarzan lighted a torch which had been brought for the purpose and preceding his warriors led the way toward the temple.

That he could accomplish much once he reached the inner chambers of the temple with his little band of picked warriors the ape-man was confident since an attack at this point would bring confusion and consternation to the easily overpowered priests, and permit Tarzan to attack the palace forces in the rear at the same time that Ja-don engaged them at the palace gates, while Ta-den and his forces swarmed the northern walls. Great value had been placed by Ja-don on the moral effect of the Dor-ul-Otho's mysterious appearance in the heart of the temple and he had urged Tarzan to take every advantage of the old chieftain's belief that many of Lu-don's warriors still wavered in their allegiance between the high priest and the Dor-ul-Otho, being held to the former more by the fear which he engendered in the breasts of all his followers than by any love or loyalty they might feel toward him.

There is a Pal-ul-donian proverb setting forth a truth similar to that contained in the old Scotch adage that "The best laid schemes o' mice and men gang aft a-gley." Freely translated it might read, "He who follows the right trail sometimes reaches the wrong destination," and such apparently was the fate that lay in the footsteps of the great chieftain of the north and his godlike ally.

Tarzan, more familiar with the windings of the corridors than his fellows and having the advantage of the full light of the torch, which at best was but a dim and flickering affair, was some distance ahead of the others, and in his keen anxiety to close with the enemy he gave too little thought to those who were to support him. Nor is this strange, since from childhood the ape-man had been accustomed to fight the battles of life single-handed so that it had become habitual for him to depend solely upon his own cunning and prowess.

And so it was that he came into the upper corridor from which opened the chambers of Lu-don and the lesser priests.

far in advance of his warriors, and as he turned into this corridor with its dim cressets flickering somberly, he saw another
enter it from a corridor before him—a warrior half carrying,
half dragging the figure of a woman. Instantly Tarzan recognized the gagged and fettered captive whom he had thought
safe in the palace of Ja-don at Ja-lur.

The warrior with the woman had seen Tarzan at the same
instant that the latter had discovered him. He heard the low
beastlike growl that broke from the ape-man's lips as he sprang
forward to wrest his mate from her captor and wreak upon him
the vengeance that was in the Tarmangani's savage heart.
Across the corridor from Pan-sat was the entrance to a smaller
chamber. Into this he leaped carrying the woman with
him.

Close behind came Tarzan of the Apes. He had cast aside
his torch and drawn the long knife that had been his father's.
With the impetuosity of a charging bull he rushed into the
chamber in pursuit of Pan-sat to find himself, when the hangings dropped behind him, in utter darkness. Almost immediately there was a crash of stone on stone before him followed
a moment later by a similar crash behind. No other evidence
was necessary to announce to the ape-man that he was again a
prisoner in Lu-don's temple.

He stood perfectly still where he had halted at the first sound
of the descending stone door. Not again would he easily be
precipitated to the *gryf* pit, or some similar danger, as had occurred when Lu-don had trapped him in the Temple of the
Gryf. As he stood there his eyes slowly grew accustomed to
the darkness and he became aware that a dim light was entering the chamber through some opening, though it was several
minutes before he discovered its source. In the roof of the
chamber he finally discerned a small aperture, possibly three
feet in diameter and it was through this that what was really

only a lesser darkness rather than a light was penetrating its Stygian blackness of the chamber in which he was imprisoned.

Since the doors had fallen he had heard no sound though his keen ears were constantly strained in an effort to discover a clue to the direction taken by the abductor of his mate. Presently he could discern the outlines of his prison cell. It was a small room, not over fifteen feet across. On hands and knees, with the utmost caution, he examined the entire area of the floor. In the exact center, directly beneath the opening in the roof, was a trap, but otherwise the floor was solid. With this knowledge it was only necessary to avoid this spot in so far as the floor was concerned. The walls next received his attention. There were only two openings. One the doorway through which he had entered, and upon the opposite side that through which the warrior had borne Jane Clayton. These were both closed by the slabs of stone which the fleeing warrior had released as he departed.

Lu-don, the high priest, licked his thin lips and rubbed his bony white hands together in gratification as Pan-sat bore Jane Clayton into his presence and laid her on the floor of the chamber before him.

"Good, Pan-sat!" he exclaimed. "You shall be well rewarded for this service. Now, if we but had the false Dor-ul-Otho in our power all Pal-ul-don would be at our feet."

"Master, I have him!" cried Pan-sat.

"What!" exclaimed Lu-don, "you have Tarzan-jad-guru? You have slain him perhaps. Tell me, my wonderful Pan-sat, tell me quickly. My breast is bursting with a desire to know."

"I have taken him alive, Lu-don, my master," replied Pan-sat. "He is in the little chamber that the ancients built to trap those who were too powerful to take alive in personal encounter."

"You have done well, Pan-sat, I——"

A frightened priest burst into the apartment. "Quick, master, quick," he cried, "the corridors are filled with the warriors of Ja-don."

"You are mad," cried the high priest. "My warriors hold the palace and the temple."

"I speak the truth, master," replied the priest, "there are warriors in the corridor approaching this very chamber, and they come from the direction of the secret passage which leads hither from the city."

"It may be even as he says," exclaimed Pan-sat. "It was from that direction that Tarzan-jad-guru was coming when I discovered and trapped him. He was leading his warriors to the very holy of holies."

Lu-don ran quickly to the doorway and looked out into the corridor. At a glance he saw that the fears of the frightened priest were well founded. A dozen warriors were moving along the corridor toward him but they seemed confused and far from sure of themselves. The high priest guessed that deprived of the leadership of Tarzan they were little better than lost in the unknown mazes of the subterranean precincts of the temple.

Stepping back into the apartment he seized a leathern thong that depended from the ceiling. He pulled upon it sharply and through the temple boomed the deep tones of a metal gong. Five times the clanging notes rang through the corridors, then he turned toward the two priests. "Bring the woman and follow me," he directed.

Crossing the chamber he passed through a small doorway, the others lifting Jane Clayton from the floor and following him. Through a narrow corridor and up a flight of steps they went, turning to right and left and doubling back through a maze of winding passageways which terminated in a spiral staircase that gave forth at the surface of the ground within the largest of the inner altar courts close beside the eastern altar.

From all directions now, in the corridors below and the grounds above, came the sound of hurrying footsteps. The five strokes of the great gong had summoned the faithful to the defense of Lu-don in his private chambers. The priests who knew the way led the less familiar warriors to the spot and presently those who had accompanied Tarzan found themselves not only leaderless but facing a vastly superior force. They were brave men but under the circumstances they were helpless and so they fell back the way they had come, and when they reached the narrow confines of the smaller passageway their safety was assured since only one foeman could attack them at a time. But their plans were frustrated and possibly also their entire cause lost, so heavily had Ja-don banked upon the success of their venture.

With the clanging of the temple gong Ja-don assumed that Tarzan and his party had struck their initial blow and so he launched his attack upon the palace gate. To the ears of Lu-don in the inner temple court came the savage war cries that announced the beginning of the battle. Leaving Pan-sat and the other priest to guard the woman he hastened toward the palace personally to direct his force and as he passed through the temple grounds he dispatched a messenger to learn the outcome of the fight in the corridors below, and other messengers to spread the news among his followers that the false Dor-ul-Otho was a prisoner in the temple.

As the din of battle rose above A-lur, Lieutenant Erich Obergatz turned upon his bed of soft hides and sat up. He rubbed his eyes and looked about him. It was still dark without.

"I am Jad-ben-Otho," he cried, "who dares disturb my slumber?"

A slave squatting upon the floor at the foot of his couch shuddered and touched her forehead to the floor. "It must be that the enemy have come, O Jad-ben-Otho." She spoke soothingly for she had reason to know the terrors of the mad frenzy into which trivial things sometimes threw the Great God.

A priest burst suddenly through the hangings of the doorway and falling upon his hands and knees rubbed his forehead against the stone flagging. "O Jad-ben-Otho," he cried, "the warriors of Ja-don have attacked the palace and the temple. Even now they are fighting in the corridors near the quarters of Lu-don, and the high priest begs that you come to the palace and encourage your faithful warriors by your presence."

Obergatz sprang to his feet. "I am Jad-ben-Otho," he screamed. "With lightning I will blast the blasphemers who dare attack the holy city of A-lur."

For a moment he rushed aimlessly and madly about the room, while the priest and the slave remained upon hands and knees with their foreheads against the floor.

"Come," cried Obergatz, planting a vicious kick in the side of the slave girl. "Come! Would you wait here all day while the forces of darkness overwhelm the City of Light?"

Thoroughly frightened as were all those who were forced to serve the Great God, the two arose and followed Obergatz towards the palace.

Above the shouting of the warriors rose constantly the cries of the temple priests: "Jad-ben-Otho is here and the false Dor-ul-Otho is a prisoner in the temple." The persistent cries reached even to the ears of the enemy as it was intended that they should.

CHAPTER XXIV

The Messenger of Death

THE sun rose to see the forces of Ja-don still held at the palace gate. The old warrior had seized the tall structure that
stood just beyond the palace and at the summit of this he kept
a warrior stationed to look toward the northern wall of the palace where Ta-den was to make his attack; but as the minutes
wore into hours no sign of the other force appeared, and now
in the full light of the new sun upon the roof of one of the
palace buildings appeared Lu-don, the high priest, Mo-sar, the
pretender, and the strange, naked figure of a man, into whose
long hair and beard were woven fresh ferns and flowers. Behind them were banked a score of lesser priests who chanted
in unison: "This is Jad-ben-Otho. Lay down your arms and
surrender." This they repeated again and again, alternating it
with the cry: "The false Dor-ul-Otho is a prisoner."

In one of those lulls which are common in battles between
forces armed with weapons that require great physical effort
in their use, a voice suddenly arose from among the followers
of Ja-don: "Show us the Dor-ul-Otho. We do not believe you!"

"Wait," cried Lu-don. "If I do not produce him before the
sun has moved his own width, the gates of the palace shall be
opened to you and my warriors will lay down their arms."

He turned to one of his priests and issued brief instructions.

The ape-man paced the confines of his narrow cell. Bitterly

he reproached himself for the stupidity which had led him into this trap, and yet was it stupidity? What else might he have done other than rush to the succor of his mate? He wondered how they had stolen her from Ja-lur, and then suddenly there flashed to his mind the features of the warrior whom he had just seen with her. They were strangely familiar. He racked his brain to recall where he had seen the man before and then it came to him. He was the strange warrior who had joined Ja-don's forces outside of Ja-lur the day that Tarzan had ridden upon the great *gryf* from the uninhabited gorge next to the Kor-ul-ja down to the capital city of the chieftain of the north. But who could the man be? Tarzan knew that never before that other day had he seen him.

Presently he heard the clanging of a gong from the corridor without and very faintly the rush of feet, and shouts. He guessed that his warriors had been discovered and a fight was in progress. He fretted and chafed at the chance that had denied him participation in it.

Again and again he tried the doors of his prison and the trap in the center of the floor, but none would give to his utmost endeavors. He strained his eyes toward the aperture above but he could see nothing, and then he continued his futile pacing to and fro like a caged lion behind its bars.

The minutes dragged slowly into hours. Faintly sounds came to him as of shouting men at a great distance. The battle was in progress. He wondered if Ja-don would be victorious and should he be, would his friends ever discover him in this hidden chamber in the bowels of the hill? He doubted it.

And now as he looked again toward the aperture in the roof there appeared to be something depending through its center. He came closer and strained his eyes to see. Yes, there was something there. It appeared to be a rope. Tarzan wondered if it had been there all the time. It must have, he reasoned, since he had heard no sound from above and it was so dark within the chamber that he might easily have overlooked it.

He raised his hand toward it. The end of it was just within his reach. He bore his weight upon it to see if it would hold him. Then he released it and backed away, still watching it, as you have seen an animal do after investigating some unfamiliar object, one of the little traits that differentiated Tarzan from other men, accentuating his similarity to the savage beasts of his native jungle. Again and again he touched and tested the braided leather rope, and always he listened for any warning sound from above.

He was very careful not to step upon the trap at any time and when finally he bore all his weight upon the rope and took his feet from the floor he spread them wide apart so that if he fell he would fall astride the trap. The rope held him. There was no sound from above, nor any from the trap below.

Slowly and cautiously he drew himself upward, hand over hand. Nearer and nearer the roof he came. In a moment his eyes would be above the level of the floor above. Already his extended arms projected into the upper chamber and then something closed suddenly upon both his forearms, pinioning them tightly and leaving him hanging in mid-air unable to advance or retreat.

Immediately a light appeared in the room above him and presently he saw the hiedous mask of a priest peering down upon him. In the priest's hands were leathern thongs and these he tied about Tarzan's wrists and forearms until they were completely bound together from his elbows almost to his fingers. Behind this priest Tarzan presently saw others and soon several lay hold of him and pulled him up through the hole.

Almost instantly his eyes were above the level of the floor he understood how they had trapped him. Two nooses had lain encircling the aperture into the cell below. A priest had waited at the end of each of these ropes and at opposite sides of the chamber. When he had climbed to a sufficient height upon the rope that had dangled into his prison below and his arms were well within the encircling snares the two priests had pulled quickly upon their ropes and he had been made an easy captive without any opportunity of defending himself or inflicting injury upon his captors.

And now they bound his legs from his ankles to his knees and picking him up carried him from the chamber. No word did they speak to him as they bore him upward to the temple yard.

The din of battle had risen again as Ja-don had urged his forces to renewed efforts. Ta-den had not arrived and the forces of the old chieftain were revealing in their lessened efforts their increasing demoralization, and then it was that the priests carried Tarzan-jad-guru to the roof of the palace and exhibited him in the sight of the warriors of both factions.

"Here is the false Dor-ul-Otho," screamed Lu-don.

Obergatz, his shattered mentality having never grasped fully the meaning of much that was going on about him, cast a casual glance at the bound and helpless prisoner, and as his eyes fell upon the noble features of the ape-man, they went wide in astonishment and fright, and his pasty countenance turned a sickly blue. Once before had he seen Tarzan of the Apes, but many times had he dreamed that he had seen him and always was the giant ape-man avenging the wrongs that had been committed upon him and his by the ruthless hands of the three German officers who had led their native troops in the ravishing of Tarzan's peaceful home. Hauptmann Fritz Schneider had paid the penalty of his needless cruelties; Unter-lieutenant von Goss, too, had paid; and now Obergatz, the last of the three, stood face to face with the Nemesis that had trailed him through his dreams for long, weary months. That he was bound and helpless lessened not the German's terror—he seemed not to realize that the man could not harm him. He but stood cringing and jibbering and Lu-don saw and was filled with apprehension that others might see and seeing realize that this bewhiskered idiot was no god—that of the two Tarzan-jad-guru was the more godly figure. Already the high priest noted that some of the palace warriors standing near were whispering together and pointing. He stepped closer to Obergatz. "You are Jad-ben-Otho," he whispered, "denounce him!"

The German shook himself. His mind cleared of all but his great terror and the words of the high priest gave him the clue to safety.

"I am Jad-ben-Otho!" he screamed.

Tarzan looked him straight in the eye. "You are Lieutenant Obergatz of the German Army," he said in excellent German. "You are the last of the three I have sought so long and in your putrid heart you know that God has not brought us together at last for nothing."

The mind of Lieutenant Obergatz was functioning clearly and rapidly at last. He too saw the questioning looks upon the faces of some of those around them. He saw the opposing warriors of both cities standing by the gate inactive, every eye turned upon him, and the trussed figure of the ape-man. He realized that indecision now meant ruin, and ruin, death. He raised his voice in the sharp barking tones of a Prussian officer, so unlike his former maniacal screaming as to quickly arouse the attention of every ear and to cause an expression of puzzlement to cross the crafty face of Lu-don.

"I am Jad-ben-Otho," snapped Obergatz. "This creature is no son of mine. As a lesson to all blasphemers he shall die upon the altar at the hand of the god he has profaned. Take him from my sight, and when the sun stands at zenith let the faithful congregate in the temple court and witness the wrath of this divine hand," and he held aloft his right palm.

Those who had brought Tarzan took him away then as Obergatz had directed, and the German turned once more to the warriors by the gate. "Throw down your arms, warriors of Ja-don," he cried, "lest I call down my lightnings to blast you where you stand. Those who do as I bid shall be forgiven. Come! Throw down your arms."

The warriors of Ja-don moved uneasily, casting looks of appeal at their leader and of apprehension toward the figures upon the palace roof. Ja-don sprang forward among his men. "Let the cowards and knaves throw down their arms and enter the palace," he cried, "but never will Ja-don and the warriors of Ja-lur touch their foreheads to the feet of Lu-don and his false god. Make your decision now," he cried to his followers.

A few threw down their arms and with sheepish looks passed through the gateway into the palace, and with the example of these to bolster their courage others joined in the desertion from the old chieftain of the north, but staunch and true around him stood the majority of his warriors and when the last weakling had left their ranks Ja-don voiced the savage cry with which he led his followers to the attack, and once again the battle raged about the palace gate.

At times Ja-don's forces pushed the defenders far into the palace ground and then the wave of combat would recede and pass out into the city again. And still Ta-den and the reinforcements did not come. It was drawing close to noon. Lu-don had mustered every available man that was not actually needed for the defense of the gate within the temple, and these he sent, under the leadership of Pan-sat, out into the city through the secret passageway and there they fell upon Ja-don's forces from the rear while those at the gate hammered them in front.

Attacked on two sides by a vastly superior force the result was inevitable and finally the last remnants of Ja-don's little army capitulated and the old chief was taken a prisoner before Lu-don. "Take him to the temple court," cried the high priest. "He shall witness the death of his accomplice and perhaps Jad-ben-Otho shall pass a similar sentence upon him as well."

The inner temple court was packed with humanity. At either end of the western altar stood Tarzan and his mate, bound and helpless. The sounds of battle had ceased and presently the ape-man saw Ja-don being led into the inner court, his wrists bound tightly together before him. Tarzan turned his eyes toward Jane and nodded in the direction of Ja-don. "This looks like the end," he said quietly. "He was our last and only hope."

"We have at least found each other, John," she replied, "and our last days have been spent together. My only prayer now is that if they take you they do not leave me."

Tarzan made no reply for in his heart was the same bitter thought that her own contained—not the fear that they would kill him but the fear that they would not kill her. The ape-man strained at his bonds but they were too many and too strong. A priest near him saw and with a jeering laugh struck the defenseless ape-man in the face.

"The brute!" cried Jane Clayton.

Tarzan smiled. "I have been struck thus before, Jane," he said, "and always has the striker died."

"You still have hope?" he asked.

"I am still alive," he said as though that were sufficient answer. She was a woman and she did not have the courage of this man who knew no fear. In her heart of hearts she knew that he would die upon the altar at high noon for he had told her, after he had been brought to the inner court, of the sentence of death that Obergatz had pronounced upon him, and she knew too that Tarzan knew that he would die, but that he was too courageous to admit it even to himself.

As she looked upon him standing there so straight and wonderful and brave among his savage captors her heart cried out against the cruelty of the fate that had overtaken him. It seemed a gross and hideous wrong that that wonderful creature, now so quick with exuberant life and strength and purpose should be presently naught but a bleeding lump of clay— and all so uselessly and wantonly. Gladly would she have offered her life for his but she knew that it was a waste of words since their captors would work upon them whatever it was their will to do—for him, death; for her—she shuddered at the thought.

And now came Lu-don and the naked Obergatz, and the high priest led the German to his place behind the altar, himself standing upon the other's left. Lu-don whispered a word to Obergatz, at the same time nodding in the direction of Ja-don. The Hun cast a scowling look upon the old warrior.

"And after the false god," he cried, "the false prophet," and he pointed an accusing finger at Ja-don. Then his eyes wandered to the form of Jane Clayton.

"And the woman, too?" asked Lu-don.

"The case of the woman I will attend to later," replied Obergatz. "I will talk with her tonight after she has had a chance to meditate upon the consequences of arousing the wrath of Jad-ben-Otho."

He cast his eyes upward at the sun. "The time approaches," he said to Lu-don. "Prepare the sacrifice."

Lu-don nodded to the priests who were gathered about Tarzan. They seized the ape-man and lifted him bodily to the altar where they laid him upon his back with his head at the south end of the monolith, but a few feet from where Jane Clayton stood. Impulsively and before they could restrain her the woman rushed forward and bending quickly kissed her mate upon the forehead. "Good-bye, John," she whispered.

"Good-bye," he answered, smiling.

The priests seized her and dragged her away. Lu-don handed the sacrificial knife to Obergatz. "I am the Great God," cried the German, "thus falleth the divine wrath upon all my enemies!" He looked up at the sun and then raised the knife high above his head.

"Thus die the blasphemers of God!" he screamed, and at the same instant a sharp staccato note rang out above the silent, spell-bound multitude. There was a screaming whistle in the air and Jad-ben-Otho crumpled forward across the body of his intended victim. Again the same alarming noise and Lu-don fell, a third and Mo-sar crumbled to the ground. And now the warriors and the people, locating the direction of this new and unknown sound turned toward the western end of the court.

Upon the summit of the temple wall they saw two figures—a Ho-don warrior and beside him an almost naked creature of the race of Tarzan-jad-guru, across his shoulders and about his hips were strange broad belts studded with beautiful cylinders that glinted in the mid-day sun, and in his hands a shining thing of wood and metal from the end of which rose a thin wreath of blue-gray smoke.

And then the voice of the Ho-don warrior rang clear upon the ears of the silent throng. "Thus speaks the true Jad-ben-Otho," he cried, "through this his Messenger of Death. Cut the bonds of the prisoners. Cut the bonds of the Dor-ul-Otho and of Ja-don, King of Pal-ul-don, and of the woman who is the mate of the son of god."

Pan-sat, filled with the frenzy of fanaticism saw the power and the glory of the régime he had served crumpled and gone. To one and only one did he attribute the blame for the disaster that had but just overwhelmed him. It was the creature who lay upon the sacrificial altar who had brought Lu-don to his death and toppled the dreams of power that day by day had been growing in the brain of the under priest.

The sacrificial knife lay upon the altar where it had fallen from the dead fingers of Obergatz. Pan-sat crept closer and then with a sudden lunge he reached forth to seize the handle of the blade, and even as his clutching fingers were poised above it, the strange thing in the hands of the strange creature upon the temple wall cried out its crashing word of doom and Pan-sat the under priest, screaming, fell back upon the dead body of his master.

"Seize all the priests," cried Ta-den to the warriors, "and let none hesitate lest Jad-ben-Otho's messenger send forth still other bolts of lightning."

The warriors and the people had now witnessed such an exhibition of divine power as might have convinced an even less superstitious and more enlightened people, and since many of them had but lately wavered between the Jad-ben-Otho of Lu-don and the Dor-ul-Otho of Ja-don it was not difficult for them to swing quickly back to the latter, especially in view of the unanswerable argument in the hands of him whom Ta-den had described as the Messenger of the Great God.

And so the warriors sprang forward now with alacrity and surrounded the priests, and when they looked again at the western wall of the temple court they saw pouring over it a great force of warriors. And the thing that startled and appalled them was the fact that many of these were black and hairy Waz-don.

At their head came the stranger with the shiny weapon and

on his right was Ta-den, the Ho-don, and on his left Om-at, the black *gund* of Kor-ul-ja.

A warrior near the altar had seized the sacrificial knife and cut Tarzan's bonds and also those of Ja-don and Jane Clayton, and now the three stood together beside the altar and as the newcomers from the western end of the temple court pushed their way toward them the eyes of the woman went wide in mingled astonishment, incredulity, and hope. And the stranger, slinging his weapon across his back by a leather strap, rushed forward and took her in his arms.

"Jack!" she cried, sobbing on his shoulder. "Jack, my son!"

And Tarzan of the Apes came then and put his arms around them both, and the King of Pal-ul-don and the warriors and the people kneeled in the temple court and placed their foreheads to the ground before the altar where the three stood.

CHAPTER XXV

Home

WITHIN an hour of the fall of Lu-don and Mo-sar, the chiefs and principal warriors of Pal-ul-don gathered in the great throneroom of the palace at A-lur upon the steps of the lofty pyramid and placing Ja-don at the apex proclaimed him king. Upon one side of the old chieftain stood Tarzan of the Apes, and upon the other Korak, the Killer, worthy son of the mighty ape-man.

And when the brief ceremony was over and the warriors with upraised clubs had sworn fealty to their new ruler, Ja-don dispatched a trusted company to fetch O-lo-a and Pan-at-lee and the women of his own household from Ja-lur.

And then the warriors discussed the future of Pal-ul-don and the question arose as to the administration of the temples and the fate of the priests, who practically without exception had been disloyal to the government of the king, seeking always only their own power and comfort and aggrandizement. And then it was that Ja-don turned to Tarzan. "Let the Dor-ul-Otho transmit to his people the wishes of his father," he said.

"Your problem is a simple one," said the ape-man, "if you but wish to do that which shall be pleasing in the eyes of god. Your priests, to increase their power, have taught you that Jad-ben-Otho is a cruel god; that his eyes love to dwell upon blood and upon suffering. But the falsity of their teachings has been demonstrated to you today in the utter defeat of the priesthood.

"Take then the temples from the men and give them instead to the women that they may be administered in kindness and charity and love. Wash the blood from your eastern altar and drain forever the water from the western.

"Once I gave Lu-don the opportunity to do these things but he ignored my commands, and again is the corridor of sacrifice filled with its victims. Liberate these from every temple in Pal-ul-don. Bring offerings of such gifts as your people like and place them upon the altars of your god. And there he will bless them and the priestesses of Jad-ben-Otho can distribute them among those who need them most."

As he ceased speaking a murmur of evident approval ran through the throng. Long had they been weary of the avarice and cruelty of the priests and now that authority had come from a high source with a feasible plan for ridding themselves of the old religious order without necessitating any change in the faith of the people they welcomed it.

"And the priests," cried one. "We shall put them to death upon their own altars if it pleases the Dor-ul-Otho to give the word."

"No," cried Tarzan. "Let no more blood be spilled. Give them their freedom and the right to take up such occupations as they choose."

That night a great feast was spread in the *pal-e-don-so* and for the first time in the history of ancient Pal-ul-don black warriors sat in peace and friendship with white. And a pact was sealed between Ja-don and Om-at that would ever make his tribe and the Ho-don allies and friends.

It was here that Tarzan learned the cause of Ta-den's failure to attack at the stipulated time. A messenger had come from Ja-don carrying instructions to delay the attack until noon, nor had they discovered until almost too late that the messenger was a disguised priest of Lu-don. And they had put him to death and scaled the walls and come to the inner temple court with not a moment to spare.

The following day O-lo-a and Pan-at-lee and the women of Ja-don's family arrived at the palace at A-lur and in the great throneroom Ta-den and O-lo-a were wed, and Om-at and Pan-at-lee.

For a week Tarzan and Jane and Korak remained the guests of Ja-don, as did Om-at and his black warriors. And then the ape-man announced that he would depart from Pal-ul-don. Hazy in the minds of their hosts was the location of heaven and equally so the means by which the gods traveled between their celestial homes and the haunts of men and so no questionings arose when it was found that the Dor-ul-Otho with his mate and son would travel overland across the mountains and out of Pal-ul-don toward the north.

They went by way of the Kor-ul-ja accompanied by the warriors of that tribe and a great contingent of Ho-don warriors under Ta-den. The king and many warriors and a multitude of people accompanied them beyond the limits of A-lur and after they had bid them good-bye and Tarzan had invoked the blessings of God upon them the three Europeans saw their simple, loyal friends prostrate in the dust behind them until the cavalcade had wound out of the city and disappeared among the trees of the nearby forest.

They rested for a day among the Kor-ul-ja while Jane investigated the ancient caves of these strange people and then they moved on, avoiding the rugged shoulder of Pastar-ul-ved and winding down the opposite slope toward the great morass. They moved in comfort and in safety, surrounded by their escort of Ho-don and Waz-don.

In the minds of many there was doubtless a question as to how the three would cross the great morass but least of all was Tarzan worried by the problem. In the course of his life he had been confronted by many obstacles only to learn that he who will may always pass. In his mind lurked an easy solution of the passage but it was one which depended wholly upon chance.

It was the morning of the last day that, as they were break-ing camp to take up the march, a deep bellow thundered from a nearby grove. The ape-man smiled. The chance had come. Fittingly then would the Dor-ul-Otho and his mate and their son depart from unmapped Pal-ul-don.

He still carried the spear that Jane had made, which he had prized so highly because it was her handiwork that he had caused a search to be made for it through the temple in A-lur after his release, and it had been found and brought to him. He had told her laughingly that it should have the place of honor above their hearth as the ancient flintlock of her Puritan grandsire had held a similar place of honor above the fireplace of Professor Porter, her father.

At the sound of the bellowing the Ho-don warriors, some of whom had accompanied Tarzan from Ja-don's camp to Ja-lur, looked questioningly at the ape-man while Om-at's Waz-don looked for trees, since the *gryf* was the one creature of Pal-ul-don which might not be safely encountered even by a great multitude of warriors. Its tough, armored hide was impreg-nable to their knife thrusts while their thrown clubs rattled from it as futilely as if hurled at the rocky shoulder of Pastar-ul-ved.

"Wait," said the ape-man, and with his spear in hand he ad-vanced toward the *gryf*, voicing the weird cry of the Tor-o-don. The bellowing ceased and turned to low rumblings and pres-ently the huge beast appeared. What followed was but a repetition of the ape-man's previous experience with these huge and ferocious creatures.

And so it was that Jane and Korak and Tarzan rode through

the morass that hems Pal-ul-don, upon the back of a prehistoric triceratops while the lesser reptiles of the swamp fled hissing in terror. Upon the opposite shore they turned and called back their farewells to Ta-den and Om-at and the brave warriors they had learned to admire and respect. And then Tarzan urged their titanic mount onward toward the north, abandoning him only when he was assured that the Waz-don and the Ho-don had had time to reach a point of comparative safety among the craggy ravines of the foothills.

Turning the beast's head again toward Pal-ul-don the three dismounted and a sharp blow upon the thick hide sent the creature lumbering majestically back in the direction of its native haunts. For a time they stood looking back upon the land they had just quit—the land of Tor-o-don and *gryf*; of *ja* and *jato*; of Waz-don and Ho-don; a primitive land of terror and sudden death and peace and beauty; a land that they all had learned to love.

And then they turned once more toward the north and with light hearts and brave hearts took up their long journey toward the land that is best of all—home.

THE END

Glossary

From conversations with Lord Greystoke and from his notes and map, there have been gleaned a number of interesting items relative to the language and customs of the inhabitants of Pal-ul-don that are not brought out in the story. For the benefit of those who may care to delve into the derivation of the proper names used in the text, and thus obtain some slight insight into the language of the race, there is appended a rough copy taken from some of Lord Greystoke's notes, together with an incomplete glossary.

A point of particular interest hinges upon the fact that the names of all male hairless pithecanthropi begin with a consonant, have an even number of syllables, and end with a consonant, while the names of the females of the same species begin with a vowel, have an odd number of syllables, and end with a vowel. On the contrary, the names of the male hairy black pithecanthropi while having an even number of syllables begin with a vowel and end with a consonant; while the females of this species have an odd number of syllables in their names which begin always with a consonant and end with a vowel.

A. Light.
Ab. Boy.
Ab-on. Acting *gund* of Kor-ul-ja.
Ad. Three.
Adad. Six.
Adadad. Nine.
Adaden. Seven.
Aden. Four.
Adenaden. Eight.
Adenen. Five.
A-lur. City of light.
An. Spear.
An-un. Father of Pan-at-lee.
As. The sun.
At. Tail.
Bal. Gold or golden.
Bar. Battle.

Ben. Great.
Bu. Moon.
Bu-lot (moon face). Son of chief Mo-sar.
Bu-lur (moon city). The city of the Waz-ho-don.
Dak. Fat.
Dak-at (fat tail). Chief of a Ho-don village.
Dak-lot. One of Ko-tan's palace warriors.
Dan. Rock.
Den. Tree.
Don. Man.
Dor. Son.
Dor-ul-Otho (son of god). Tarzan.

E. Where.

Ed. Seventy.

El. Grace or graceful.

En. One.

Enen. Two.

Es. Rough.

Es-sat (rough skin). Chief of Om-at's tribe of hairy blacks.

Et. Eighty.

Fur. Thirty.

Ged. Forty.

Go. Clear.

Gryf. "Triceratops. A genus of huge herbivorous dinosaurs of the group Ceratopsia. The skull had two large horns above the eyes, a median horn on the nose, a horny beak, and a great bony hood or transverse crest over the neck. Their toes, five in front and three behind, were provided with hoofs, and the tail was large and strong." Webster's Dict. The *gryf* of Pal-ul-don is similar except that it is omniverous, has strong, powerfully armed jaws and talons instead of hoofs. Coloration: face yellow with blue bands encircling the eyes; hood red on top, yellow underneath; belly yellow; body a dirty, slate blue; legs same. Bony protuberances yellow except along the spine—these are red. Tail conforms with body and belly. Horns, ivory.

Gund. Chief.

Guru. Terrible.

Het. Fifty.

Ho. White.

Ho-don. The hairless white men of Pal-ul-don.

Id. Silver.

Id-an. One of Pan-at-lee's two brothers.

In. Dark.

In-sad. Kor-ul-ja warrior accompanying Tarzan, Om-at, and Ta-den in search of Pan-at-lee.

In-tan. Kor-ul-lul left to guard Tarzan.

Ja. Lion.

Jad. The.

Jad-bal-lul. The golden lake.

Jad-ben-lul. The big lake.

Jab-ben-Otho. The Great God.

Jad-guru-don. The terrible man.

Jad-in-lul. The dark lake.

Ja-don (the lion-man). Chief of a Ho-don village and father of Ta-den.

Jad Pele ul Jad-ben-Otho. The valley of the Great God.

Ja-lur (lion city). Ja-don's capital.

Jar. Strange.

Jar-don. Name given Korak by Om-at.

Jato. Saber-tooth hybrid.

Ko. Mighty.

Kor. Gorge.

Kor-ul-gryf. Gorge of the *gryf*.

Kor-ul-ja. Name of Es-sat's gorge and tribe.

Kor-ul-lul. Name of another Waz-don gorge and tribe.

Ko-tan. King of the Ho-don.

Lav. Run or running.

Lee. Doe.

Lo. Star.

Lot. Face.

Lu. Fierce.

Lu-don (fierce man). High priest of A-lur.

Lul. Water.

Lur. City.

Ma. Child.

Mo. Short.

Mo-sar (short nose). Chief and pretender.

Mu. Strong.

No. Brook.

O. Like or similar.

Od. Ninety.

O-dan. Kor-ul-ja warrior accompanying Tarzan, Om-at, and Ta-den in search of Pan-at-lee.

Og. Sixty.

O-lo-a (like-star-light). Ko-tan's daughter.

Om. Long.

Om-at (long tail). A black.

On. Ten.

Otho. God.

Pal. - Place; land; country.

Pal-e-don-so (place where men eat). Banquet hall.

Pal-ul-don (land of man). Name of the country.

Pal-ul-ja. Place of lions.

Pan. Soft.

Pan-at-lee. Om-at's sweetheart.

Pan-sat (soft skin). A priest.

Pastar. Father.

Pastar-ul-ved. Father of Mountains.

Pele. Valley.

Ro. Flower.

Sad. Forest.

San. One hundred.

Sar. Nose.

Sat. Skin.

So. Eat.

Sod. Eaten.

Sog. Eating.

Son. Ate.

Ta. Tall.

Ta-den (tall tree). A white.

Tan. Warrior.

Tarzan-jad-guru. Tarzan the Terrible.

To. Purple.

Ton. Twenty.

Tor. Beast.

Tor-o-don. Beastlike man.

Tu. Bright.

Tu-lur (bright city). Mo-sar's city.

Ul. Of.

Un. Eye.

Ut. Corn.

Ved. Mountain.

Waz. Black.

Waz-don. The hairy black men of Pal-ul-don.

Waz-ho-don (black white men). A mixed race.

Xot. One thousand.

Yo. Friend.

Za. Girl.

COSIMO is a specialty publisher of books and publications that inspire, inform, and engage readers. Our mission is to offer unique books to niche audiences around the world.

COSIMO BOOKS publishes books and publications for innovative authors, nonprofit organizations, and businesses. **COSIMO BOOKS** specializes in bringing books back into print, publishing new books quickly and effectively, and making these publications available to readers around the world.

COSIMO CLASSICS offers a collection of distinctive titles by the great authors and thinkers throughout the ages. At **COSIMO CLASSICS** timeless works find new life as affordable books, covering a variety of subjects including: Business, Economics, History, Personal Development, Philosophy, Religion & Spirituality, and much more!

COSIMO REPORTS publishes public reports that affect your world, from global trends to the economy, and from health to geopolitics.

FOR MORE INFORMATION CONTACT US AT
INFO@COSIMOBOOKS.COM

❋ if you are a book lover interested in our current catalog of books

❋ if you represent a bookstore, book club, or anyone else interested in special discounts for bulk purchases

❋ if you are an author who wants to get published

❋ if you represent an organization or business seeking to publish books and other publications for your members, donors, or customers.

COSIMO BOOKS ARE ALWAYS AVAILABLE AT ONLINE BOOKSTORES

VISIT COSIMOBOOKS.COM
BE INSPIRED, BE INFORMED